AF779120

Thermal Flows

Thermal Flows

Editor

Marcello Lappa

MDPI • Basel • Beijing • Wuhan • Barcelona • Belgrade • Manchester • Tokyo • Cluj • Tianjin

Editor
Marcello Lappa
Department of Mechanical and
Aerospace Engineering
University of Strathclyde
Glasgow
United Kingdom

Editorial Office
MDPI
St. Alban-Anlage 66
4052 Basel, Switzerland

This is a reprint of articles from the Special Issue published online in the open access journal *Fluids* (ISSN 2311-5521) (available at: www.mdpi.com/journal/fluids/special_issues/thermal_flows).

For citation purposes, cite each article independently as indicated on the article page online and as indicated below:

LastName, A.A.; LastName, B.B.; LastName, C.C. Article Title. *Journal Name* **Year**, *Volume Number*, Page Range.

ISBN 978-3-0365-2228-9 (Hbk)
ISBN 978-3-0365-2227-2 (PDF)

© 2022 by the authors. Articles in this book are Open Access and distributed under the Creative Commons Attribution (CC BY) license, which allows users to download, copy and build upon published articles, as long as the author and publisher are properly credited, which ensures maximum dissemination and a wider impact of our publications.
The book as a whole is distributed by MDPI under the terms and conditions of the Creative Commons license CC BY-NC-ND.

Contents

About the Editor

Marcello Lappa

Marcello Lappa attained a qualification to the rank of Full Professor in Italy in 2013. He joined the Department of Aerospace and Mechanical Engineering of the University of Strathclyde (UK, Glasgow) in 2015. On Oct 2017 he took on the role of Director (Programme Advisor of Studies) of the MSc course in Mechanical Engineering. He is listed in the "World Ranking of Top 2% Scientists in 2021" Stanford University database in the subfield Fluids & Plasmas. Over the last 25 years, he has authored 3 international books, more than 120 publications in peer reviewed journals or as book chapters and many other conference papers. Since 2005 he serves as the Editor-in-Chief of the International Scientific Journal "Fluid Dynamics Materials Processing". His research focuses on fluid motion and stability behavior, computational fluid dynamics, organic and inorganic materials and crystal growth, multiphase flows, solidification, high-temperature gas-dynamics, particle dynamics and microgravity science. At the University of Strathclyde, he has secured over 2 million € of external funding in a position of PI. He acts as a Reviewer for several funding bodies (EPSRC-UK, DFG-Germany, FNRS-Belgium, GIF-Israel, ANVUR-Italy, NSERC-Canada, NVSTE-Kazakhstan and European Space Agency).

fluids MDPI

Editorial
Thermal Flows

Marcello Lappa

Department of Mechanical and Aerospace Engineering, University of Strathclyde, James Weir Building, 75 Montrose Street, Glasgow G1 1XJ, UK; marcello.lappa@strath.ac.uk

check for updates

Citation: Lappa, M. Thermal Flows. *Fluids* **2021**, *6*, 227. https://doi.org/10.3390/fluids6060227

Received: 2 June 2021
Accepted: 7 June 2021
Published: 18 June 2021

Publisher's Note: MDPI stays neutral with regard to jurisdictional claims in published maps and institutional affiliations.

Copyright: © 2021 by the author. Licensee MDPI, Basel, Switzerland. This article is an open access article distributed under the terms and conditions of the Creative Commons Attribution (CC BY) license (https://creativecommons.org/licenses/by/4.0/).

Flows of thermal origin and heat transfer problems are central in a variety of disciplines and industrial applications. The present Special Issue entitled "Thermal Flows" consists of a collection of studies by distinct investigators and research groups, dealing with different types of flows relevant to both natural and technological contexts. It comprises 11 papers covering some of the latest advances in these fields. Both reviews of the state-of-the-art and new theoretical, numerical and experimental investigations are presented, which illustrate the structure of these flows, their stability behavior and the possible bifurcations to different patterns of symmetry and/or spatiotemporal regimes. Moreover, different categories of fluids are considered (liquid metals, gases, common fluids such as water and silicone oils, organic and inorganic transparent liquids and nanofluids).

The first part of this collection [1–3] focuses on a very classical problem, generally falling under the heading of Rayleigh–Bénard convection (RB), that is, the behavior of fluid systems heated from below and cooled from above. In particular, the first study [1] finds its motivations in the field of thermal comfort in the building heating and ventilation sector, where thermal control is traditionally implemented via floor heating and/or ceiling cooling; interestingly, the response of RB convection to the application of time-periodic cooling is assessed by means of sophisticated LBM simulations. Thermogravitational convection is still at the root of the following two numerical investigations [2,3], where, however, a shift is implemented in the intended outcomes. In particular, the context is changed from the design of heating/cooling systems to the much more fundamental area relating to natural phenomena of a geophysical or astrophysical interest. Indeed, flows of the RB type are also known for their remarkable ability to produce important large-scale features (such as a self-sustained magnetic field) in rotating cosmic objects, relevant exemplars being the sun [2] and terrestrial or gas giant planets [3]. Rotating thermal convection is yet the main subject of analysis in [4], where the non-linear dynamics of such flow are considered on the surface of a hemisphere. It is shown that, just like a half soap bubble located on a heated plate, a hemisphere undergoing rigid or differential rotation can mimic the behavior of large-scale fluid systems (the numerous atmospheric and oceanic flows, which can effectively be modeled as two-dimensional thermal convection on a curved surface).

The next paper on the list [5] is still somehow relevant to typical natural phenomena, as it is concerned with the behavior of a viscous non-isothermal mixing layer of water. In isothermal conditions, such a kind of flow is the simplest configuration allowing for the well-known Kelvin–Helmholtz (or KH) instability that takes place when two parallel currents with different velocities meet. It typically develops in the form of a wave. This study examines the changes that are introduced in these dynamics when the fluid displays a stable thermal stratification.

Paper [6] is a review of the landmark experimental results obtained with regard to a completely different mechanism of convection, which does not require the presence of buoyancy or gravity, namely Marangoni (or thermocapillary) convection. This genus of fluid motion is relevant to a variety of technological sectors (including, but not limited to, soldering processes, film and spray coating and several techniques for the production of crystals of different types). The authors of paper [6] offer a masterful exposition on the state of the art for Marangoni flows in "liquid bridges" (LBs) made of silicone oil and other transparent organic liquids (an LB is a column of liquid held by surface tension between two supporting disks at different temperatures). When this type of convection, initially steady and axisymmetric, becomes unstable, it can produce waves travelling in the azimuthal direction that, under certain conditions, can cause the accumulation of tracer particles with various sizes and densities into esthetically appealing structures known as "PAS". Surface tension effects are still central in the next work [7], where the planar laminar free-surface flow of a Newtonian fluid (as typically encountered in slot and blade coating applications) is examined analytically at moderate Reynolds and capillary numbers under adverse or favorable pressure gradients.

Following the treatment of thermogravitational and thermocapillary convection, two studies [8,9] are also included, which concentrate on a third fundamental variant of thermally induced flows, namely fluid motion produced in non-isothermal systems by the application of "vibrations" (thermovibrational convection). Vibrations enable a time-varying acceleration, which formally makes the emerging convective structures a variant of buoyancy convection. As in the parent phenomena, these processes are extremely sensitive to the relative directions of the acceleration and the prevailing temperature gradient. In particular, [8] focuses on a case where these vectors are parallel. The complexity of this specific arrangement (with respect to the companion situation in which vibrations and temperature difference are perpendicular) stems from the properties that are inherited from the corresponding case with steady gravity, i.e., the standard RB. The need to overcome a threshold to induce convection from an initial quiescent state, together with the opposite tendency of acceleration to damp fluid motion when its sign is reversed, causes a variety of possible solutions that can display synchronous, non-synchronous, time-periodic and multi-frequency responses. A very special case is represented by "circular translational vibrations", i.e., a situation where the entire fluid domain moves along a circular path [9]. The peculiarity of this flow originates from the absence of a preferred direction for the vibrations and the existence of a rotating force field in the non-inertial reference frame moving with the fluid container.

The "rotating fluid" theme, already addressed to a certain extent in [9] and in [2–4], returns (although under different perspectives) in [10], where the main purpose is the presentation of an analytical dependence on the critical Reynolds number for a flow near a rotating disk (such a subject being essential for the analysis of the thermal response of disks and blades in relation to the design of gas and steam turbines).

The last work of the sequence [11] finally addresses another important long-standing issue in engineering, that is, the thermal performances of circular pipes with inserts. Such a subject is examined in combination with the use of a nanofluid (water and titanium dioxide) for different values of the Reynolds number. The resulting problem aligns with current efforts aimed to design more efficient heat-exchange technologies to be used in a variety of fields (processor and electronic chip cooling, nuclear reactor cooling, food processing, cooling systems in commercial buildings, just to cite a few).

As witnessed by all of these studies, problems concerned with heat transfer and related (forced or natural) flows continue to burgeon and bring surprises to this day. Along these lines, the focused papers included in this Special Issue have been collected under the optimistic hope that they will serve as a new important resource for physicists, engineers and advanced students interested in the physics of non-isothermal fluid systems, fluid mechanics, environmental phenomena, meteorology, geophysics, thermal, mechanical and materials engineering.

Conflicts of Interest: The author declares no conflict of interest.

References

1. Nasseri, L.; Himrane, N.; Ameziani, D.E.; Bourada, A.; Bennacer, R. Time-periodic cooling of Rayleigh–Bénard convection. *Fluids* **2021**, *6*, 87. [CrossRef]
2. Silva, L.; Gupta, P.; MacTaggart, D.; Simitev, R.D. Effects of shell thickness on cross-helicity generation in convection-driven spherical dynamos. *Fluids* **2020**, *5*, 245. [CrossRef]

3. Simitev, R.D.; Busse, F.H. Onset of inertial magnetoconvection in rotating fluid spheres. *Fluids* **2021**, *6*, 41. [CrossRef]
4. Fischer, P.; Bruneau, C.H.; Kellay, H. Numerical study of rotating thermal convection on a hemisphere. *Fluids* **2020**, *5*, 185. [CrossRef]
5. Vitoshkin, H.; Gelfgat, A. Non-modal three-dimensional optimal perturbation growth in thermally stratified mixing layers. *Fluids* **2021**, *6*, 37. [CrossRef]
6. Ueno, I. Experimental study on coherent structures by particles suspended in half-zone thermocapillary liquid bridges: Review. *Fluids* **2021**, *6*, 105. [CrossRef]
7. Khayat, R.E.; Hossain, M.T. Coating flow near channel exit. A theoretical perspective. *Fluids* **2020**, *5*, 180. [CrossRef]
8. Crewdson, G.; Lappa, M. The zoo of modes of convection in liquids vibrated along the direction of the temperature gradient. *Fluids* **2021**, *6*, 30. [CrossRef]
9. Kozlov, V.; Rysin, K.; Vjatkin, A. Vibroconvective patterns in a layer under translational vibrations of circular polarization. *Fluids* **2021**, *6*, 108. [CrossRef]
10. Dmitrenko, A.V. Determination of critical Reynolds number for the flow near a rotating disk on the basis of the theory of stochastic equations and equivalence of measures. *Fluids* **2021**, *6*, 5. [CrossRef]
11. Papazian, K.; Al Hajaj, Z.; Saghir, M.Z. Thermal performance of a heated pipe in the presence of a metal foam and twisted tape inserts. *Fluids* **2020**, *5*, 195. [CrossRef]

fluids

MDPI

Article

Time-Periodic Cooling of Rayleigh–Bénard Convection

Lyes Nasseri [1], Nabil Himrane [2], Djamel Eddine Ameziani [1], Abderrahmane Bourada [3] and Rachid Bennacer [4,*]

1 LTPMP (Laboratoire de Transports Polyphasiques et Milieux Poreux), Faculty of Mechanical and Proceeding Engineering, USTHB (Université des Sciences et de la Technologie Houari Boumedienne), Algiers 16111, Algeria; lyes.nasseri@gmail.com (L.N.); deameziani@usthb.dz (D.E.A.)

2 Labo of Energy and Mechanical Engineering (LEMI), Faculty of Technology, UMBB (Université M'hamed Bougara-Boumerdes, Boumerdes 35000, Algeria; n.himrane@univ-boumerdes.dz

3 Laboratory of Transfer Phenomena, RSNE (Rhéologie et Simulation Numérique des Ecoulements) Team, FGMGP (Faculté de génie Mécaniques et de Génie des Procédés Engineering), USTHB (Université des Sciences et de la Technologie Houari Boumedienne), Bab Ezzouar, Algiers 16111, Algeria; abderrahmanebourada@gmail.com

4 CNRS (Centre National de la Recherche Scientifique), LMT (Laboratoire de Mécanique et Technologie—Labo. Méca. Tech.), Université Paris-Saclay, ENS (Ecole National Supérieure) Paris-Saclay, 91190 Gif-sur-Yvette, France

* Correspondence: rachid.bennacer@ens-paris-saclay.fr

Abstract: The problem of Rayleigh–Bénard's natural convection subjected to a temporally periodic cooling condition is solved numerically by the Lattice Boltzmann method with multiple relaxation time (LBM-MRT). The study finds its interest in the field of thermal comfort where current knowledge has gaps in the fundamental phenomena requiring their exploration. The Boussinesq approximation is considered in the resolution of the physical problem studied for a Rayleigh number taken in the range $10^3 \leq Ra \leq 10^6$ with a Prandtl number equal to 0.71 (air as working fluid). The physical phenomenon is also controlled by the amplitude of periodic cooling where, for small values of the latter, the results obtained follow a periodic evolution around an average corresponding to the formulation at a constant cold temperature. When the heating amplitude increases, the physical phenomenon is disturbed, the stream functions become mainly multicellular and an aperiodic evolution is obtained for the heat transfer illustrated by the average Nusselt number.

Keywords: Rayleigh–Bénard convection; time periodical cooling; Lattice Boltzmann method

check for updates

Citation: Nasseri, L.; Himrane, N.; Ameziani, D.E.; Bourada, A.; Bennacer, R. Time-Periodic Cooling of Rayleigh–Bénard Convection. *Fluids* **2021**, *6*, 87. https://doi.org/10.3390/fluids6020087

Academic Editor: Marcello Lappa

Received: 19 January 2021
Accepted: 12 February 2021
Published: 16 February 2021

Publisher's Note: MDPI stays neutral with regard to jurisdictional claims in published maps and institutional affiliations.

Copyright: © 2021 by the authors. Licensee MDPI, Basel, Switzerland. This article is an open access article distributed under the terms and conditions of the Creative Commons Attribution (CC BY) license (https://creativecommons.org/licenses/by/4.0/).

1. Introduction

The building sector is one of the largest energy consuming sectors. It represents a large proportion of total energy consumption, much higher than industry and transport in many countries [1–5]. The major challenge to guarantee good comfort is that energy consumption and the level of comfort are often in conflict in a room [6].

In a confined space, thermal comfort depends on the operation of controlled installations (i.e., heating, ventilation and air conditioning) [7]. Therefore, to assess the energy performance of the building, a thermal analysis is essential in order to predict the thermal responses and calculate building loads (heating/cooling). The key of good thermal comfort is the temperature control. To assess it, two models exist: physical or white-box models based on the energy and mass balance equations, and data-driven or black box models based on artificial neural network and developed after sufficient data are available.

In order to propose an adequate design of the indoor environment where people work and live, several papers have been published on the thermal comfort of buildings [8–10]. Rayleigh–Bénard (RB) convection is a classical problem of natural convection. It appears in several applications of engineering and building, such as thermal comfort by using floor heating or ceiling cooling. The other challenge is the energy storing, so either phase change materials or liquid storing are integrated to enhance the thermal building inertia.

Chandrasekhar [11] and Drazin and Reid [12] implemented a full report on the linearization theory. Moreover, many numerical studies of RB natural convection in rectangular enclosures have been carried out for Newtonian fluids [13–15]. Other researchers were interested in the Rayleigh-Bénard convection in viscoplastic fluids leading to numerical and experimental investigations [14–17]. The Rayleigh number (*Ra*) is the parameter that quantifies the intensity of the thermal driving in convection. For sufficiently large *Ra*, Rayleigh–Bénard convection flow becomes turbulent. Significant progress in our understanding of turbulent convection has been obtained by both experimental and numerical studies [18–20]. The Rayleigh–Bénard convection with all the issues illustrated above becomes more complex in the case of time-dependent boundary conditions [21]. The Rayleigh–Bénard problem, with linear temperature increase, was studied by Kaviany [22] and extended later on by Kaviany and Vogel [23], with the inclusion of solute concentration gradients. The solute gradient represents either the phase diagram solute redistribution near phase change interface or the stabilized long-term energy storing solar pound. According to the results of the Rayleigh–Bénard convection studies, a sufficient condition has been found to control the frequency of heat pulsation in order to initiate convection in a periodically heated and cooled cavity from top wall [24].

Aniss et al. [25] have studied the influence of the gravitational modulation on the stability threshold in the case of a Newtonian fluid confined in Hele-Shaw cell and subjected to vertical periodic motion. A time-dependent perturbation expressed in Fourier series has been applied to the wall temperature according to Bhadauria and Bhatia [26]. They found that it is possible to advance or delay the onset of convection. Umavathi [27] also investigated the effect of external modulation on the thermal convection in a porous medium saturated by a nanofluid. It was found in this last study that the low frequency symmetric thermal modulation is destabilizing while moderate and high frequency symmetric modulation is always stabilizing. Recently, Himraneet et al. [28] have studied numerically the Rayleigh–Bénard convection using the Lattice Boltzmann method. The authors considered periodic heating at the lower wall of the cavity. For high values of Rayleigh number, they obtained an unsteady regime in the form of temporal evolution with several frequencies. Abourida et al. [29] have examined Rayleigh–Bénard convection in a square enclosure with a top wall submitted to constant or sinusoidal cold temperature and sinusoidally heated bottom wall. It was observed that by varying the two imposed temperatures, basic differences were noted in comparison to the case of variable hot temperature and that of constant boundary temperature conditions. Raji et al. [30] have presented Rayleigh–Bénard convection inside square enclosure with a time-periodic cold temperature in the top wall. They proved that the variable cooling can lead to a significant improvement in heat transfer compared to constant cooling, particularly at certain low periods. The influence of thermoelectric effect on the Rayleigh–Bénard instability has been well investigated by researchers [31–33].

The literature review has shown that Rayleigh–Bénard's convection problem is still relevant and often considered with constant boundary conditions whereas reality suggests that heating and cooling conditions are modulated over time (i.e., scrolling of the days). In the present study, the problem of time-periodic cooling applied to Rayleigh–Bénard convection is investigated using Lattice Boltzmann method with multiple relaxation time (LBM-MRT). The study finds its interest in the field of thermal comfort where current knowledge has gaps in the fundamental phenomena (theoretical funds) requiring their exploration.

2. Mathematical Formulation

Lattice Boltzmann method was adopted for the resolution of fluid flow and transports phenomena. This method has non-linearity in its mathematical formulation and therefore approximates the temporal variation wall temperature (totally explicitly unsteady). Fluid flow is described as a movement of particles. In two-dimensional domain, the particles' moving is done by distribution functions with DnQm model, where "n" denote dimensions and "m" discrete velocities [34]. In our study, we used D2Q9 model for dynamic field and D2Q5 for thermal model. Finally, collisions and advection terms are computed by the Boltzmann equation of the distribution functions [35,36]:

$$f_i(x + e_i\delta_t, t + \delta_t) - f_i(x, t) = \Omega(f_i) + \delta_t F_i \tag{1}$$

where f_i is the distribution function with velocity e_i at lattice node x at time t, δ_t is the discrete time step, $\Omega(f_i)$ is the collision operator and F_i is the implemented external forces term. Then, the collision operator in indicial form is as follows:

$$f_i(x + e_i\delta_t, t + \delta_t) - f_i(x, t) = -1/\tau\left[f_i(x, t) - f_i^{eq}(x, t)\right] + \delta_t F_i \tag{2}$$

where f_i^{eq} is the equilibrium function which is expressed by:

$$f_i^{eq} = w_i\rho\left[1 + 3e_i v + 9(e_i v)^2/2 - 3v^2/2\right] \tag{3}$$

The factors $\mathrm{w_i}$ are given as: $\left\{\frac{4}{9}, \frac{1}{9}, 0, 0, \frac{1}{9}, \frac{1}{36}, 0, 0, \frac{1}{36}\right\}$, τ is the relaxation time without dimension. The nine discrete velocities are defined as follows:

$$e_i = \begin{cases} (0,0);\ i = 0 \\ c[\cos((i-1)\pi/2), \sin((i-1)\pi/2)];\ i = 1,2,3,4 \\ \sqrt{2c}[\cos((2i-9)\pi/2), \sin((2i-9)\pi/2)];\ i = 5,6,7,8 \end{cases} \tag{4}$$

The collision term is expressed in the Multiple Relaxation Time model (MRT) D'Humières [37], where better stability is observed with a wide range of Prandtl Number values [38].

$$\Omega = -M^{-1}C[m_i(x, t) - m_i{}^{eq}(x, t)] \tag{5}$$

The flow field formulation becomes:

$$f_i(x + e_i\delta_t, t + \delta_t) - f_i(x, t) = -M^{-1}C[m_i(x, t) - m_i{}^{eq}(x, t)] + M^{-1}\delta_t(1 - C/2)D \tag{6}$$

where M is the projection matrix of f_i and f_i^{eq} into the moment space. So, the expression of $m = Mf$ and $m^{eq} = Mf^{eq}$ are given by:

$$\begin{pmatrix} \rho \\ e \\ \phi \\ j_x - (\delta_t/2)\rho F_x \\ q_x \\ j_y - (\delta_t/2)\rho F_y \\ q_y \\ p_{xx} \\ p_{xy} \end{pmatrix} = \begin{bmatrix} 1 & 1 & 1 & 1 & 1 & 1 & 1 & 1 & 1 \\ -4 & -1 & -1 & -1 & -1 & 2 & 2 & 2 & 2 \\ 4 & -2 & -2 & -2 & -2 & 1 & 1 & 1 & 1 \\ 0 & 1 & 0 & -1 & 0 & 1 & -1 & -1 & 1 \\ 0 & -2 & 0 & 2 & 0 & 1 & -1 & -1 & 1 \\ 0 & 0 & 1 & 0 & -1 & 1 & 1 & -1 & -1 \\ 0 & 0 & -2 & 0 & 2 & 1 & 1 & -1 & -1 \\ 0 & 1 & -1 & 1 & -1 & 0 & 0 & 0 & 0 \\ 0 & 0 & 0 & 0 & 0 & 1 & -1 & 1 & -1 \end{bmatrix} \begin{pmatrix} f_0 \\ f_1 \\ f_2 \\ f_3 \\ f_4 \\ f_5 \\ f_6 \\ f_7 \\ f_8 \end{pmatrix}. \tag{7}$$

The fluid density ρ, components moment j_x and j_y are the conserved quantities. The six other moments are non-conserved ones and are relaxed linearly in time namely: energy e, energy squared ϕ, energy flux in the two directions q_x, q_Y and diagonal/off-diagonal component of the strain-rate tensor p_{xx}, p_{xy}. The collision operator is carried out in the moment space and in indicial form:

$$m_i^*(x, t) = m_i(x, t) - C[m_i(x, t) - m_i{}^{eq}(x, t)] \tag{8}$$

For the thermal model, the two-dimensional D2Q5 model with five velocities is used in this work. This model was chosen and validated by several authors in the literature [30] due to its simplicity and accuracy. The Boltzmann equation with multi-relaxation time can be written as:

$$g_i(x + e_i\delta_t, t + \delta_t) - g_i(x,t) = -N^{-1}E[n_i(x,t) - n_i^{eq}(x,t)] \tag{9}$$

where $g_i(x,t)$ is distribution function of temperature, N is projection matrix of g_i and g_i^{eq} into the moment space, and with the same procedure of temperature (first population) $n = Ng$. The transformation matrix N is given by:

$$\begin{pmatrix} n_0 \\ n_1 \\ n_2 \\ n_3 \\ n_4 \end{pmatrix} = \begin{bmatrix} 1 & 1 & 1 & 1 & 1 \\ 0 & 1 & 0 & -1 & 0 \\ 0 & 0 & 1 & 0 & -1 \\ -4 & 1 & 1 & 1 & 1 \\ 0 & 1 & -1 & 1 & -1 \end{bmatrix} \begin{pmatrix} g_0 \\ g_1 \\ g_2 \\ g_3 \\ g_4 \end{pmatrix} \tag{10}$$

The boundary conditions, according to their macroscopic mathematical formulation are illustrated in Figure 1 for velocities:

$$U(0,Y,t) = U(1,Y,t) = U(X,0,t) = U(X,1,t) = 0 \tag{11}$$

$$V(0,Y,t) = V(1,Y,t) = V(X,0,t) = V(X,1,t) = 0 \tag{12}$$

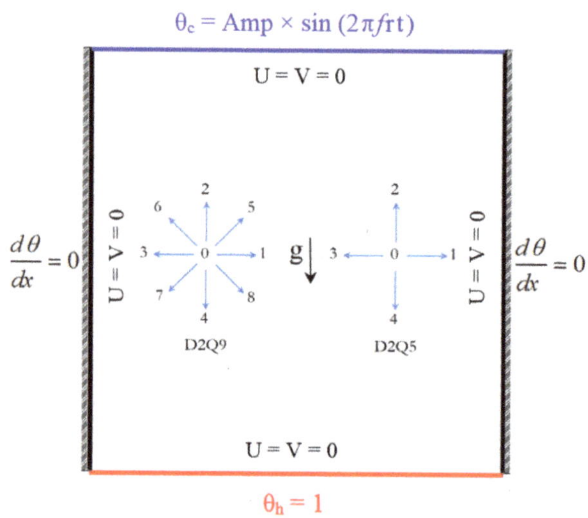

Figure 1. Physical model.

Temperature boundary conditions:

$$\theta(X,1,t) = Amp \cdot \sin(2\pi f t) \tag{13}$$

$$\theta(X,0,t) = 1 \tag{14}$$

$$\left.\frac{\partial\theta(X,Y,t)}{\partial X}\right|_{X=0} = 0 \; and \; \left.\frac{\partial\theta(X,Y,t)}{\partial X}\right|_{X=1} = 0 \tag{15}$$

The instantaneous and average Nusselt number *Nu(t)* at hot and cold wall (*Y* = *0* and *Y* = *1*) is obtained as:

$$Nu_h(X,0,t) = -\left.\frac{\partial\theta(X,Y,t)}{\partial Y}\right|_{Y=0} \; and \; Nu_{h_{avg}}(t) = \int_0^1 Nu(X,t)dX \tag{16}$$

$$Nu_c(X,1,t) = -\frac{\partial\theta(X,Y,t)}{\partial Y}\bigg|_{Y=1} \quad and\ Nu_{c_{avg}}(t) = \int_0^1 Nu(X,t)dX \tag{17}$$

Before presenting the different results obtained, it is necessary to test the validity of our digital code. Thus, we compared the results of our simulations with those of the various studies carried out on the analysis of Rayleigh–Bénard convection in square cavities filled with air. Different discretization methods have been adopted by these reference studies. Table 1 quantitatively summarizes the values of the average Nusselt number obtained as a function of thermal gradient intensity imposed and characterized by the Rayleigh number. We note that our results show good agreement with those of the literature. The code is also successfully validated for the case of time-periodic temperature condition. Figure 2 shows a good agreement between the present hot and cold Nusselt numbers as function of time and those of Wang et al. [39].

Table 1. Nusselt numbers' validation.

References	Ra			
	10^3	10^4	10^5	10^6
Present Work	1.0035	2.1502	3.912	6.321
Ourtatani et al. [13]	1.0004	2.158	3.910	6.309
Turan [14]	1.0000	2.154	3.907	6.309
Bouabdallah et al. [40]	1.0000	2.2000	3.900	6.400

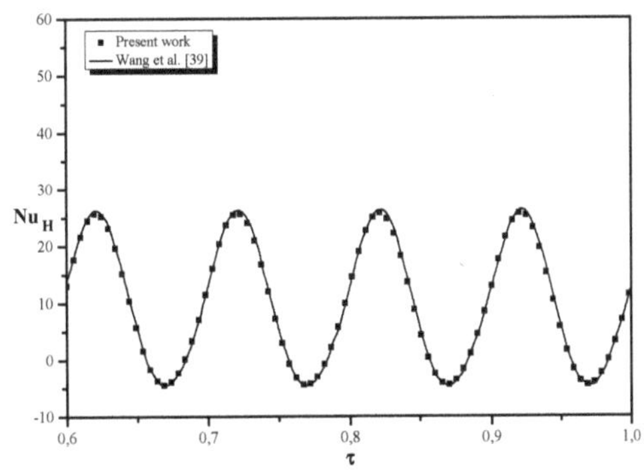

(**a**) Comparison of the average Nusselt number at the hot wall with that of Wang et al. [39].

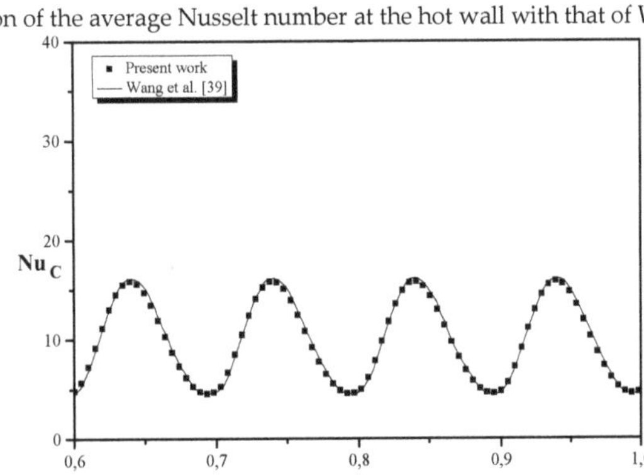

(**b**) Comparison of the average Nusselt number at the cold wall with that of Wang et al. [39].

Figure 2. Time history of Nusselt numbers. Ra = 10^6, Pr = 6.2, τ_p = 0.1 and Amp = 0.8.

In what follows, we present the influence of the various control parameters governing the natural convection problem. The different results are represented in terms of streamlines, isotherms and heat transfer rate over a time period. Finally, we analyze the periodicity of the convective regime, by means of phase portraits obtained by the heat exchange coefficients.

3. Results

The range of the Rayleigh number values is taken $10^3 \leq \text{Ra} \leq 10^6$. The amplitude of heating is taken between 0.0 and 0.8, and the working fluid is assumed to be air with $\text{Pr} = 0.71$.

In order to better study the behavior of the flow, we varied the amplitude for the three cases, 0.2, 0.5 and 0.8. Figure 3 represents the stream functions for the amplitude 0.2 as a function of the Rayleigh number for the four quarter-period $f = 1/2\pi\tau_p = 1/3 \times 10^{-4}$. Note that the stream function is obtained from the velocity integral and represents the fluid flow rate. When the Rayleigh number is equal to Ra = 10^4, an invariant cell is observed for the four quarter-period.

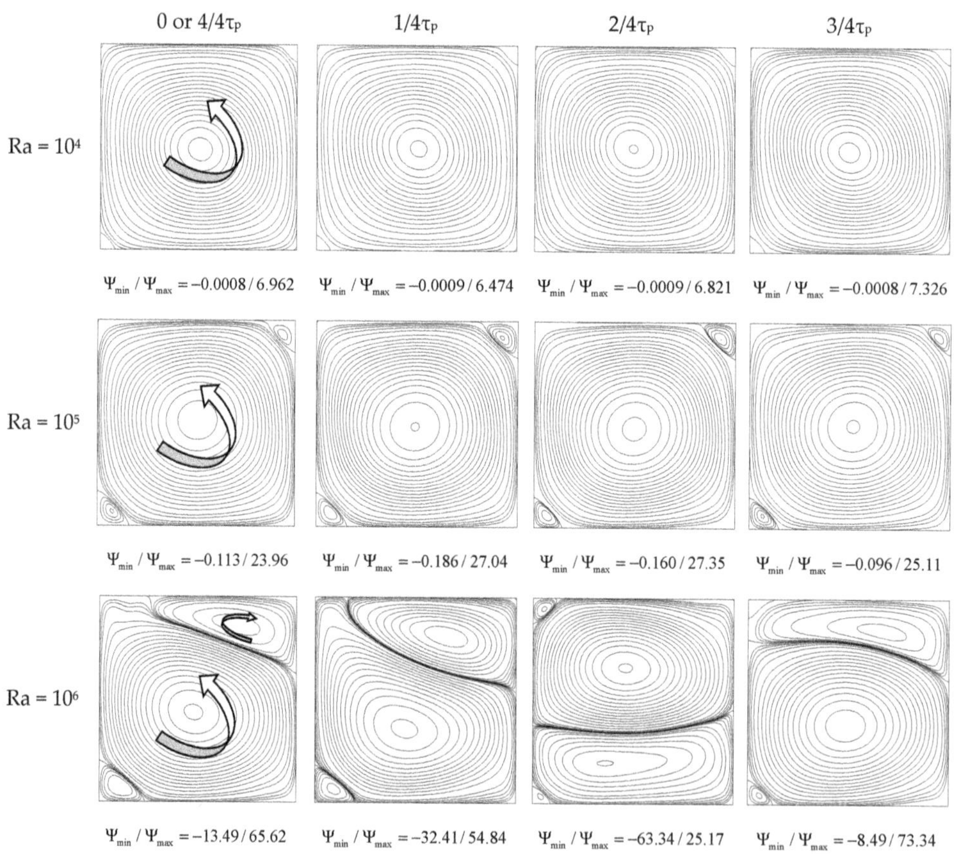

Figure 3. Evolution of streamlines over a period as a function of the Rayleigh number (Amp = 0.2 and $|f| = 0.33 \times 10^{-4}$).

This means that the modulation of cooling has little influence on the dynamic fields. The only difference is that the flow increases characterized by ψ_{max} is increasingly larger than the general temperature gradient is maximum (in the heating period). Same case for the Rayleigh Ra = 10^5 with the presence of small vortices on the upper left and lower right side. As the thermal draft increases, for a Rayleigh number 10^6, the flow loses its symmetry and subsequently becomes multicellular. For this Rayleigh value and a period of $\tau p/2$, the upper secondary cell grew further and took up most of the space of the cavity, in turn becoming counter-rotating bicellular. This spatial multi-cellular (mainly bicellular) competition is persistent throughout the period.

Figures 4 and 5 represent the isocurrents for amplitudes 0.5 and 0.8 respectively, as a function of Rayleigh for the four quarter-period (with f = $1/3 \times 10^{-4}$). The flow structure for the Rayleigh 10^4 is essentially single-cell, with vortices on the upper right and lower left sides being very small. These findings are similar to the same Rayleigh with a smaller temperature modulation (Amp = 0.2). On the other hand, for Rayleigh equal to 10^5, it is always the same type of flow (i.e., mainly single-cell) but the vortices of the corners are strongly present and influence the size of the main cell. The upper secondary cell still grows (Ra = 10^6) and the flow becomes predominantly counter-rotating bicellular and then becomes mainly monocellular again at the end of the period due to the decrease of the upper cell.

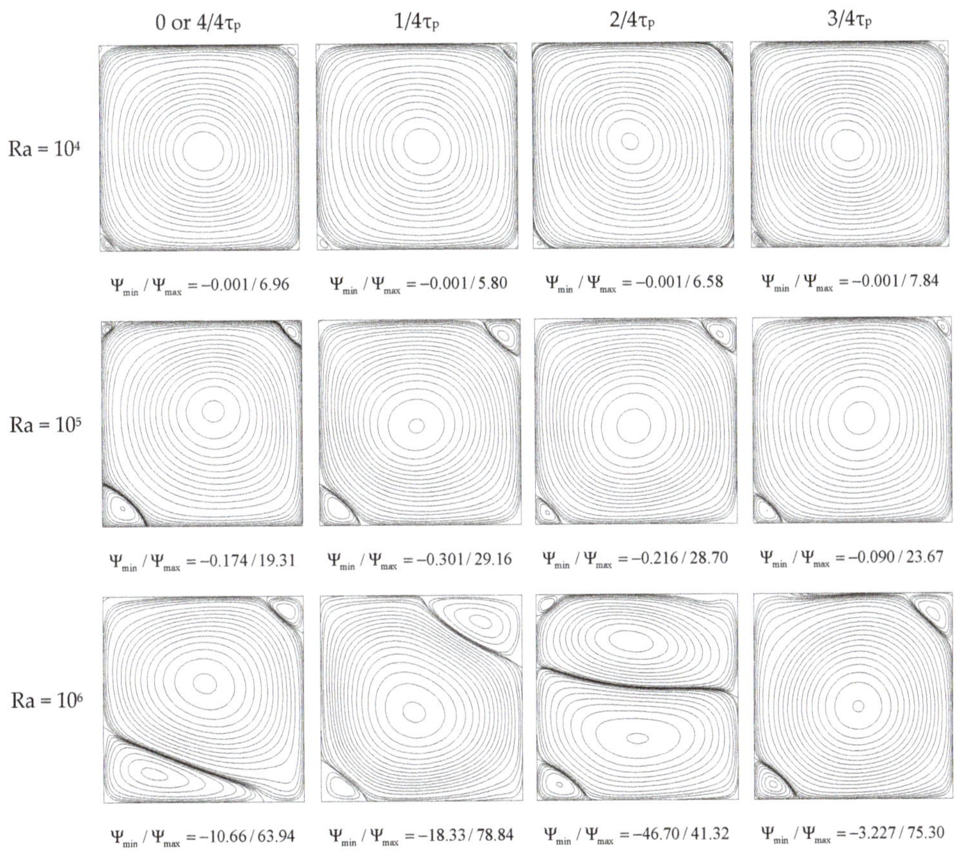

Figure 4. Evolution of streamlines over a period as a function of the Rayleigh number (Amp = 0.5 and $|f| = 0.33 \times 10^{-4}$).

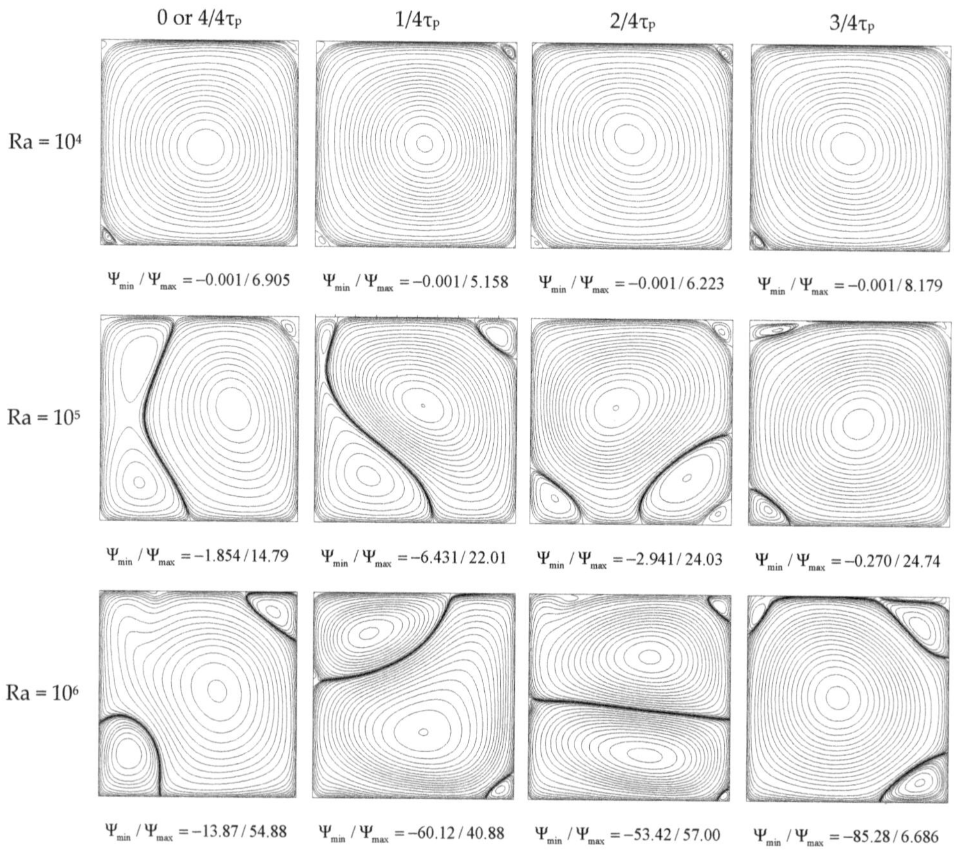

Figure 5. Evolution of streamlines "Ψ" over a period as a function of the Rayleigh number (Amp = 0.8 and $|f| = 0 \times 33\ 10^{-4}$).

Finally, for the temporal step τp, this main cell will be oriented on the right side with one other cell on the lower left side. When the amplitude increases (i.e., Amp = 0.8, Figure 5), an essentially bicellular spatial competition is observed even for Ra = 10^5 and this competition persists even for the maximum Rayleigh of 10^6.

The isotherms corresponding to the different Rayleigh number and the quarter-periods for Amp = 0.2 (Figure 6) show that the heat distribution is consistent with the circulation of the fluid revealed by the stream functions. We also note that the isothermal lines are transported by the movement of fluid. For the Rayleigh 10^3 (not presented in this figure), the isotherms are stratified for all quarter-periods. The isotherms' distortion begins around Ra = 10^4, evolving the form of a vortex. We notice that for Ra = 10^5, isotherms are concentrated on the two horizontal walls for the four beats of the period. When the Rayleigh number reaches to 10^6, we clearly see the conformity of the temperature distribution with respect to the fluid circulation, and distortion is always present for the four-quarter period.

Figure 7 provides evolution of isotherms as function of Rayleigh number and the quarter-period for Amp = 0.8. Visibly, this figure shows that the heat distribution is consistent with the circulation of the fluid revealed by the stream functions. Additionally, we note that the isothermal lines are transported by the fluid flow. For Ra = 10^3(not shown in the figure), isotherms are stratified for all quarter periods, and distortion begins around the Rayleigh value 10^4 in appearance of a vortex. We notice that for Raleigh equal to 10^5, the temperature lines are found concentrated on the two horizontal walls for the four times of the period.

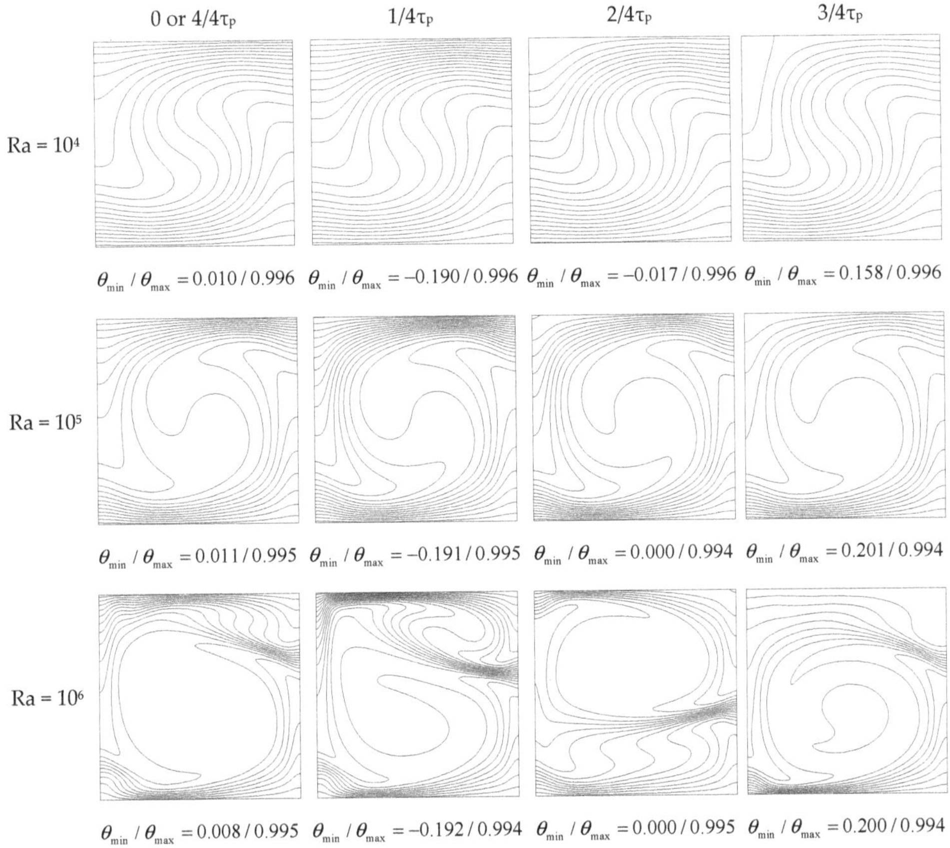

Figure 6. Evolution of isotherms over a period as a function of the Rayleigh number (Amp = 0.2 and $|f| = 0.33 \times 10^{-4}$).

When the Rayleigh reaches to 10^6, we can clearly see the conformity of the temperature distribution with respect to the circulation of the fluid, and the distortion is always present for the four quarters of a period. It can be seen, for the last quarter of a period $\tau p/2$, that the temperature lines are concentrated much more on the cold wall compared to that of the opposite side; this is due to a cooling which is more intense.

Figure 8 gives the temporal evolution of the Nusselt number at the cold wall for the last considered periods. The periodic oscillatory regime is established for most Rayleigh numbers. For high Rayleigh numbers, the cold Nusselt curves exhibit quasi-periodic shapes when the cooling modulation amplitude is small (Amp $\leq$ 0.5) and aperiodic shapes for high amplitudes (Amp > 0.5). We note that the more the amplitude increases, the more the heat transfer module intensifies. In the case of Rayleigh 10^6, the curve loses its periodicity due to the appearance of unsteady phenomena. Note that when the amplitude increases, instabilities appear even for the intermediate Rayleigh numbers and lower than Ra < 10^6. Additionally, it is apparent that for low modulation amplitudes, the curves have a quasi-symmetrical shape, whereas when the modulation increases, the physical behavior of the cavity is different between heating and cooling. It should also be noted that it is no longer the increase in the Rayleigh number that governs the growth of transfer rate.

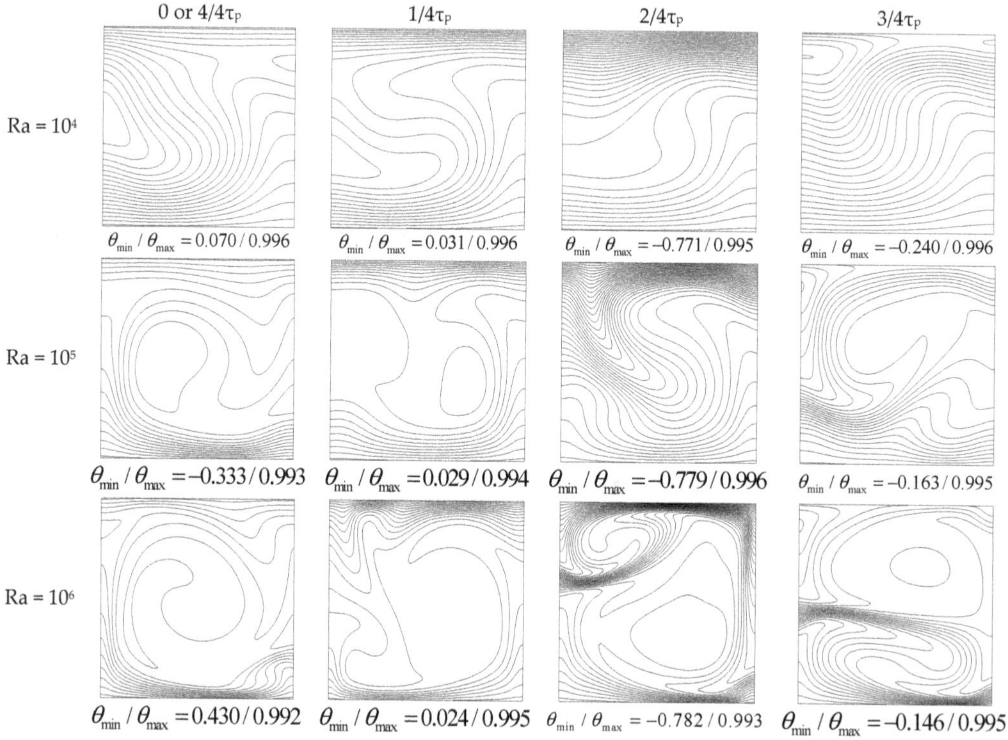

Figure 7. Evolution of isotherms over a period as a function of the Rayleigh number (Amp = 0.8 and $|f| = 0.33 \times 10^{-4}$).

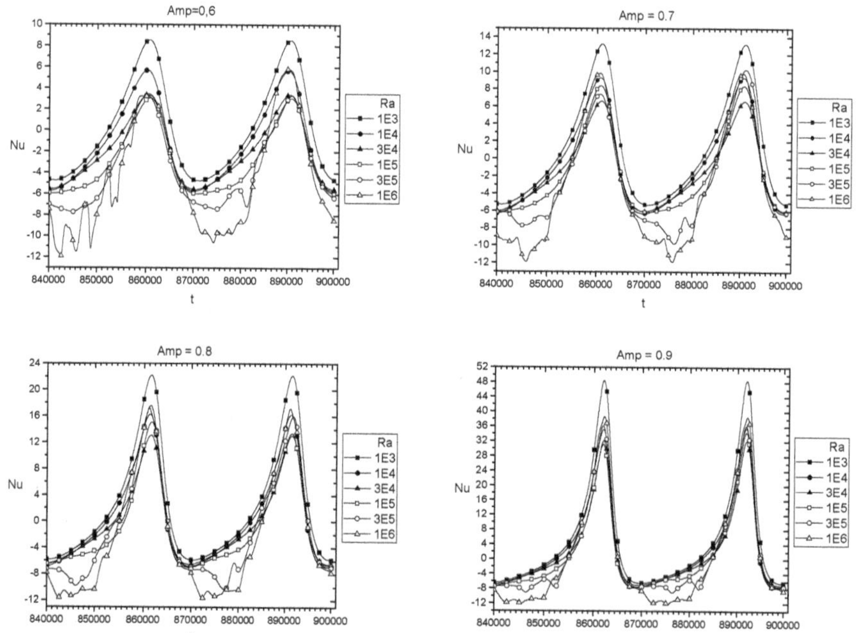

Figure 8. *Cont.*

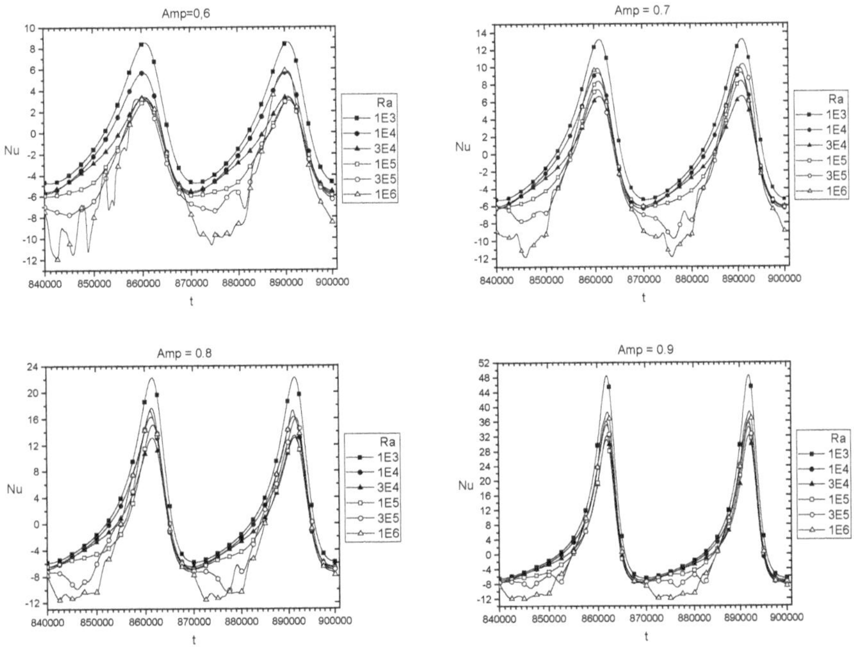

Figure 8. Evolution of the Nusselt number "Nu_c" (at the top wall) as a function of time for different Ra over the last two periods.

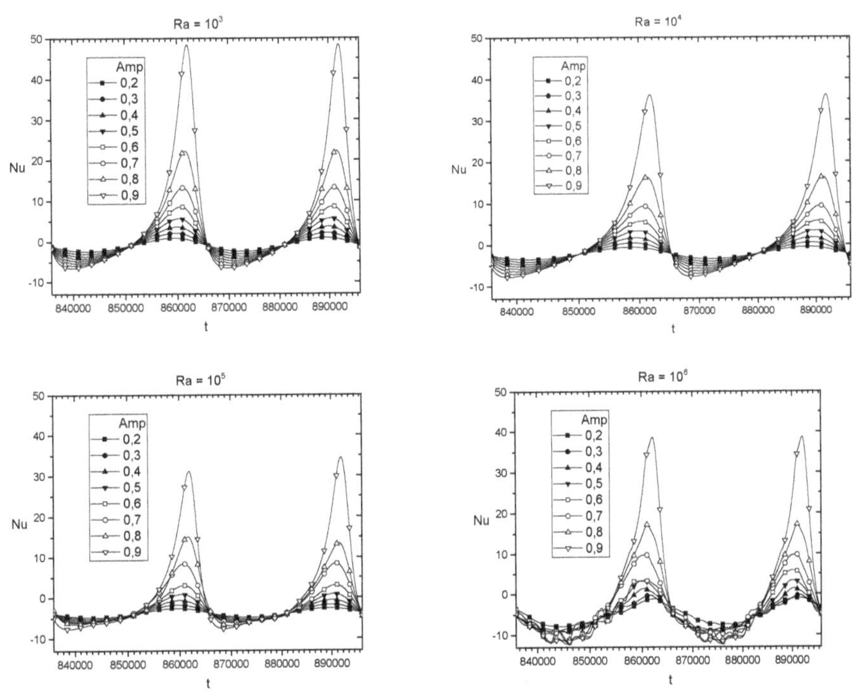

Figure 9. Evolution of the Nusselt number "Nu_c" (at the top wall) as a function of time for different Amp over the last two periods.

Figure 10 illustrates the phase portraits of the normalized Nu_c and Nu_h (cold and hot Nusselt respectively) for different Ra, and for Amp = 0.4 and 0.8, respectively. It is well known that when boundary conditions are not modulated, hot and cold Nusselt numbers must be the same to ensure energy balance. In the case where the boundary conditions are modulated, the heat transfers on the hot and cold sides follow different evolutions [41–44]. The main observation is that for a Rayleigh of 10^3, 10^4 and 10^5, the limit cycle indicates that the regime is periodic; on the other hand for the Rayleigh 10^6, the limit cycle is replaced by a cross cycle indicating the birth of natural instabilities and their addition to the pulsation imposed by the boundary condition. For Amp = 0.8, the limit and cross cycles are stretched diagonally showing that reaching the maximum value of Nu_c induces the fall of Nu_h to its minimum value and vice versa.

Amp = 0.4

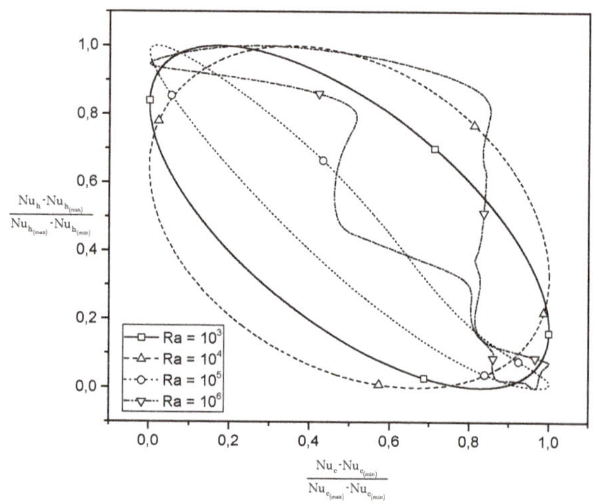

Amp = 0.8

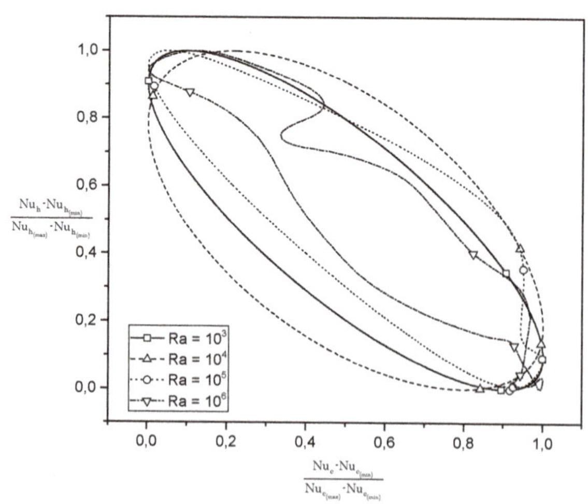

Figure 10. Phase plot for the normalized Nu_c and Nu_h for different Ra (Amp = 0.4 and 0.8).

4. Conclusions

In this study, the Lattice Boltzmann method has been used in order to investigate Rayleigh–Bénard convection in square cavity submitted to a time-periodic cooling. The Rayleigh number value considered is between 10^3 and 10^6, while the amplitude varies 0.2 to 0.8 and Prandtl value kept constant at 0.71. The flow state as well as the thermal fields depends on the values of control parameters (Ra and Amp). For the dynamic field, the obtained results show that the flow structure changes from predominantly monocellular to predominantly counter-rotating bicellular flow for Rayleigh number values Ra= 10^6 and low values of the heating amplitude. This phenomenon was obtained for lower Rayleigh values (Ra = 10^4) by increasing the value of the heating amplitude.

The analysis of the heat transfer show that for small values of amplitude heating, the averaged Nusselt curves follow a periodic evolution around an average corresponding to the formulation according to a constant cold temperature. An unsteady evolution is observed when the thermal draft increases; this unsteadiness has appeared for low Rayleigh numbers by increasing the value of the heating amplitude.

Author Contributions: Conceptualization, D.E.A.; Formal analysis, L.N., N.H., D.E.A. and R.B.; Investigation, L.N. and N.H.; Methodology, L.N. and N.H.; Supervision, R.B.; Writing—original draft, L.N.; Writing—review & editing, N.H., D.E.A. and A.B. All authors have read and agreed to the published version of the manuscript.

Funding: This research received no external funding.

Institutional Review Board Statement: Not applicable.

Informed Consent Statement: Not applicable.

Data Availability Statement: No data available.

Acknowledgments: The authors acknowledge the financial support of the General Direction of Scientific Research and Technological Development (DGRSDT), Algeria.

Conflicts of Interest: The authors declare that they have no conflict of interest.

Abbreviations

C	Lattice speed
c_s	Sound speed ($c_s = 1/\sqrt{3}$)
c_i	Micro-discrete velocities
f	Frequence
f	Distribution function of momentum equation
f^{eq}	Distribution function for equilibrium momentum equations
G	Gravitational acceleration (m/s^2)
g	Thermal distribution function
H	Convective exchange coefficient
$\overline{H}$	Correction term
J	Components of momentum
L	Cavity length (m)
m	Moments
m^{eq}	Equilibrium moments
M	Transformation matrix
N	Thermal transformation matrix
Nu	Average Nusselt number
Nu_x	Local Nusselt number
Pr	Prandtl number $Pr = \upsilon/\alpha$
r_j	Position node
Ra	Rayleigh number $Ra = \left(g\,\beta\,\Delta T_{ref} L^3\right)/(\upsilon\,\alpha)$
S	Matrix with S_j diagonal relaxation rates elements
t	Lattice time
t_k	Time at a step k
T	Dimensional temperature
U	Dimensionless horizontal velocity component
V	Dimensionless vertical velocity component
x	Dimensional longitudinal coordinate
y	Dimensional vertical coordinate

X	Dimensionless longitudinal coordinate
Y	Dimensionless vertical coordinate
Subscript	
c	Cold
h	Hot
t	Time derivative
Greek symbols	
α	Thermal diffusivity coefficient (m^2/s)
β	Coefficient of thermal expansion (K^{-1})
$\bar{\varepsilon}$	The second order energy
μ	Dynamic viscosity (kg/m/s)
v	Kinematic viscosity (m^2/s)
δ_t	Time step
Ψ	Dimensionless stream function
τ	Relaxation time
ρ	Density (Kg/m^3)
θ	Dimensionless temperature
Θ	Diagonal relaxation matrix of σ_i
Ψ	Thermal source term vector
φ	Energy flux

References

1. Fiaschi, D.; Bandinelli, R.; Conti, S. A case study for energy issues of public buildings and utilities in a small municipality: Investigation of possible improvements and integration with renewables. *Appl. Energy* **2012**, *97*, 101–114. [CrossRef]
2. Pérez-Lombard, L.; Ortiz, J.; Pout, C. A review on buildings energy consumption information. *Energy Build.* **2008**, *40*, 394–398. [CrossRef]
3. Yang, L.; Yan, H.; Lam, J.C. Thermal comfort and building energy consumption implications—A review. *Appl. Energy* **2014**, *115*, 164–173. [CrossRef]
4. Ameziani, D.; Bennacer, R.; Bouhadef, K.; Azzi, A. Effect of the days scrolling on the natural convection in an open ended storage silo. *Int. J. Therm. Sci.* **2009**, *48*, 2255–2263. [CrossRef]
5. Himrane, N.; Ameziani, D.E.; Bouhadef, K.; Bennacer, R. Storage Silos Self Ventilation: Interlinked Heat and Mass Transfer Phenomenon. *Numer. Heat Transf. Part A Appl.* **2014**, *66*, 379–401. [CrossRef]
6. Yang, R.; Wang, L. Multi-objective optimization for decision-making of energy and comfort management in building automation and control. *Sustain. Cities Soc.* **2012**, *2*, 1–7. [CrossRef]
7. Enescu, D. A review of thermal comfort models and indicators for indoor environments. *Renew. Sustain. Energy Rev.* **2017**, *79*, 1353–1379. [CrossRef]
8. Manz, H.; Schaelin, A.; Simmler, H. Airflow patterns and thermal behavior of mechanically ventilated glass double façades. *Build. Environ.* **2004**, *39*, 1023–1033. [CrossRef]
9. Dimoudi, A.; Androutsopoulos, A.; Lykoudis, S. Experimental work on a linked, dynamic and ventilated, wall component. *Energy Build.* **2004**, *36*, 443–453. [CrossRef]
10. De Dear, R.J.; Akimoto, T.; Arens, E.A.; Brager, G.; Candido, C.; Cheong, K.W.D.; Li, B.; Nishihara, N.; Sekhar, S.C.; Tanabe, S.; et al. Progress in thermal comfort research over the last twenty years. *Indoor Air* **2013**, *23*, 442–461. [CrossRef]
11. Chandrasekhar, S. *Hydrodynamic and Hydromagnetic Stability*; Oxford University Press: London, UK, 1961.
12. Drazin, P.; Reid, W.; Busse, F.H. Hydrodynamic Stability. *J. Appl. Mech.* **1982**, *49*, 467–468. [CrossRef]
13. Ouertatani, N.; Cheikh, N.B.; Beya, B.B.; Lili, T. Numerical simulation of two-dimensional Rayleigh–Bénard convection in an enclosure. *Comptes Rendus Mécanique* **2008**, *336*, 464–470. [CrossRef]
14. Turan, O.; Chakraborty, N.; Poole, R.J. Laminar Rayleigh-Bénard convection of yield stress fluids in a square enclosure. *J. Non-Newtonian Fluid Mech.* **2012**, *171*, 83–96. [CrossRef]
15. Park, H. Rayleigh–Bénard convection of nanofluids based on the pseudo-single-phase continuum model. *Int. J. Therm. Sci.* **2015**, *90*, 267–278. [CrossRef]
16. Aghighi, M.S.; Ammar, A.; Metivier, C.; Gharagozlu, M. Rayleigh-Bénard convection of Casson fluids. *Int. J. Therm. Sci.* **2018**, *127*, 79–90. [CrossRef]
17. Kebiche, Z.; Castelain, C.; Burghelea, T. Experimental investigation of the Rayleigh–Bénard convection in a yield stress fluid. *J. Non-Newton. Fluid Mech.* **2014**, *203*, 9–23. [CrossRef]
18. Chavanne, X.; Chilla, F.; Chabaud, B.; Castaing, B.; Hebral, B. Turbulent Rayleigh–Bénard convection in gaseous and liquid He. *Phys. Fluids* **2001**, *13*, 1300–1320. [CrossRef]
19. Chillà, F.; Schumacher, J. New perspectives in turbulent Rayleigh-Bénard convection. *Eur. Phys. J. E* **2012**, *35*, 58. [CrossRef]
20. Zhu, X.; Mathai, V.; Stevens, R.J.A.M.; Verzicco, R.; Lohse, D. Transition to the Ultimate Regime in Two-Dimensional Rayleigh-Bénard Convection. *Phys. Rev. Lett.* **2018**, *120*, 144502. [CrossRef]

21. Bhadauria, B.S. Time-periodic heating of Rayleigh–Benard convection in a vertical magnetic field. *Phys. Scr.* **2006**, *73*, 296–302. [CrossRef]
22. Kaviany, M. Onset of Thermal Convection in a Fluid Layer Subjected to Transient Heating From Below. *J. Heat Transf.* **1984**, *106*, 817–823. [CrossRef]
23. Kaviany, M.; Vogel, M. Effect of Solute Concentration Gradients on the Onset of Convection: Uniform and Nonuniform Initial Gradients. *J. Heat Transf.* **1986**, *108*, 776–782. [CrossRef]
24. Lage, J. Convective currents induced by periodic time-dependent vertical density gradient. *Int. J. Heat Fluid Flow* **1994**, *15*, 233–240. [CrossRef]
25. Aniss, S.; Souhar, M.; Belhaq, M. Asymptotic study of the convective parametric instability in Hele-Shaw cell. *Phys. Fluids* **2000**, *12*, 262–268. [CrossRef]
26. Bhadauria, B.S.; Bhatia, P.K. Time-periodic Heating of Rayleigh-Benard Convection. *Phys. Scr.* **2002**, *66*, 59–65. [CrossRef]
27. Umavathi, J.C. Rayleigh–Benard convection subject to time dependent wall temperature in a porous medium layer saturated by a nanofluid. *Meccanica* **2015**, *50*, 981–994. [CrossRef]
28. Himrane, N.; Ameziani, D.E.; Nasseri, L. Study of thermal comfort: Numerical simulation in a closed cavity using the lattice Boltzmann method. *SN Appl. Sci.* **2020**, *2*, 1–7. [CrossRef]
29. Abourida, B.; Hasnaoui, M.; Douamna, S. Transient Natural Convection in a Square Enclosure with Horizontal Walls Submitted to Periodic Temperatures. *Numer. Heat Transf. Part A Appl.* **1999**, *36*, 737–750. [CrossRef]
30. Raji, A.; Hasnaoui, M.; Firdaouss, M.; Ouardi, C. Natural Convection Heat Transfer Enhancement in a Square Cavity Periodically Cooled from Above. *Numer. Heat Transf. Part A Appl.* **2013**, *63*, 511–533. [CrossRef]
31. Eidel'Man, E.D. Excitation of an electric instability by heating. *Physics-Uspekhi* **1995**, *38*, 1231–1246. [CrossRef]
32. Éidel'Man, E.D. Influence of the thickness of the liquid layer on the ratio of the dimensions of a convection cell. *Tech. Phys.* **1998**, *43*, 1275–1279. [CrossRef]
33. Nevskii, S.; Sarychev, V.; Konovalov, S.; Granovskii, A.; Gromov, V. Formation Mechanism of Micro- and Nanocrystalline Surface Layers in Titanium and Aluminum Alloys in Electron Beam Irradiation. *Metals* **2020**, *10*, 1399. [CrossRef]
34. Mohamad, A. *Lattice Boltzmann Method*; Springer: London, UK, 2011; Volume 70.
35. Guo, Y.; Bennacer, R.; Shen, S.; Ameziani, D.; Bouzidi, M. Simulation of mixed convection in slender rectangular cavity with lattice Boltzmann method. *Int. J. Numer. Methods Heat Fluid Flow* **2010**, *20*, 130–148. [CrossRef]
36. Khali, S.; Nebbali, R.; Ameziani, D.E.; Bouhadef, K. Numerical investigation of non-Newtonian fluids in annular ducts with finite aspect ratio using lattice Boltzmann method. *Phys. Rev. E* **2013**, *87*, 053002. [CrossRef] [PubMed]
37. D'Humieres, D. Generalized Lattice-Boltzmann Equations. In Proceedings of the 18th International Symposium, Rarefied Gas Dynamics, Vancouver, BC, Canada, 26–30 July 1994; pp. 450–458.
38. Lallemand, P.; Luo, L.-S. Theory of the lattice Boltzmann method: Dispersion, dissipation, isotropy, Galilean invariance, and stability. *Phys. Rev. E* **2000**, *61*, 6546–6562. [CrossRef]
39. Wang, G.; Meng, X.; Zeng, M.; Ozoe, H.; Wang, Q.W. Natural Convection Heat Transfer of Copper–Water Nanofluid in a Square Cavity With Time-Periodic Boundary Temperature. *Heat Transf. Eng.* **2013**, *35*, 630–640. [CrossRef]
40. Bouabdallah, S.; Ghernaout, B.; Teggar, M.; Benchatti, A.; Benarab, F.-Z. Onset of Natural Convection and Transition Laminar-Oscillatory Convection Flow in Rayleigh-Bénard Configuration. *Int. J. Heat Technol.* **2016**, *34*, 151–157. [CrossRef]
41. Kwak, H.S.; Hyun, J.M. Natural convection in an enclosure having a vertical sidewall with time-varying temperature. *J. Fluid Mech.* **1996**, *329*, 65–88. [CrossRef]
42. Soong, C.; Tzeng, P.; Hsieh, C. Numerical study of bottom-wall temperature modulation effects on thermal instability and oscillatory cellular convection in a rectangular enclosure. *Int. J. Heat Mass Transf.* **2001**, *44*, 3855–3868. [CrossRef]
43. Kazmierczak, M.; Chinoda, Z. Buoyancy-driven flow in an enclosure with time periodic boundary conditions. *Int. J. Heat Mass Transf.* **1992**, *35*, 1507–1518. [CrossRef]
44. Wang, Q.-W.; Wang, G.; Zeng, M.; Ozoe, H. Upward Heat Flux through the Horizontal Fluid Layer of Water with Sinusoidal Wall Temperature at the Top or Bottom Boundary. *Numer. Heat Transf. Part A Appl.* **2007**, *52*, 817–829. [CrossRef]

fluids

MDPI

Article

Effects of Shell Thickness on Cross-Helicity Generation in Convection-Driven Spherical Dynamos

Luis Silva, **Parag Gupta**, **David MacTaggart** and **Radostin D. Simitev** *

School of Mathematics and Statistics, University of Glasgow, Glasgow G12 8QQ, UK; lacsilva@gmail.com (L.S.); p.gupta.1@research.gla.ac.uk (P.G.); david.mactaggart@glasgow.ac.uk (D.M.)

* Correspondence: Radostin.Simitev@glasgow.ac.uk

Received: 6 November 2020; Accepted: 12 December 2020; Published: 16 December 2020

check for updates

Abstract: The relative importance of the helicity and cross-helicity electromotive dynamo effects for self-sustained magnetic field generation by chaotic thermal convection in rotating spherical shells is investigated as a function of shell thickness. Two distinct branches of dynamo solutions are found to coexist in direct numerical simulations for shell aspect ratios between 0.25 and 0.6—a mean-field dipolar regime and a fluctuating dipolar regime. The properties characterising the coexisting dynamo attractors are compared and contrasted, including differences in temporal behaviour and spatial structures of both magnetic fields and rotating thermal convection. The helicity α-effect and the cross-helicity γ-effect are found to be comparable in intensity within the fluctuating dipolar dynamo regime, where their ratio does not vary significantly with the shell thickness. In contrast, within the mean-field dipolar dynamo regime the helicity α-effect dominates by approximately two orders of magnitude and becomes stronger with decreasing shell thickness.

Keywords: rotating thermal convection; convection-driven dynamos; numerical simulations; bistability; mean-field magnetohydrodynamics; spherical shells

1. Introduction

Thermal flows give rise to some of the most characteristic large-scale features of cosmic objects—their self-sustained magnetic fields [1,2]. For instance, the Sun and several of the planets in the Solar System display substantial magnetic fields [3,4]. The solar magnetic field drives solar activity and strongly affects planetary atmospheres [5,6]. Earth's field shields life from solar radiation [7]. Farther out, the gas giants, the ice giants, and the Jovian moons all have significant magnetic fields [8]. These fields are sustained by dynamo processes in the interiors or the atmospheres of their celestial hosts where vigorous convective motions of electrically conductive fluids generate large-scale electric currents [9–11]. The convective flows are driven primarily by thermal buoyancy forces due to thermonuclear fusion in stellar interiors and secular cooling in planetary interiors, respectively. Thermal convection in celestial bodies is highly turbulent in nature and, at the same time, strongly influenced both by rotation and by the self-generated magnetic fields. Considerable attention has therefore been devoted to this fascinating and important subject, and for topical reviews we refer to the papers by Busse and Simitev [12], Jones [13], Wicht and Sanchez [14] and references within.

Conceptually, dynamo generation of large-scale magnetic fields is understood on the basis of mean-field dynamo theory [15–17], a well-established theory of magnetohydrodynamic turbulence. A cornerstone of the theory is the turbulence modelling of the mean electromotive force—the sole source term arising in the Reynolds-averaged magnetic induction equation governing the evolution of the large-scale field, see Section 3.5 further below. The electromotive force is usually approximated by an expansion in terms of the mean field and its spatial derivatives where the expansion coefficients are known informally as "mean-field effects". The turbulent helicity effect, also called α-effect (in this work, when we refer to "helicity" without further qualification, we intend the helicity associated with the α-effect—this shorthand should not be confused with other helicities, such as "magnetic helicity"), has been studied extensively in the research literature on mean-field dynamo theory, for example, see [16,18] and references therein. In contrast, the cross-helicity effect, also known as γ-effect [19], has been a subject to a rather small number of studies, for example, [20,21] and works cited therein. This is due to the currently prevailing treatment of turbulence where large-scale velocity is neglected because of the Galilean invariance of the momentum equation. However, such treatment leads to the neglect of the large-scale shear effects, which are, in fact, significant. For example, large-scale rotation is ubiquitous in astro/geophysical objects, for example, the Solar internal differential rotation is substantial and well measured [22,23] while numerical simulations suggest it is an essential ingredient of the dynamo process and likely to be responsible for the regular oscillations of convection-driven spherical dynamos [24,25]. Similarly, a number of studies of plane-parallel flows confirm that cross-helicity effects are not small compared to helicity effects [26,27]. Apart from its role in dynamo generation, cross-helicity is an important Solar observable. For instance, measurements of the cross-helicity component $\langle u_z b_z \rangle$ at the Solar surface are available from the Swedish 1-m Solar Telescope and can be used to calculate the magnetic eddy diffusivity of the quiet Sun by quasilinear mean-field theory [28].

Cross-helicity has not been explored in models of self-consistent dynamos driven by thermal convection in rotating spherical shells and this paper aims to contribute in this direction. The main goal of this work is to investigate the relative importance of the helicity and cross-helicity effects as a function of the thickness of the convective shell. Intuitive arguments suggest that the α-effect is important in the case of the geodynamo and the cross-helicity effect is important in the case of the global solar dynamo. Indeed, the geodynamo operates in the relatively thick fluid outer code of the Earth where large-scale columnar structures are believed to develop. The coherent columnar structures are characterised by relatively large-scale vorticity and generate a strong helicity α-effect. In contrast, the global solar dynamo operates in the thinner solar convection zone where columnar structures are thought difficult to maintain and so vorticity may have a less regular structure, thus increasing the relative importance of the cross-helicity effect. To assess this hypothesis, we present a set of dynamo simulations that differ mainly in their shell thickness aspect ratio $\eta = r_i/r_o$, see Figure 1, while other governing parameters are kept fixed. Along with estimates of the relative strength of the helicity and cross-helicity effects, we report on the mechanisms of electromotive force generation and its spatial distribution. Variation of shell thickness is also relevant to the case of the geodynamo as the inner core did not exist at the time of formation of the Earth, but nucleated sometime later in the geological history of the planet and continues to grow in size.

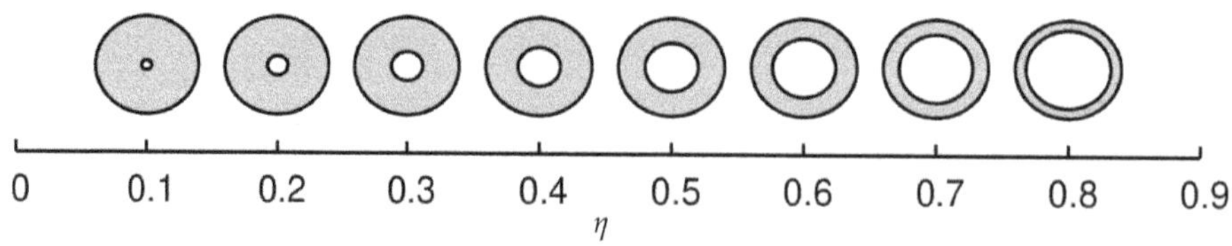

Figure 1. Illustration of shell thickness aspect ratio variation.

The geodynamo and the solar global dynamo are also different in that the former has a dominant and rarely reversing dipole, while the latter exhibits a regular periodic cycle. To capture this essential difference while comparing quid pro quo, we have performed this study at parameter values where two distinct dynamo branches are known to coexist [29–31]. These branches have rather different magnetic field properties, in particular one branch is non-reversing while the other branch is cyclic, and also display significant differences in zonal flow intensity and profile. It is reasonable to expect that the two branches will offer different mechanisms of helicity and cross-helicity generation and thus in this paper we proceed to study both branches. Bistability, in itself, may play a role in aperiodic magnetic field polarity reversals, a notable feature of the geodynamo [32], as well as in the regular cycle of the solar dynamo [33]. We have previously investigated the hysteretic transitions between the coexisting dynamo branches with variation of the Rayleigh, Prandtl and Coriolis numbers (defined further below). In addition, in this paper we demonstrate for the first time that the distinct dynamo branches coexist also when the shell thickness η is varied. The discussion of this dichotomous behaviour runs as a secondary theme of the article.

The paper is structured as follows. Details of the mathematical model and the numerical methods for solution are given in Section 2. In Section 3, we describe the set of dynamo simulations performed in the context of this work. We pay particular attention to the description of the two coexisting dynamo branches, which are studied for the first time here as a function of the thickness of the convective shell. In the process, we describe the typical morphology and time dependent behaviour of thermal convection flows. In Section 3.5, we briefly summarise the mean field arguments related to the helicity and cross-helicity mechanisms for the generation of a large-scale magnetic field. In Section 3.6, the cross-helicity properties of our dynamo solutions and the relative contributions of the α-and γ-effects are assessed. Section 4 is devoted to concluding remarks.

2. Materials and Methods

This section describes a standard mathematical formulation of the problem of thermal convection and magnetic field generation in rotating spherical fluid shells. A set of transformations used to recast the problem in a scalar stream-function form and a pseudo-spectral algorithm used for the numerical solution of the equations is presented. The exposition in this section is standard and follows our previous articles, for example, [34,35]. This section also serves as an introduction and a review of the typical approach to the formulation and solution of this important problem.

2.1. Mathematical Formulation

We consider a spherical shell full of electrically conducting fluid as shown in Figure 2. The shell rotates with a constant angular velocity $\boldsymbol{\Omega}$ about the vertical coordinate axis. We assume that a static state exists with the temperature distribution

$$T_S = T_0 - \beta d^2 r^2/2, \tag{1a}$$

$$\beta = q/(3\kappa c_p), \tag{1b}$$

$$T_0 = T_1 - (T_2 - T_1)/(1-\eta). \tag{1c}$$

The evolution of the system is governed by the equations of momentum, heat and magnetic induction, along with solenoidality conditions for the velocity and magnetic fields,

$$\nabla \cdot \mathbf{u} = 0, \tag{2a}$$

$$(\partial_t + \mathbf{u}\cdot\nabla)\mathbf{u} = -\nabla\pi - \tau\mathbf{k}\times\mathbf{u} + \Theta\mathbf{r} + \nabla^2\mathbf{u} + \mathbf{B}\cdot\nabla\mathbf{B}, \tag{2b}$$

$$P(\partial_t + \mathbf{u}\cdot\nabla)\Theta = R\mathbf{r}\cdot\mathbf{u} + \nabla^2\Theta, \tag{2c}$$

$$\nabla\cdot\mathbf{B} = 0, \tag{2d}$$

$$P_m(\partial_t + \mathbf{u}\cdot\nabla)\mathbf{B} = P_m\mathbf{B}\cdot\nabla\mathbf{u} + \nabla^2\mathbf{B}, \tag{2e}$$

written for the perturbations from the static reference state and with notations defined in Table 1. In this formulation, the Boussinesq approximation is used with the density ϱ having a constant value ϱ_0 except in the gravity term where

$$\varrho = \varrho_0(1-\alpha\Theta), \tag{3}$$

and α is the specific thermal expansion coefficient $\alpha \equiv -(\mathrm{d}\varrho/\mathrm{d}T)/\varrho = \text{const}$. With the units of Table 2, five dimensionless parameters appear in the governing equations, namely the shell radius ratio η, the Rayleigh number R, the Coriolis number τ, the Prandtl number P and the magnetic Prandtl number P_m defined by

$$\eta = \frac{r_i}{r_o}, \ R = \frac{\alpha\gamma\beta d^6}{\nu\kappa}, \ \tau = \frac{2\Omega d^2}{\nu}, \ P = \frac{\nu}{\kappa}, \ P_m = \frac{\nu}{\lambda}, \tag{4}$$

where λ is the magnetic diffusivity. Since the velocity $\mathbf{u}$ and the magnetic flux density $\mathbf{B}$ are solenoidal vector fields, the general representation in terms of poloidal and toroidal components is used

$$\mathbf{u} = \nabla \times (\nabla v \times \mathbf{r}) + \nabla w \times \mathbf{r}, \tag{5a}$$
$$\mathbf{B} = \nabla \times (\nabla h \times \mathbf{r}) + \nabla g \times \mathbf{r}. \tag{5b}$$

Taking $\mathbf{r} \cdot \nabla\times$ and $\mathbf{r} \cdot \nabla \times \nabla\times$ of the momentum Equation (2b), two equations for w and v are obtained

$$[(\nabla^2 - \partial_t)\mathcal{L}_2 + \tau\partial_\varphi]w - \tau\mathcal{Q}v = \mathbf{r} \cdot \nabla \times (\mathbf{u} \cdot \nabla\mathbf{u} - \mathbf{B} \cdot \nabla\mathbf{B}), \tag{6a}$$
$$[(\nabla^2 - \partial_t)\mathcal{L}_2 + \tau\partial_\varphi]\nabla^2 v + \tau\mathcal{Q}w - \mathcal{L}_2\Theta = -\mathbf{r} \cdot \nabla \times [\nabla \times (\mathbf{u} \cdot \nabla\mathbf{u} - \mathbf{B} \cdot \nabla\mathbf{B})], \tag{6b}$$

where ∂_φ denotes the partial derivative with respect to the angle φ of a spherical system of coordinates (r, θ, φ) and where the operators $\mathcal{L}_2$ and $\mathcal{Q}$ are defined as

$$\mathcal{L}_2 \equiv -r^2\nabla^2 + \partial_r(r^2\partial_r),$$
$$\mathcal{Q} \equiv r\cos\theta\nabla^2 - (\mathcal{L}_2 + r\partial_r)(\cos\theta\partial_r - r^{-1}\sin\theta\partial_\theta).$$

The heat equation for the dimensionless deviation Θ from the static temperature distribution can be written in the form

$$\nabla^2\Theta + R\mathcal{L}_2 v = P(\partial_t + \mathbf{u} \cdot \nabla)\Theta, \tag{6c}$$

and the equations for h and g are obtained by taking $\mathbf{r}\cdot$ and $\mathbf{r} \cdot \nabla\times$ of the dynamo Equation (2e)

$$\nabla^2\mathcal{L}_2 h = P_m[\partial_t\mathcal{L}_2 h - \mathbf{r} \cdot \nabla \times (\mathbf{u} \times \mathbf{B})], \tag{6d}$$
$$\nabla^2\mathcal{L}_2 g = P_m[\partial_t\mathcal{L}_2 g - \mathbf{r} \cdot \nabla \times (\nabla \times (\mathbf{u} \times \mathbf{B}))]. \tag{6e}$$

For the flow we assume stress-free boundaries with fixed temperatures

$$v = \partial^2_{rr}v = \partial_r(w/r) = \Theta = 0 \quad \text{at } r = r_i \text{ and } r = r_o. \tag{7a}$$

For the magnetic field we assume electrically insulating boundaries such that the poloidal function h must be matched to the function $h^{(e)}$, which describes the potential fields outside the fluid shell

$$g = h - h^{(e)} = \partial_r(h - h^{(e)}) = 0 \quad \text{at } r = r_i \text{ and } r = r_o. \tag{7b}$$

This is a standard formulation of the spherical convection-driven dynamo problem [13,36–38] for which an extensive collection of results already exists [24,34,39,40]. The results reported below are not strongly model dependent as confirmed by simulations of convection driven by differential heating [41], for cases with no-slip conditions at the inner boundary and an electrical conductivity of the exterior equal to that of the fluid [25,42], and for thermo-compositional driving [35]. Thus, aiming to retain a general physical perspective, we intentionally use here a generic model formulation with a minimal number of physical parameters including only those of first-order importance for stellar and planetary applications.

Table 1. Notation used in Section 2.1, where not defined in the main text.

Notation	Quantity	Notation	Quantity
(r, θ, φ)	Spherical polar coordinates	T_S	Background temperature distribution
t	Time	T_1, T_2	Temperature inner, outer boundary
$\mathbf{r}$	Position vector wrt centre of sphere	q	Density of uniformly distributed heat sources
d	Thickness of the spherical shell	κ	Thermal diffusivity
r_i, r_o	Inner and outer radii of the shell	ν	Kinematic viscosity
$\mathbf{u}$	Velocity field perturbation	μ	Magnetic permeability
$\mathbf{B}$	Magnetic flux density perturbation	c_p	Specific heat at constant pressure
Θ	Temperature perturbation from the background state	γ	Gravitational acceleration magnitude
π	Effective pressure	∂	Partial derivative notation

Table 2. Units of non-dimensionalisation.

Quantity	Unit
Length	d
Time	d^2/ν
Temperature	$\nu^2/\gamma\alpha d^4$
Magnetic flux density	$\nu(\mu\varrho)^{1/2}/d$

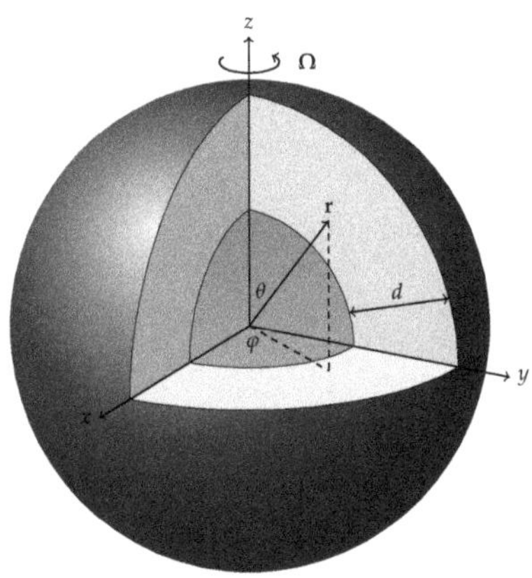

Figure 2. Schematic illustration of the three-dimensional region considered in this study, the associated spherical coordinate system and the position of the axis of rotation. The region is assumed full of electrically conducting fluid.

2.2. Numerical Methods

For the direct numerical integration of the convection-driven dynamo problem specified by the scalar Equation (6) and the boundary conditions (7) we use a pseudo-spectral method described by [43]. The code has been benchmarked for accuracy, most recently in [44,45], and has been made open source [46]. All dependent variables in the code are spatially discretised by means of spherical harmonics Y_l^m and Chebychev polynomials T_n, for example,

$$v(r,\theta,\varphi) = \sum_{l,m,n}^{N_l,N_m,N_n} V_{l,n}^m(t) T_n\big(2(r-r_i)-1\big) Y_l^m(\theta,\varphi), \tag{8}$$

and similarly for the other unknown scalars, w, h, g and Θ. The nonlinear terms in the equations are computed in physical space and then projected onto spectral space at every time step. Time integration makes use of an IMEX combination of the Crank–Nicolson scheme for the diffusion terms and the Adams–Bashforth scheme for the nonlinear terms with both schemes of second order accuracy.

When the spectral powers of the kinetic and magnetic energies drop by more than three orders of magnitude from the spectral maximum to the cut-off wavelength, we consider the simulations to be reasonably resolved [47]. In all the cases reported here, a minimum of 41 collocation points in the radial direction has been considered, together with spherical harmonics up to order 96. These numbers provide sufficient resolution, as demonstrated in Figure 7 for two typical dynamo solutions.

2.3. Diagnostics

It is convenient to characterise the non-magnetic convection and the convection-driven dynamo solutions using their energy densities. To understand the interactions between various components of the flow, we decompose the kinetic energy density into mean poloidal, mean toroidal, fluctuating poloidal and fluctuating toroidal parts as follows

$$\overline{E}_p = \frac{1}{2}\langle \mid \nabla \times (\nabla \overline{v} \times \mathbf{r}) \mid^2 \rangle, \quad \overline{E}_t = \frac{1}{2}\langle \mid \nabla \overline{w} \times \mathbf{r} \mid^2 \rangle, \tag{9a}$$

$$\widetilde{E}_p = \frac{1}{2}\langle \mid \nabla \times (\nabla \widetilde{v} \times \mathbf{r}) \mid^2 \rangle, \quad \widetilde{E}_t = \frac{1}{2}\langle \mid \nabla \widetilde{w} \times \mathbf{r} \mid^2 \rangle, \tag{9b}$$

where $\langle \cdot \rangle$ indicates the average over the fluid shell and time as described in Section 3.5 and $\overline{v}$ refers to the axisymmetric component of the poloidal scalar field v, while $\widetilde{v}$ is defined as $\widetilde{v} = v - \overline{v}$. The corresponding magnetic energy densities $\overline{M}_p$, $\overline{M}_t$, $\widetilde{M}_p$ and $\widetilde{M}_t$ are defined analogously with the scalar fields h and g for the magnetic field replacing v and w.

To assess the predominant configuration of the magnetic field, we define the dipolarity ratio

$$\mathcal{D} = \overline{M}_p / \widetilde{M}_p. \tag{10}$$

When $\overline{M}_p > \widetilde{M}_p$ then $\mathcal{D} > 1$ and the corresponding solutions will be referred to as "Mean Dipolar", for reasons to be explained below, and denoted by **MD** following [29]. When $\overline{M}_p < \widetilde{M}_p$ then $\mathcal{D} < 1$ and the corresponding solutions will be referred to as "Fluctuating Dipolar" and denoted by **FD**.

To quantify heat transport by convection the Nusselt numbers at the inner and outer spherical boundaries Nu_i and Nu_o are used. These are defined by

$$Nu_i = 1 - \frac{P}{r_i R} \frac{d\overline{\overline{\Theta}}}{dr}\bigg|_{r=r_i}, \qquad Nu_o = 1 - \frac{P}{r_o R} \frac{d\overline{\overline{\Theta}}}{dr}\bigg|_{r=r_o}, \tag{11}$$

where the double bar indicates the average over the spherical surface.

Other quantities are defined in the text as required.

3. Results

3.1. Parameter Values Used

In order to investigate the effects of the shell thickness on the properties of non-magnetic convection and on dynamo solutions we perform a suite of numerical simulations varying the shell aspect ratio between $\eta = 0.1$ and $\eta = 0.7$. To compare the simulations on an equal footing, as well as to keep the number of runs required to a manageable level, all parameters except those depending on the aspect ratio are kept at fixed values. The value of the Prandtl number is set to $P = 0.75$ allowing us to use a relatively low value of the magnetic Prandtl number $P_m = 1.5$ as appropriate for natural dynamos. The Coriolis number is fixed to $\tau = 2 \times 10^4$ representing a compromise between the fast rotation rate appropriate for the geodynamo and the relatively slow rotation rate appropriate for the solar dynamo. To ensure that dynamos are driven equally strongly, we fix the value of the Rayleigh number at 3.8 times the critical value R_c for the onset of convection for each shell thickness aspect ratio as shown in Figure 3 below. The required values of the critical Rayleigh number are determined as explained in the next section where we also discuss general features of the onset of thermal convection.

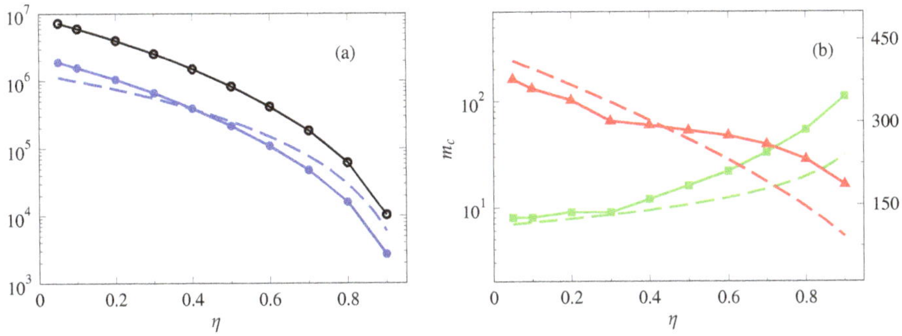

Figure 3. Critical parameter values for the onset of convection and values of the Rayleigh number used in this work as a function of the shell thickness aspect ratio η in the case $P = 0.75$, and $\tau = 2 \times 10^4$. (**a**) The critical Rayleigh number R_c for the linear onset of convection is plotted in solid blue curve marked by full circles. The values used in the simulations are given by $R = 3.8R_c$; they are plotted in solid black curve marked by empty circles. (**b**) The critical wave number m_c (left y-axis) and the critical frequency ω_c (right y-axis) for the onset of convection are denoted by green squares and red triangles, respectively. Local asymptotic approximations (12) are shown by correspondingly coloured dashed curves in all panels. (Colour online).

3.2. Linear Onset of Thermal Convection

The onset of thermal convection in rapidly rotating spherical shells has been extensively studied, for example, most recently as a special case of the onset of thermo-compositional convection [48]. In general, two major regimes are found at onset–columnar convection and equatorially-attached convection. The equatorially-attached regime occurs at small values of the Prandtl number P and consists of flows that take the form of non-spiralling rolls trapped near the equator with a relatively large azimuthal length scale. This regime can be understood as a form of inertial oscillations, for example, [49]. The columnar regime is realised at moderate and large values of P and features elongated rolls parallel to axis of rotation that are spiralling strongly and have a relatively short azimuthal length scale. At the selected values of the Prandtl and the Coriolis numbers, the simulations reported in this study belong to the columnar regime of rapidly rotating convection.

To determine accurate values for the critical parameters at onset we use our open source numerical code [50]. The code implements a Galerkin spectral projection method due to Zhang and Busse [51] to solve the linearised versions of Equations (6a) and (6c). The method leads to a generalised eigenvalue problem for the critical Rayleigh number R_c and frequency ω_c of the most unstable mode of thermal convection at specified other parameter values and at specified azimuthal wave number m of the convective perturbation. Numerical extremisation and continuation problems then are tackled in order to follow the marginal stability curve in the parameter space as detailed in [48]. The critical values thus obtained are shown in Figure 3. The critical Rayleigh number R_c and drift frequency ω_c decrease with decreasing shell thickness while the critical azimuthal wave number m_c increases.

It is interesting to compare and validate these results against theoretical results for the onset convection in rapidly rotating systems. The asymptotic analysis of this problem has a long and distinguished history of local and global linear stability analysis [52–56], see also [48] for a brief overview. Converting results of Yano [57] to our dimensionless parameters, length and time scales, we obtain

$$R_c = 7.252 \left(\frac{P\tau}{1+P} \right)^{4/3} (1-\eta)^{7/3}, \tag{12a}$$

$$m_c = 0.328 \left(\frac{P\tau}{1+P} \right)^{1/3} (1-\eta)^{-2/3}, \tag{12b}$$

$$\omega_c = 0.762 \left(\frac{\tau^2}{P(1+P)^2} \right)^{1/3} (1-\eta)^{2/3}, \tag{12c}$$

for the critical parameters of viscous columnar convection in an internally heated spherical shell. While expressions (12) are not strictly valid asymptotic results for the spherical shell configuration studied here, they provide a reasonable agreement with the numerical results plotted in Figure 3.

3.3. *Finite-Amplitude Convection and Dynamo Features*

As the value of the Rayleigh number is increased away from the onset, rotating columnar convection undergoes a sequence of transitions from steady flow patterns drifting with constant angular velocity to increasingly chaotic states as described in detail in [40]. When the amplitude of convection becomes sufficiently large so that the magnetic Reynolds number defined as $Rm = Pm\sqrt{2E}$ reaches values of the order 10^2, onset of dynamo action is typically observed [34]. Three examples of dynamo solutions are shown in Figure 4 to (i) illustrate typical spatial features of chaotic thermal convection in rotating shells and the associated magnetic field morphology and (ii) to reveal how these features vary with decreasing shell thickness. Outside of the tangent cylinder the flow consists of pairs of adjacent spiralling convection columns as seen in the second row of Figure 4. Within the columns the fluid particles travel in clockwise and anticlockwise directions parallel to the equatorial plane and up towards the poles or down towards the equatorial plane as columns extend through the height of the convective shell. In agreement with the linear analysis, as the shell thickness is decreased the azimuthal wave number rapidly increases with the thin shell solution $\eta = 0.7$ showing a cartridge of fine scale columns closely adjacent to each other and exhibiting much weaker spiralling and slower drift than in the thick shell cases. These convective patterns strongly influence the structure and the morphology of magnetic fields as illustrated by the first row of Figure 4 where magnetic fieldlines of the three dynamo solutions are shown. The fieldlines are intricately knotted and exhibit a rather complicated structure within the convective domain in all three cases. The imprint of the convective columns is visible in the thick shell cases $\eta = 0.2$ and $\eta = 0.4$ where the magnetic fieldlines are coiled around the convective columnar structures indicating the presence of toroidal field and poloidal field feedback and amplification processes. Outside of the convective domain, the magnetic field of the thickest shell case $\eta = 0.2$ is well organised and emerges from the polar regions of the domain in the form of big bundles of opposite polarities with fieldlines proceeding to close and forming extensive overarching loops that are characteristic of a strong dipolar field symmetry. A similar picture is seen in the mid-thickness case $\eta = 0.4$ although in this case there appear to be several magnetic "poles" where strong bundles of vertical fieldlines emerge at the surface of the spherical domain. In the thin shell case $\eta = 0.7$ the magnetic field is much less organised with numerous fieldline coils inside the convective domain and barely visible but still dominant dipolar structure outside.

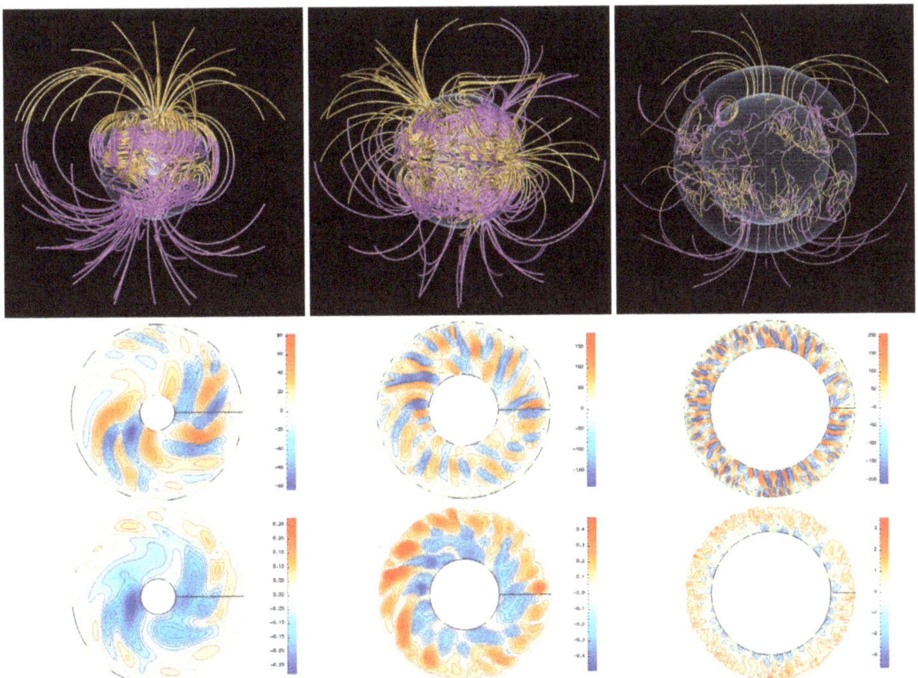

Figure 4. Snapshots of spatial structures of dynamo solutions with increasing shell thickness aspect ratio η and with $R = 3.8 \times R_c$, $\tau = 2 \times 10^4$, $P = 0.75$ and $P_m = 1.5$. Three cases are shown as follows: $\eta = 0.2$, R =4,000,000 (left column); $\eta = 0.4$, R =1,500,000 (middle column); and $\eta = 0.7$, R =180,000 (right column). Magnetic poloidal fieldlines are plotted in the top row, contours of the radial velocity u_r in the equatorial plane are plotted in the middle row, and contours of the temperature perturbation Θ in the equatorial plane are plotted in the bottom row. (Colour online).

While typical, the spatial structures described in relation to Figure 4 are only snapshots of the three dynamo solutions at fixed moments in time. An illustration of the temporal behaviour exhibited in our dynamo simulations is shown in Figure 5. The main magnetic and kinetic energy density components of two distinct dynamo cases are plotted as functions of time, and the chaotic nature of the solutions is clearly visible. The time dependence of the time series consists of continual oscillations around the mean values of the respective densities with periods much shorter than the viscous diffusion time. Kinetic energy densities are displayed in the second row of the figure and show that the fluctuating components of motion dominate the flow with the fluctuating toroidal velocity being the strongest. The mean poloidal component of motion is negligible in both cases in agreement with the constraint of the Proudman–Taylor theorem on motions parallel to the axis of rotation. The mean toroidal component, representing differential rotation, appears to be weak in both cases plotted in Figure 5 more so in the case to the left marked **MD** for reasons we will discuss further below. The differential rotation, however, is known to be the component most strongly impaired in the presence of the magnetic field [34]. This leads us to a discussion of the features of the magnetic energy densities plotted in the first row of Figure 5. Here, the differences between the two cases illustrated are rather more pronounced. The total magnetic energy density of the case in Figure 5a is approximately six times larger than that in Figure 5d. More significant is the essential qualitative difference in the balance of magnetic energy components. The axisymmetric poloidal component $\overline{M}_p$ is dominant in the case shown in Figure 5a while it has a relatively small contribution in the case of Figure 5d. The axial dipole coefficient H_1^0 and the axial quadrupole coefficient H_2^0 in Figure 5c,f reveal that this difference is due to the fact that the case to the left is dominated by a strong dipole and the case to the right is less strongly dipolar and the time series suggests the presence of magnetic field oscillations.

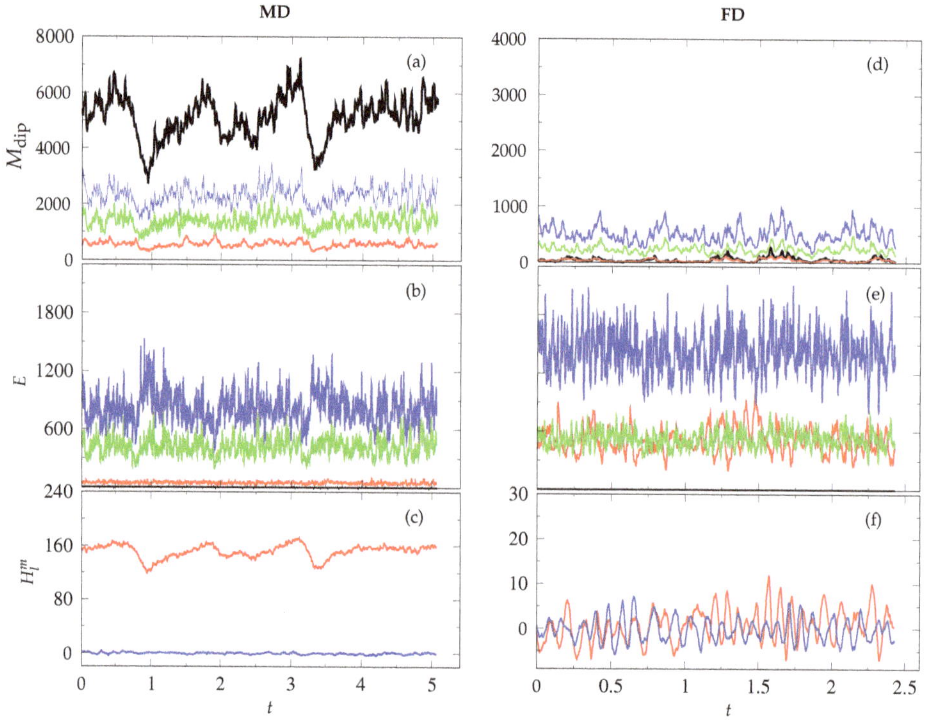

Figure 5. Chaotic dynamo attractors at identical parameter values—a Mean Dipolar (MD) dynamo (left column (**a–c**)) and a Fluctuating Dipolar (FD) dynamo (right column (**d–f**)) both at $\eta = 0.5$, $R = 8.2 \times 10^5$, $\tau = 2 \times 10^4$, $P = 0.75$ and $P_m = 1.5$. Panels (**a**,**d**) show time series of magnetic dipolar energy densities and panels (**b**,**e**) show kinetic energy densities. The component $\overline{X}_p$ is shown by solid black line, while $\overline{X}_t$, $\widetilde{X}_p$, and $\widetilde{X}_t$ are shown by red, green and blue lines, respectively. X stands for either M or E. Panels (**c**,**f**) show the axial dipolar H_1^0 and the axial quadrupolar H_2^0 coefficients at midshell $r = (r_i + r_o)/2$ by red and blue lines, respectively. Note the very different ordinate scales between panels (**a**,**c**,**d**,**f**). The ordinate scales of panels (**b**,**e**) are identical. (Colour online).

The solutions plotted in Figure 5a–f are examples of two types of dipolar dynamos that have been observed in numerical simulations [29,34,58,59], namely those with $\mathcal{D} > 1$ to which we will refer to as "Mean Dipolar" (**MD**) and those with $\mathcal{D} \leq 1$ that we will call "Fluctuating Dipolar" (**FD**). The typical spatial structures of the **MD** and **FD** dynamos are illustrated in Figure 6. The radial magnetic field plotted in the second column of Figure 6 shows the predominant dipolar symmetry of the dynamos, particularly clearly in the **MD** case where the north and the south hemispheres have opposite polarities entirely. The **FD** case displays a band of reversed polarity in a belt near the equator. In time this band propagates towards the poles and replaces the initial polarity leading to periodically occurring reversals. The stationary dipole of the **MD** case is stronger in intensity and inhibits differential rotation. This is confirmed by the profiles of the differential rotation plotted in the left part of the third column of Figure 6 that are markedly different. The **FD** case is characterised with a stronger geostrophic rotation largely aligned with the tangent cylinder while the mean zonal flow of the **MD** is weaker and exhibits a non-geostrophic rotation that is retrograde near the equator. The columnar convective structure of the solutions remains similar in the **MD** and the **FD** case. Time-averaged kinetic and magnetic energy power spectra are shown in Figure 7.

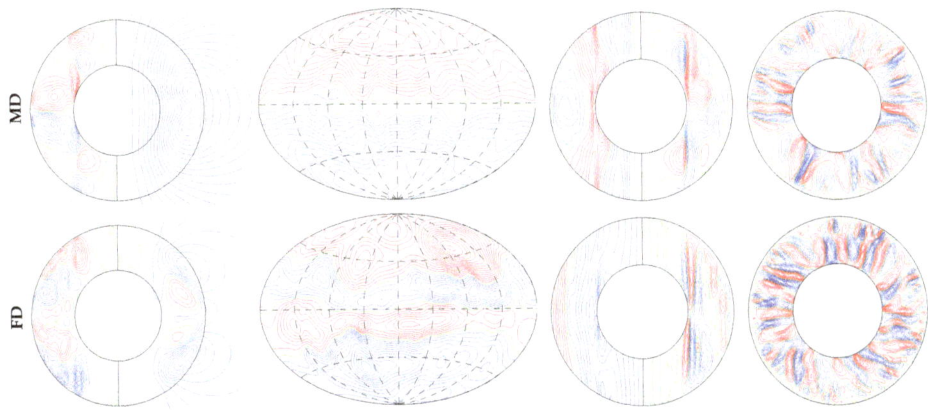

Figure 6. A **MD** (top row) and a **FD** (bottom row) dynamo solutions at $\eta = 0.5$, $R = 8.2 \times 10^5$, $\tau = 2 \times 10^4$, $P = 0.75$ and $P_m = 1.5$ corresponding to the cases shown in Figure 5. The first column shows meridional lines of constant $\overline{B_\varphi}$ in the left half and of $r \sin\theta\partial_\theta\overline{h} = const.$ in the right half. The second column shows lines of constant B_r at $r = 1.675r_o$. The third column shows meridional lines of constant $\overline{u}_\varphi$ in the left half and of $r\sin\theta\partial_\theta\overline{v}$ in the right half. The fourth column shows contours of the radial flow u_r on the equatorial plane. Positive values are shown in red; negative values are shown in blue, and the zeroth contour line is shown in green. (Colour online).

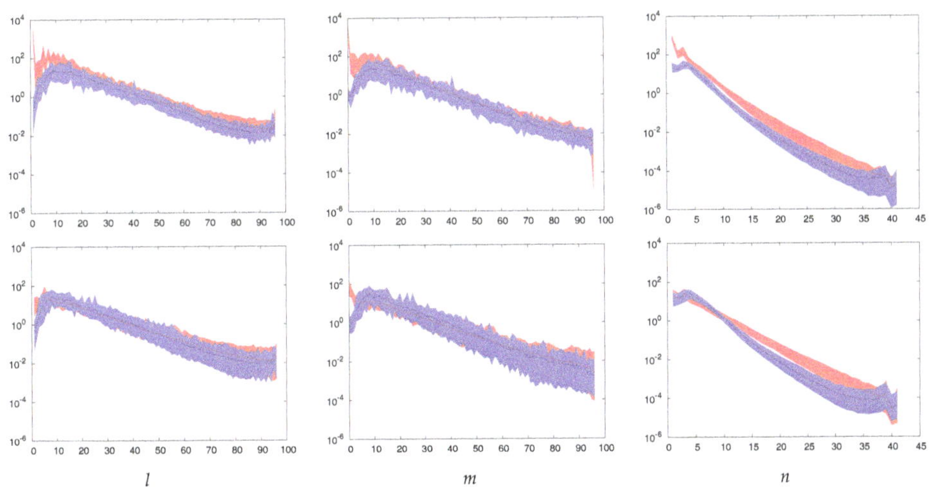

Figure 7. Typical power spectra of velocity (blue) and magnetic field (red). The top row shows an **MD** dynamo solution whereas the bottom row shows an **FD** dynamo solution both at $\eta = 0.4$, R =1,500,000, $\tau = 2 \times 10^4$, $P = 0.75$ and $P_m = 1.5$. From left to right, power spectra as a function of the spherical harmonic degree l, order m, and Chebychev polynomial degree n are shown, respectively. Lines represent the average spectra and shaded areas go from the minimum to the maximum values for each mode in the averaging period. A period of one viscous-diffusion time unit is used for the time-averaging period in both cases. (Colour online).

3.4. Bistability and General Effects of Shell Thickness Variation

One of the most remarkable features of **MD** and **FD** dynamos introduced above is that these two very distinct types can coexist at identical parameter values. Coexistence was first reported in [29]. Indeed, in each of the Figures 5–7, two different cases obtained at the same parameter values are shown. Within the parameter range of coexistence it is the initial conditions that determine which of the two chaotic attractors will be realised. Figure 8 shows the dipolarity ratio $\mathcal{D}$ as a function of the shell thickness aspect ratio η. Several observations can be made immediately. First, bistability only seems to occur for aspect ratios between $\eta = 0.25$ and $\eta = 0.6$ and both to the left and to the right of this interval **FD** dynamos are found. In contrast, alternating regimes appeared on each side of the hysteresis loop in previous studies [29,42] where continuation as a function of all remaining parameters R, P, P_m and τ was performed. A further observation is that the **FD** dynamos have a decreasing dipolarity with increasing aspect ratio, that is, dipolarity seems to decrease with shell thickness. The **MD** dynamos, on the other hand, show little variation of dipolarity with aspect ratio but can still be separated into two groups, one for thin shells and another for thick shells. In this respect, it is apparent that thinner shells result in dynamos that are more dipole-dominated.

It is also interesting to note that there is a clear division between **MD** and **FD** dynamos also in the energy density space. Figure 9 shows a compilation of plots of magnetic energy density as a function of kinetic energy density. Dots represent instantaneous values; circles/triangles are mean values over time. The aspect ratio, η, increases from darker to lighter colours. Blue dots and circles represent simulations that started off as fluctuating dipolar dynamos whereas warm colours and greens represent simulations starting off as mean dipolar dynamos. Green symbols and dots represent simulations starting off as mean dipolar dynamos at $\eta = 0.6$ and $\eta = 0.7$ which were repeated starting from a higher magnetic energy and lower kinetic energy (triangles) relative to the original simulations (circles). Three regions can be clearly identified that correspond to simulations that finished as high and low dipolarity **MD** dynamos (regions I and II in Figure 9), and to simulations that finished as **FD** dynamos (region III in Figure 9). It is evident that dipolarity is preserved throughout the computations (most warm coloured dots and circles end up in region I and II; all blue dots and symbols end up in region III). The exception to this rule happens when the magnetic energy density of the initial **MD** condition is not big enough or its ration to the kinetic energy density is small (green circles). In this case the solutions drift to an **FD** state and remain there. If, on the other hand, the initial **MD** condition sees its magnetic energy density scaled up sufficiently, the solution will remain an **MD** dynamo (green dots and triangles).

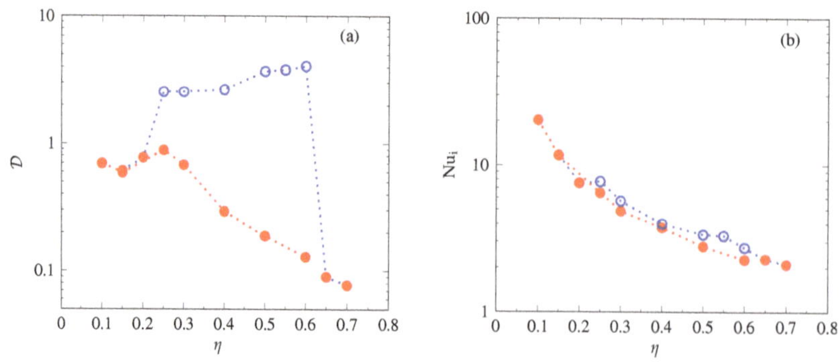

Figure 8. Bistability as a function of the shell thickness aspect ratio η. (**a**) The dipolarity ratio $\mathcal{D} = \overline{M}_p/\widetilde{M}_p$ and (**b**) the Nusselt number at $r = r_i$ in the cases $R = 3.8 \times R_c$, $\tau = 2 \times 10^4$, $P = 0.75$ and $P_m = 1.5$. Full red and empty blue circles indicate **FD** and **MD** dynamos, respectively. Red dotted lines and blue dotted lines connect dynamos that were started from **FD** and **MD** initial conditions, respectively. (Colour online).

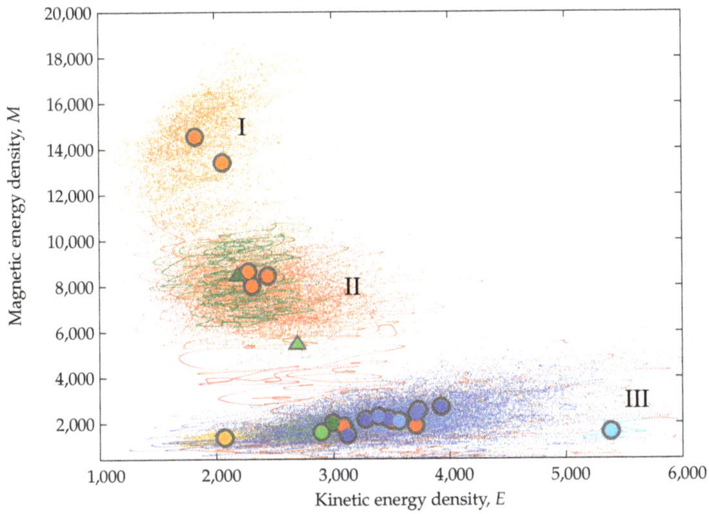

Figure 9. A "phase portrait" of magnetic vs. kinetic energy density values for dynamos with $R = 3.8 \times R_c$, $\tau = 2 \times 10^4$, $P = 0.75$ and $P_m = 1.5$. Dots are instantaneous values; large markers are time-averaged values. The aspect ratio η increases from darker to lighter colours (blue to orange). Blue dots and points represent dynamos that were started from **FD** initial conditions. Warm colours and greens represent simulations that were started from **MD** initial conditions. Green symbols and dots represent simulations that were started as from **MD** initial conditions at $\eta = 0.6$ and $\eta = 0.7$ and that were repeated starting from a higher magnetic energy and lower kinetic energy (triangles) relative to the original simulations (circles). (Colour online).

3.5. The Cross-Helicity Effect

In order to model the effect of turbulence (or, at least, small-scale chaotic motion) on dynamo action, we consider a separation of scales. This approach is justified as dynamos tend to exhibit long-lasting large-scale structures (for example, the Earth's dipolar field) together with complex turbulent motions at smaller scales. We perform an averaging approach where, for the velocity field $\mathbf{u}$ and the magnetic field $\mathbf{b}$, we write

$$\mathbf{u} = \mathbf{U} + \mathbf{u}', \tag{13a}$$

$$\mathbf{b} = \mathbf{B} + \mathbf{b}'. \tag{13b}$$

Capital letters represent large-scale components of each field, and will be referred to as the "mean" components within this and the following section. As described in the literature [15,16,20], there are several ways to perform this scale separation. Here, we perform the scale separation by assuming that the steady large-scale components of the flow and magnetic field can be identified with their respective time-averaged zonal components. The mean flow is then described as

$$\mathbf{U} = \langle \mathbf{u} \rangle = \frac{1}{2\pi\tau} \iint \mathbf{u} \, \mathrm{d}\varphi \, \mathrm{d}t, \tag{14}$$

for a suitable time scale τ, and a similar expression can be constructed for the mean magnetic field. In principle, we can apply this separation of scales to all the main dynamical variables and all the model equations. Here, however, we only focus on the induction equation in order to gauge the effect of turbulent transport on the generation of the magnetic field through dynamo action.

Applying the above scale separation to the induction equation

$$\partial_t \mathbf{b} = \nabla \times (\mathbf{u} \times \mathbf{b}) + \lambda \nabla^2 \mathbf{b}, \tag{15}$$

where λ is the magnetic diffusivity (note that Equation (16) is an alternative formulation of Equation (2e)), we find the induction equation for the mean magnetic field to be

$$\partial_t \mathbf{B} = \nabla \times (\mathbf{U} \times \mathbf{B}) + \nabla \times \mathbf{E}_\mathrm{M} + \lambda \nabla^2 \mathbf{B}, \tag{16}$$

where the turbulent electromotive force, $\mathbf{E}_\mathrm{M}$, is defined as

$$\mathbf{E}_\mathrm{M} = \langle \mathbf{u}' \times \mathbf{b}' \rangle. \tag{17}$$

Through an application of the two-scale direct-interaction approximation (TSDIA) of inhomogeneous MHD turbulence (see [60] and references therein), the turbulent electromotive force can be written, in terms of mean variables, as

$$\mathbf{E}_\mathrm{M} = \alpha \mathbf{B} - \beta \mathbf{J} + \gamma \boldsymbol{\Omega}. \tag{18}$$

Here, $\mathbf{J} = \nabla \times \mathbf{B}$ and $\boldsymbol{\Omega} = \nabla \times \mathbf{U}$. The coefficients α, β and γ can be expressed in terms of the turbulent residual helicity, $H = \langle \mathbf{b}' \cdot \mathbf{j}' - \mathbf{u}' \cdot \boldsymbol{\omega}' \rangle$, the turbulent MHD energy, $K = \langle \mathbf{u}'^2 + \mathbf{b}'^2 \rangle / 2$, and the turbulent cross-helicity $W = \langle \mathbf{u}' \cdot \mathbf{b}' \rangle$, respectively [15,61]. Following [20], they are modelled as

$$\alpha = C_\alpha \tau \langle \mathbf{b}' \cdot \mathbf{j}' - \mathbf{u}' \cdot \boldsymbol{\omega}' \rangle = C_\alpha \tau H, \tag{19a}$$

$$\beta = C_\beta \tau \langle \mathbf{u}'^2 + \mathbf{b}'^2 \rangle = C_\beta \tau K, \tag{19b}$$

$$\gamma = C_\gamma \tau \langle \mathbf{u}' \cdot \mathbf{b}' \rangle = C_\gamma \tau W, \tag{19c}$$

with C_α, C_β and C_γ being model constants. Here, τ is the characteristic time of turbulence, which is often expressed as

$$\tau = K/\epsilon, \tag{20}$$

with the dissipation rate of the turbulent MHD energy, ϵ, defined by

$$\epsilon = \nu \left\langle \frac{\partial u'_a}{\partial x_b} \frac{\partial u'_a}{\partial x_b} \right\rangle + \lambda \left\langle \frac{\partial b'_a}{\partial x_b} \frac{\partial b'_a}{\partial x_b} \right\rangle. \tag{21}$$

Substituting (18) into the mean induction equation (16), we have

$$\partial_t \mathbf{B} = \nabla \times (\mathbf{U} \times \mathbf{B}) + \nabla \times (\alpha \mathbf{B} + \gamma \boldsymbol{\Omega}) - \nabla \times [(\lambda + \beta) \nabla \times \mathbf{B}]. \tag{22}$$

Thus, in addition to the transport enhancement or structure destruction due to turbulence through the enhanced diffusion $\lambda + \beta$, there is also transport suppression or structure formation due to turbulence represented by the helicities α and γ [60].

In the classical mean field theory of dynamos [10,15], the turbulent electromotive force is composed of the first two terms on the right-hand side of Equation (18), namely $\alpha \mathbf{B} - \beta \mathbf{J}$. Dynamos resulting from this model are known as "α dynamos", where the turbulent diffusion is balanced by an α-effect. The properties of these terms have been discussed widely in the literature, and so we do not repeat this discussion here. Instead, let us now consider the final term on the right-hand side of Equation (18), $\gamma \boldsymbol{\Omega}$. Unlike the other terms describing the electromotive force, the mean variable in this term depends on the mean velocity and not the mean magnetic field. Yokoi [20] describes how a fluid element subject to a Coriolis-like force (a mean vorticity field) can contribute to the turbulent electromotive force through γ, a measure of the turbulent cross helicity. Dynamos in which the main balance is between $-\beta \mathbf{J}$ and $\gamma \boldsymbol{\Omega}$ are known as "cross-helicity dynamos", where the cross-helicity term replaces the α-effect term in balancing the turbulent diffusion.

Cross-helicity dynamos have been studied much less than α dynamos, and this study represents an initial step in addressing this potentially important imbalance. In particular in Figure 10, we calculate all three contributions to the turbulent electromotive force in our dynamo simulations in order to determine their relative importance. These results are discussed below.

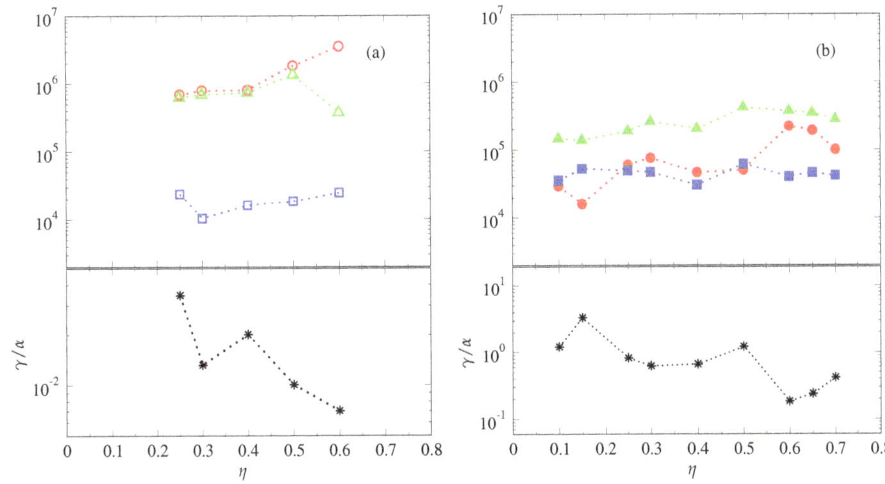

Figure 10. Magnitude of α-, β-, and γ-effects with increasing shell thickness aspect ratio η for dynamo solutions with $R = 3.8 \times R_c$, $\tau = 2 \times 10^4$, $P = 0.75$ and $P_m = 1.5$. The upper panels show root-mean squared time-averaged values of the α-effect (red circles), β-effect (green triangles up) and γ-effect (blue squares). The lower panels show the ratio of γ-to α-effects. Column (**a**) contains **MD** dynamo solutions (empty symbols) while column (**b**) contains **FD** dynamo solutions (full symbols) as shown in Figure 8. (Colour online).

3.6. *Properties and Relative Importance of Cross-Helicity*

The variation of the turbulent transport coefficients α, β, and γ as a function of shell thickness is displayed in Figure 10. For simplicity, in this initial investigation, we take $C_A\tau = 1$, where $A = \alpha$, β, or γ. Thus, the three effects are represented by the turbulent residual helicity H, the turbulent MHD energy K and the turbulent cross-helicity W, respectively. For **MD** dynamo solutions, there is a clear disparity between the α- and β-effects, and the γ-effect. The γ-effect is, for the range of η considered, about two orders of magnitude smaller than the other effects. Thus, across a wide range of shell thickness aspect ratios, **MD** dynamos can be considered to be operating predominantly as α dynamos. In contrast, for **FD** dynamo solutions, a different picture emerges. Across the range of η considered, the α- and γ-effects are of a similar magnitude. Thus, both these effects are potentially important in balancing the β-effect. Therefore, **FD** dynamo solutions represent a "mixture" of an α dynamo and a cross-helicity dynamo.

Figure 11 displays z-projections of the azimuthally-averaged components of the electromotive force. For the **MD** dynamo solutions, shown in (a), the γ-effect follows an antisymmetric pattern about the equator, just like the other effects. This behaviour is expected from the pseudoscalar nature of γ and the symmetry of magnetic fields in **MD** dynamos [20]. For **FD** dynamo solutions, such as those displayed in (b), the components of the electromotive force no longer exhibit antisymmetry about the equator. This behaviour is, in part, due to the more complex spatial structure of the magnetic fields of **FD** dynamos compared to **MD** dynamos. This feature, combined with generally weaker magnetic field strengths and different flow profiles (see Figures 5 and 6, for example), results in the α-effect being weaker for **FD** dynamos. Thus, both the α- and γ-effects become of comparable importance in sustaining dynamo action.

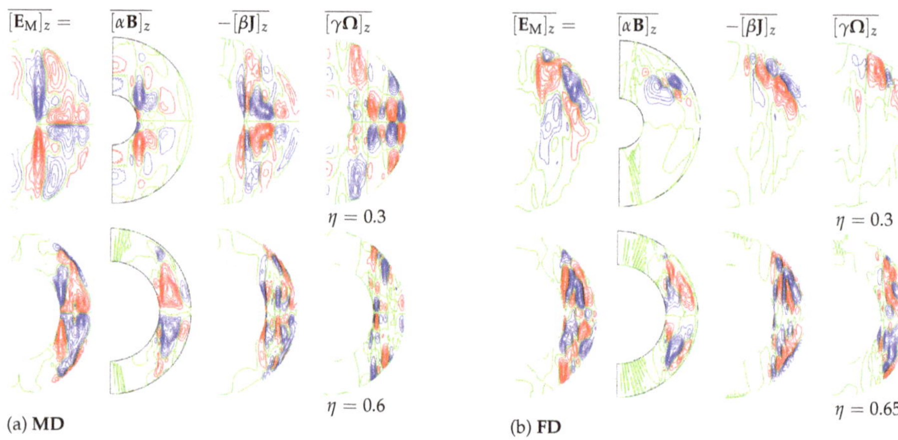

Figure 11. Spacial structures of the azimuthally-averaged z-component of the electromotive force $\mathbf{E}_M$ and its α-, β- and γ-effect constituents as given by Equation (18). Four dipolar dynamo solutions are plotted as follows. (**a**) **MD** dynamo solutions with $\eta = 0.3$, $P = 0.75$, $\tau = 2 \times 10^4$, R =2,500,000, $P_m = 1.5$ (**top row**) and $\eta = 0.6$, $P = 0.75$, $\tau = 2 \times 10^4$, R =410,000, $P_m = 1.5$ (**bottom row**). (**b**) **FD** dynamo solutions with $\eta = 0.3$, $P = 0.75$, $\tau = 2 \times 10^4$, R =2,500,000, $P_m = 1.5$ (**top row**) and $\eta = 0.65$, $P = 0.75$, $\tau = 2 \times 10^4$, R =300,000, $P_m = 1.5$ (**bottom row**). In each column contour lines of the quantities denoted at the column heading are plotted with positive contours shown in red, negative contours shown in blue, and the zeroth contour shown in green. (Colour online).

4. Summary and Discussion

Rotating thermal convection is ubiquitous within the interiors and the atmospheres of celestial bodies. These fluid regions usually contain plasmas or metallic components so vigorous convection drives large-scale electric currents and generates the self-sustained magnetic fields characteristic of these cosmic objects. In this article the relative importance of two main mechanisms for magnetic field generation and amplification is assessed, namely the helicity- and the cross-helicity effects of mean-field dynamo theory. The motivation for this study was to test the hypothesis that the turbulent helicity effect, also known as the α-effect, is more important in the case of the geodynamo, while the cross-helicity effect, also known as the γ-effect, is more significant in the case of the solar global dynamo, due to differences between the shell aspect ratio of the solar convection zone and that of Earth's inner core. The following novel results are reported in the article.

(a) Critical parameter values for onset of convection determined numerically as functions of the shell radius ratio, η.
(b) Bistability and coexistence of two distinct dynamo attractors found as a function of the shell radius ratio, η.
(c) Spatial distributions and time-averaged values of turbulent helicity and cross-helicity EMF effects obtained (1) for both types of dynamo attractors, as well as (2) as functions of the shell radius ratio, η.

Further details and a discussion of these results follows.

To assess α- and γ- electromotive effects, we performed, and report here, an extensive suite of over 40 direct numerical simulations of self-sustained dynamo action driven by thermal convection in rotating spherical fluid shells, where the shell thickness aspect ratio η is varied at fixed values of the other parameters. The simulations are based on the Boussinesq approximation of the governing nonlinear magnetohydrodynamic equations with stress-free velocity boundary conditions. While the use of fully compressible equations is desirable, it is not feasible for global dynamo simulations.

Indeed, the fully compressible MHD equations allow sound wave solutions with periods many orders of magnitude shorter than the convective turnover time and the magnetic diffusion timescales that are of primary interest. The Boussinesq approximation is justified and generally used for modelling convection in Earth's inner core where density variation between the inner–outer core boundary and the core mantle boundary is small [13,14,38,47]. The density contrast between the bottom (ρ_i) and the top (ρ_o) of the Solar convection zone is five orders of magnitude giving a density scale number of $\log(\rho_i/\rho_o) \approx 12$ [62], and the anelastic approximation is more appropriate and commonly used in global solar convection models, for example, [11,33,63]. However, anelastic and Boussinesq simulations show many similarities [63], with Boussinesq models able to mimic solar periodicity and active longitude phenomena [25,42]. Thus, in this work the Boussinesq approximation is used for uniformity across various shell radius ratios and to focus on the effects of shell thickness in isolation from effects of density stratification.

Coexistence of distinct chaotic dynamo states has been reported to occur in this problem in terms of certain governing parameters in [29,31]. In this study, we establish that two essentially different nonlinear dynamo attractors coexist also for an extensive range of shell thickness aspect ratios $\eta \in [0.25, 0.6]$. Since this is precisely the range of values where most celestial dynamos operate, this result is significant as it demonstrates that field morphologies may be dependent on the initial state of a dynamo. We proceed to discuss in detail the contrasting properties characterising the coexisting dynamo regimes (mean-field dipolar (**MD**) dynamos and fluctuating dipolar (**FD**) dynamos) including differences in temporal behaviour and spatial structures of both the magnetic field and rotating thermal convection. We find that the relative importance of the electromotive dynamo effects is different in the cases of mean-field dipolar dynamos and fluctuating dipolar dynamos. The helicity α-effect and the cross-helicity γ-effect are comparable in intensity in the case of fluctuating dipolar dynamos and their ratio does not vary significantly with shell thickness. In contrast, in the case of mean-field dipolar dynamos the helicity α-effect dominates by approximately two orders of magnitude and becomes even stronger with decreasing shell thickness. Our results, therefore, indicate that both dynamo mechanisms are important for solar global magnetic field generation as the solar dynamo is of a fluctuating dipolar type. Our results also indicate that the cross-helicity effect may be important in understanding dynamo mechanisms in stellar dynamos. The latter may also be of fluctuating dipolar type and markedly different from the solar dynamo, for example, having large-scale magnetic structures being dominant in only one hemisphere [64]. Since the geodynamo is of a mean-field dipolar type, the helicity effect appears, indeed, to be more significant in this case and our results show this effect will become even stronger as the inner solid core grows in size by iron freezing. Simulations of the geodynamo with nucleation and growth of the inner core have been recently reported by Driscoll [65] and Landeau et al. [66]. These authors find that pre-inner core nucleation dynamos exhibit weak thermal convection, low magnetic intensity and non-dipolar field morphology, while post-inner core nucleation and with increasing inner core size their solutions have stronger axial dipole morphology. Our results similarly demonstrate that **FD** and multipolar dynamos occur when the value of the shell radius ratio η is smaller than 0.25. However, our **FD** solutions exhibit vigorous convection and can be described as strong-field dynamos even though they are of lower magnetic field intensity than corresponding **MD** dynamos. A further discrepancy is that for $\eta > 0.25$ we find that **MD** and **FD** dynamos coexist. These discrepancies can be attributed to significant differences in thermal and velocity boundary conditions between our model and the models of [65,66]. Most importantly, the governing parameters values in [65,66] are controlled by thermochemical evolution models and vary with inner core size (age), while in our study all parameter values apart from η are kept fixed.

It will be of interest to revisit the analysis of helicity and cross-helicity effects using the more general anelastic approximation of the governing equations. Further, there are many questions that remain to be answered on how the dynamic balance between the components of the electromotive force affects different aspects of dynamo action, including how to switch between **MD** and **FD** dynamos.

Author Contributions: Conceptualization, R.D.S.; methodology, R.D.S.; software, R.D.S.; validation, L.S.; formal analysis, R.D.S. and D.M.; investigation, L.S. and P.G.; resources, R.D.S.; data curation, L.S. and P.G.; writing—original draft preparation, R.D.S.; writing—review and editing, R.D.S. and D.M.; visualization, L.S., R.D.S. and P.G.; supervision, R.D.S.; funding acquisition, R.S. All authors have read and agreed to the published version of the manuscript.

Funding: This research was funded by the Leverhulme Trust grant number RPG-2012-600.

Acknowledgments: Numerical simulations were carried out in part at the DiRAC Data Centric system at Durham University, operated by the Institute for Computational Cosmology on behalf of the STFC DiRAC HPC Facility (www.dirac.ac.uk). This equipment was funded by BIS National E-infrastructure capital grant ST/K00042X/1, STFC capital grants ST/H008519/1 and ST/K00087X/1, STFC DiRAC Operations grant ST/K003267/1 and Durham University. DiRAC is part of the National E-Infrastructure.

Conflicts of Interest: The authors declare no conflict of interest.

Abbreviations

The following abbreviations are used in this manuscript:

MD	Mean Dipolar Dynamo
FD	Fluctuating Dipolar Dynamo

References

1. Parker, E.N. *Cosmical Magnetic Fields. Their Origin and Their Activity*; OUP: London, UK, 1979.
2. Brun, A.S.; Browning, M.K. Magnetism, dynamo action and the solar-stellar connection. *Living Rev. Sol. Phys.* **2017**, *14*. [CrossRef]
3. Zwaan, C. Elements and Patterns in the Solar Magnetic Field. *Annu. Rev. Astron. Astrophys.* **1987**, *25*, 83–111. [CrossRef]
4. Busse, F.; Simitev, R. Planetary Dynamos. In *Treatise on Geophysics*; Elsevier: Amsterdam, The Netherlands, 2015; pp. 239–254.
5. Usoskin, I.G. A history of solar activity over millennia. *Living Rev. Sol. Phys.* **2017**, *14*. [CrossRef]
6. Owens, M.J.; Forsyth, R.J. The Heliospheric Magnetic Field. *Living Rev. Sol. Phys.* **2013**, *10*. [CrossRef]
7. Russell, C.T. The Magnetosphere. *Annu. Rev. Earth Planet. Sci.* **1991**, *19*, 169–182. [CrossRef]
8. Busse, F.H.; Simitev, R. *Dynamos of Giant Planets*; Cambridge University Press: Cambridge, UK, 2006; pp. 467–474.
9. Larmor, J. How could a rotating body such as the Sun become a magnet? *Rep. Br. Assoc.* **1919**, *87*, 159–160.
10. Moffatt, H.K. *Magnetic Field Generation in Electrically Conducting Fluids*; Cambridge University Press: Cambridge, UK, 1978.
11. Charbonneau, P. Solar Dynamo Theory. *Annu. Rev. Astron. Astrophys.* **2014**, *52*, 251–290. [CrossRef]
12. Busse, F.H.; Simitev, R. Dynamos driven by convection in rotating spherical shells. *Astr. Nachr.* **2005**, *326*, 231. [CrossRef]
13. Jones, C.A. 8.05-Thermal and Compositional Convection in the Outer Core. In *Treatise on Geophysics*, 2nd ed.; Schubert, G., Ed.; Elsevier: Amsterdam, The Netherlands, 2015; pp. 115–159.
14. Wicht, J.; Sanchez, S. Advances in geodynamo modelling. *Geophys. Astrophys. Fluid Dyn.* **2019**, *113*, 2–50. [CrossRef]
15. Krause, F.; Raedler, K.H. *Mean-Field Magnetohydrodynamics and Dynamo Theory*; Pergamon: Oxford, UK, 1980.
16. Brandenburg, A. Advances in mean-field dynamo theory and applications to astrophysical turbulence. *J. Plasma Phys.* **2018**, *84*. [CrossRef]
17. Moffatt, K.; Dormy, E. *Self-Exciting Fluid Dynamos*; Cambridge University Press: Cambridge, UK, 2019.
18. Brandenburg, A.; Subramanian, K. Astrophysical magnetic fields and nonlinear dynamo theory. *Phys. Rep.* **2005**, *417*, 1–209. [CrossRef]
19. Yoshizawa, A.; Yokoi, N. Turbulent Magnetohydrodynamic Dynamo for Accretion Disks Using the Cross-Helicity Effect. *Astrophys. J.* **1993**, *407*, 540. [CrossRef]
20. Yokoi, N. Cross helicity and related dynamo. *Geophys. Astrophys. Fluid Dyn.* **2013**, *107*, 114–184. [CrossRef]
21. Pipin, V.V.; Yokoi, N. Generation of a Large-scale Magnetic Field in a Convective Full-sphere Cross-helicity Dynamo. *Astrophys. J.* **2018**, *859*, 18. [CrossRef]

22. Thompson, M.J.; Toomre, J.; Anderson, E.R.; Antia, H.M.; Berthomieu, G.; Burtonclay, D.; Chitre, S.M.; Christensen-Dalsgaard, J.; Corbard, T.; DeRosa, M.; et al. Differential Rotation and Dynamics of the Solar Interior. *Science* **1996**, *272*, 1300–1305. [CrossRef]
23. Schou, J.; Antia, H.M.; Basu, S.; Bogart, R.S.; Bush, R.I.; Chitre, S.M.; Christensen-Dalsgaard, J.; Mauro, M.P.D.; Dziembowski, W.A.; Eff-Darwich, A.; et al. Helioseismic Studies of Differential Rotation in the Solar Envelope by the Solar Oscillations Investigation Using the Michelson Doppler Imager. *Astrophys. J.* **1998**, *505*, 390–417. [CrossRef]
24. Busse, F.H.; Simitev, R. Parameter dependences of convection-driven dynamos in rotating spherical fluid shells. *Geophys. Astrophys. Fluid Dyn.* **2006**, *100*, 341. [CrossRef]
25. Simitev, R.D.; Busse, F.H. How far can minimal models explain the solar cycle? *Astrophys. J.* **2012**, *749*, 9. [CrossRef]
26. Hamba, F. Turbulent dynamo effect and cross helicity in magnetohydrodynamic flows. *Phys. Fluids Fluid Dyn.* **1992**, *4*, 441–450. [CrossRef]
27. Yokoi, N.; Balarac, G. Cross-helicity effects and turbulent transport in magnetohydrodynamic flow. *J. Phys. Conf. Ser.* **2011**, *318*, 072039. [CrossRef]
28. Rüdiger, G.; Küker, M.; Schnerr, R.S. Cross helicity at the solar surface by simulations and observations. *Astron. Astrophys.* **2012**, *546*, A23. [CrossRef]
29. Simitev, R.D.; Busse, F.H. Bistability and hysteresis of dipolar dynamos generated by turbulent convection in rotating spherical shells. *EPL (Europhys. Lett.)* **2009**, *85*, 19001. [CrossRef]
30. Busse, F.H.; Simitev, R. Remarks on some typical assumptions in dynamo theory. *Geophys. Astrophys. Fluid Dyn.* **2011**, *105*, 234. [CrossRef]
31. Simitev, R.D.; Busse, F.H. Bistable attractors in a model of convection-driven spherical dynamos. *Phys. Scr.* **2012**, *86*, 018409. [CrossRef]
32. Busse, F.H.; Simitev, R. Toroidal flux oscillations as possible causes of geomagnetic excursions and reversals. *Phys. Earth Planet. Inter.* **2008**, *168*, 237. [CrossRef]
33. Matilsky, L.I.; Toomre, J. Exploring Bistability in the Cycles of the Solar Dynamo through Global Simulations. *Astrophys. J.* **2020**, *892*, 106. [CrossRef]
34. Simitev, R.; Busse, F.H. Prandtl-number dependence of convection-driven dynamos in rotating spherical fluid shells. *J. Fluid Mech.* **2005**, *532*, 365. [CrossRef]
35. Mather, J.F.; Simitev, R.D. Regimes of thermo-compositional convection and related dynamos in rotating spherical shells. *Geophys. Astrophys. Fluid Dyn.* **2020**, 1–24. [CrossRef]
36. Busse, F.H. Homogeneous Dynamos in Planetary Cores and in the Laboratory. *Annu. Rev. Fluid Mech.* **2000**, *32*, 383–408. [CrossRef]
37. Dormy, E.; Simitev, R.; Busse, F.; Soward, A. Dynamics of Rotating Fluids. In *Mathematical Aspects of Natural Dynamos*; Chapman and Hall/CRC: Boca Raton, FL, USA, 2007; pp. 120–198.
38. Roberts, P.H.; King, E.M. On the genesis of the Earth's magnetism. *Rep. Prog. Phys.* **2013**, *76*, 096801. [CrossRef]
39. Grote, E.; Busse, F.; Tilgner, A. Regular and chaotic spherical dynamos. *Phys. Earth Planet. Inter.* **2000**, *117*, 259–272, [CrossRef]
40. Simitev, R.; Busse, F.H. Patterns of convection in rotating spherical shells. *New J. Phys.* **2003**, *5*, 97. [CrossRef]
41. Simitev, R.; Busse, F.; Grote, E. Convection in rotating spherical shells and its dynamo action. In *Earth's Core and Lower Mantle*; CRC Press: Boca Raton, FL, USA, 2003; pp. 130–152.
42. Simitev, R.; Busse, F.H. Solar cycle properties described by simple convection-driven dynamos. *Phys. Scr.* **2012**, *86*, 018407. [CrossRef]
43. Tilgner, A. Spectral methods for the simulation of incompressible flows in spherical shells. *Int. J. Numer. Meth. Fluids* **1999**, *30*, 713–724, [CrossRef]
44. Marti, P.; Schaeffer, N.; Hollerbach, R.; Cébron, D.; Nore, C.; Luddens, F.; Guermond, J.L.; Aubert, J.; Takehiro, S.; Sasaki, Y.; et al. Full sphere hydrodynamic and dynamo benchmarks. *Geophys. J. Int.* **2014**, *197*, 119–134, [CrossRef]
45. Matsui, H.; Heien, E.; Aubert, J.; Aurnou, J.M.; Avery, M.; Brown, B.; Buffett, B.A.; Busse, F.; Christensen, U.R.; Davies, C.J.; et al. Performance benchmarks for a next generation numerical dynamo model. *Geochem. Geophys. Geosyst.* **2016**, *17*, 1586–1607, [CrossRef]

46. Silva, L.A.C.; Simitev, R.D. *Pseudo-Spectral Code For Numerical Simulation Of Nonlinear Thermo-Compositional Convection and Dynamos in Rotating Spherical Shells*; University of Glasgow: Glasgow, UK. [CrossRef]
47. Christensen, U.; Olson, P.; Glatzmaier, G.A. Numerical modelling of the geodynamo: A systematic parameter study. *Geophys. J. Int.* **1999**, *138*, 393–409. [CrossRef]
48. Silva, L.; Mather, J.F.; Simitev, R.D. The onset of thermo-compositional convection in rotating spherical shells. *Geophys. Astrophys. Fluid Dyn.* **2019**, *113*, 377–404. [CrossRef]
49. Busse, F.H.; Simitev, R. Inertial convection in rotating fluid spheres. *J. Fluid Mech.* **2004**, *498*, 23–30. [CrossRef]
50. Silva, L.A.C.; Simitev, R.D. *Spectral Code for Linear Analysis of The Onset of Thermo-Compositional Convection in Rotating Spherical Fluid Shells*; University of Glasgow: Glasgow, UK, 2018. [CrossRef]
51. Zhang, K.K.; Busse, F.H. On the onset of convection in rotating spherical shells. *Geophys. Astrophys. Fluid Dyn.* **1987**, *39*, 119–147. [CrossRef]
52. Roberts, P.H. On the Thermal Instability of a Rotating-Fluid Sphere Containing Heat Sources. *Philos. Trans. R. Soc. Math. Phys. Eng. Sci.* **1968**, *263*, 93–117. [CrossRef]
53. Busse, F.H. Thermal instabilities in rapidly rotating systems. *J. Fluid Mech.* **1970**, *44*, 441. [CrossRef]
54. Soward, A.M. On the Finite amplitude thermal instability of a rapidly rotating fluid sphere. *Geophys. Astrophys. Fluid Dyn.* **1977**, *9*, 19–74. [CrossRef]
55. Jones, C.A.; Soward, A.M.; Mussa, A.I. The onset of thermal convection in a rapidly rotating sphere. *J. Fluid Mech.* **2000**, *405*, 157–179. [CrossRef]
56. Dormy, E.; Soward, A.M.; Jones, C.A.; Jault, D.; Cardin, P. The onset of thermal convection in rotating spherical shells. *J. Fluid Mech.* **2004**, *501*, 43–70. [CrossRef]
57. Yano, J.I. Asymptotic theory of thermal convection in rapidly rotating systems. *J. Fluid Mech.* **1992**, *243*, 103. [CrossRef]
58. Christensen, U.R.; Aubert, J. Scaling properties of convection-driven dynamos in rotating spherical shells and application to planetary magnetic fields. *Geophys. J. Int.* **2006**, *166*, 97–114. [CrossRef]
59. Olson, P.L.; Glatzmaier, G.A.; Coe, R.S. Complex polarity reversals in a geodynamo model. *Earth Planet. Sci. Lett.* **2011**, *304*, 168–179. [CrossRef]
60. Yokoi, N. Turbulence, Transport and Reconnection. In *Topics in Magnetohydrodynamic Topology, Reconnection and Stability Theory*; MacTaggart, D., Hillier, A., Eds.; Springer: Cham, Switzerland, 2020; pp. 177–265.
61. Yoshizawa, A. Self-consistent turbulent dynamo modeling of reversed field pinches and planetary magnetic fields. *Phys. Fluids Plasma Phys.* **1990**, *2*, 1589–1600. [CrossRef]
62. Christensen-Dalsgaard, J.; Dappen, W.; Ajukov, S.V.; Anderson, E.R.; Antia, H.M.; Basu, S.; Baturin, V.A.; Berthomieu, G.; Chaboyer, B.; Chitre, S.M.; et al. The Current State of Solar Modeling. *Science* **1996**, *272*, 1286–1292. [CrossRef]
63. Simitev, R.D.; Kosovichev, A.G.; Busse, F.H. Dynamo effects near the transition from solar to anti-solar differential rotation. *Astrophys. J.* **2015**, *810*, 80. [CrossRef]
64. MacTaggart, D.; Gregory, S.G.; Neukirch, T.; Donati, J.F. Magnetohydrostatic modelling of stellar coronae. *Mon. Not. RAS* **2015**, *456*, 767–774. [CrossRef]
65. Driscoll, P.E. Simulating 2 Ga of geodynamo history. *Geophys. Res. Lett.* **2016**, *43*, 5680–5687. [CrossRef]
66. Landeau, M.; Aubert, J.; Olson, P. The signature of inner-core nucleation on the geodynamo. *Earth Planet. Sci. Lett.* **2017**, *465*, 193–204. [CrossRef]

Publisher's Note: MDPI stays neutral with regard to jurisdictional claims in published maps and institutional affiliations.

© 2020 by the authors. Licensee MDPI, Basel, Switzerland. This article is an open access article distributed under the terms and conditions of the Creative Commons Attribution (CC BY) license (http://creativecommons.org/licenses/by/4.0/).

fluids

MDPI

Article

Onset of Inertial Magnetoconvection in Rotating Fluid Spheres

Radostin D. Simitev [1,*] and Friedrich H. Busse [2]

1 School of Mathematics and Statistics, University of Glasgow, Glasgow G12 8QQ, UK
2 Institute of Physics, University of Bayreuth, 95440 Bayreuth, Germany; friedrich.busse@uni-bayreuth.de
* Correspondence: Radostin.Simitev@glasgow.ac.uk

Abstract: The onset of convection in the form of magneto-inertial waves in a rotating fluid sphere permeated by a constant axial electric current is studied in this paper. Thermo-inertial convection is a distinctive flow regime on the border between rotating thermal convection and wave propagation. It occurs in astrophysical and geophysical contexts where self-sustained or external magnetic fields are commonly present. To investigate the onset of motion, a perturbation method is used here with an inviscid balance in the leading order and a buoyancy force acting against weak viscous dissipation in the next order of approximation. Analytical evaluation of constituent integral quantities is enabled by applying a Green's function method for the exact solution of the heat equation following our earlier non-magnetic analysis. Results for the case of thermally infinitely conducting boundaries and for the case of nearly thermally insulating boundaries are obtained. In both cases, explicit expressions for the dependence of the Rayleigh number on the azimuthal wavenumber are derived in the limit of high thermal diffusivity. It is found that an imposed azimuthal magnetic field exerts a stabilizing influence on the onset of inertial convection and as a consequence magneto-inertial convection with azimuthal wave number of unity is generally preferred.

Keywords: rotating thermal magnetoconvection; linear onset; sphere

check for updates

Citation: Simitev, R.D.; Busse, F.H. Onset of Inertial Magnetoconvection in Rotating Fluid Spheres. *Fluids* **2021**, 6, 41. https://doi.org/doi:10.3390/fluids6010041

Received: 15 December 2020
Accepted: 10 January 2021
Published: 13 January 2021

Publisher's Note: MDPI stays neutral with regard to jurisdictional claims in published maps and institutional affiliations.

Copyright: © 2021 by the authors. Licensee MDPI, Basel, Switzerland. This article is an open access article distributed under the terms and conditions of the Creative Commons Attribution (CC BY) license (https://creativecommons.org/licenses/by/4.0/).

1. Introduction

Buoyancy-driven motions of rotating, electrically conducting fluids in the presence of magnetic fields represent a fundamental aspect of the dynamics of stellar and planetary interiors, see, e.g., in [1–4]. The problem of magnetic field generated and sustained by convection is rather difficult to attack both analytically and numerically because of its essential nonlinearity and scale separation [5,6]. Valuable insights can be gained by studying magnetoconvection, the simpler case of an imposed magnetic field, which has received much attention ever since the early work of Chandrasekhar [7], see in [8,9]. For instance, the propagation of rotating magnetoconvection modes excited in the deep convective region of the Earth's core has been proposed as a possible mechanism for explaining features of observed longitudinal geomagnetic drifts [10,11], see also the recent review of Finlay et al. [12]. A rather detailed classification of magnetoconvection waves in a rotating cylindrical annulus has been recently attempted by Hori et al. [13] and the authors of [14] who proceeded further to make useful comparisons with nonlinear spherical dynamo simulations and to provide estimates for the strength of the "hidden" azimuthal part of the magnetic field within the core. These authors used the rotating annulus model of Busse [15,16] and only considered values of the Prandtl number of the order unity. However, both spherical geometry as well as small values of the Prandtl number are essential features of a planetary or a stellar interior [17]. At sufficiently small values of the Prandtl number, a different style of convection exists that is sometimes called inertial or equatorially-attached convection or thermo-inertial waves [18–21]. In this limit, convection oscillates so fast that the viscous force does not enter the leading-order balance. The latter is then reduced to the Poincaré equation in a rotating spherical system [22–24]. On the longer time scale of the next order of approximation the buoyancy force maintains convection against the weak viscous dissipation. This regime of convection thus represents a transition between thermal convection and wave propagation in rapidly rotating geometries. It is important to understand how this regime of inertial convection is affected by an imposed magnetic field.

With this in mind, we study in the present paper the onset of magneto-inertial-convection. In particular, we consider a rotating fluid sphere permeated by a constant axial electric current as proposed by Malkus [11] in the limit of low viscosity and high thermal diffusivity (small Prandtl number). A similarly configured problem was also investigated by Zhang and Busse [25] who derived an explicit dependence of the critical Rayleigh number on the imposed field strength but were not able to obtain an explicit dependence on the azimuthal wavenumber of the modes as this requires the evaluation of a volume integral of the temperature perturbation [25]. In an earlier paper, we proposed a Green's function method for the exact solution of the heat equation [26] which then allowed the analytical evaluation of the integral quantities needed to find a fully explicit expressions for the critical Rayleigh number and frequency for the onset of convection and to study mode competition. Here, we apply the same approach to the case of magneto-inertial convection and we consider both value and flux boundary conditions for the temperature.

In the following we start with the mathematical formulation of the problem in Section 2. The special limit of a high ratio of thermal to magnetic diffusivity will be treated in Section 3. The general case requires the symbolic evaluation of lengthy analytical expressions and will be presented in Section 4. A discussion of the results and an outlook on related problems will be given in the final Section 5 of the paper.

2. Mathematical Formulation of the Problem

We consider a homogeneously heated and self-gravitating sphere as illustrated in Figure 1. The sphere is filled with incompressible and electrically conducting fluid characterized by its magnetic diffusivity η, kinematic viscosity ν, thermal diffusivity κ, and density ϱ. The sphere is rotating with a constant angular velocity $\Omega\mathbf{k}$ where $\mathbf{k}$ is the axial unit vector. The gravity field is given by $\mathbf{g} = -gr_0\mathbf{r}$, where $\mathbf{r}$ is the position vector with respect to the center of the sphere, r is its length measured in fractions of the radius r_0 of the sphere, and g is the amplitude of the gravitational acceleration. Following Malkus [11], we assume that the fluid sphere is permeated by a toroidal magnetic field $\mathbf{B} \sim \mathbf{k} \times \mathbf{r}$. As the Lorentz force like the centrifugal force can be balanced by the pressure gradient, a static state of no motion exists with the temperature distribution $T_S = T_0 - \beta r_0^2 r^2/2$. We employ the Boussinesq approximation and assume constant material properties η, ν, κ, and ϱ everywhere except in the buoyancy term where the density is assumed to have a linear dependence on temperature with a coefficient of thermal expansion $\alpha \equiv (d\varrho/dT)/\varrho = \text{const.}$

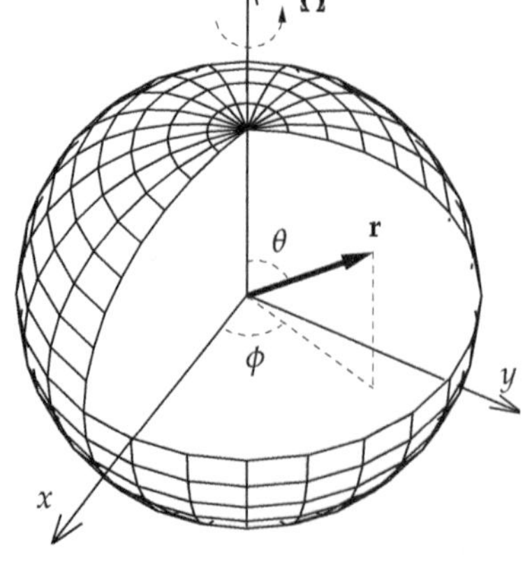

Figure 1. Geometrical configuration of the problem. A part of the outer spherical surface is removed to expose the interior of sphere to which the conducting fluid is confined.

In order to study the onset of magnetoconvection in this system, we consider the linearized momentum, magnetic induction, heat, continuity, and solenoidality equations:

$$\partial_t \tilde{\mathbf{u}} + \tau \mathbf{k} \times \tilde{\mathbf{u}} + \nabla(\pi - \tilde{\mathbf{b}} \cdot \mathbf{j} \times \mathbf{r}) + (\mathbf{j} \times \mathbf{r}) \cdot \nabla \tilde{\mathbf{b}} - \mathbf{j} \times \tilde{\mathbf{b}} = \tilde{\Theta} \mathbf{r} + P_m \nabla^2 \tilde{\mathbf{u}}, \tag{1a}$$

$$\partial_t \tilde{\mathbf{b}} - (\mathbf{j} \times \mathbf{r}) \cdot \nabla \tilde{\mathbf{u}} + \mathbf{j} \times \tilde{\mathbf{u}} = \nabla^2 \tilde{\mathbf{b}}, \tag{1b}$$

$$\hat{R}\, \mathbf{r} \cdot \tilde{\mathbf{u}} + \nabla^2 \tilde{\Theta} - (P/P_m) \partial_t \tilde{\Theta} = 0, \tag{1c}$$

$$\nabla \cdot \tilde{\mathbf{u}} = 0, \quad \nabla \cdot \tilde{\mathbf{b}} = 0, \tag{1d}$$

respectively, that govern the evolution of infinitesimal velocity perturbations $\tilde{\mathbf{u}}$, temperature perturbations $\tilde{\Theta}$, and magnetic field perturbations $\tilde{\mathbf{b}}$ away from the static state. The equations have been non-dimensionalized using the radius r_0 as a unit of length, r_0^2/η as a unit of time, $\eta^2/g\alpha r_0^4$ as a unit of temperature, and $\sqrt{\mu\varrho}\eta/r_0$ as a unit of magnetic flux density. The dimensionless magnetic field takes the form $\mathbf{j} \times \mathbf{r} + \tilde{\mathbf{b}}$, where $\mathbf{j} = j\mathbf{k}$ is the vector of the density of the imposed electric current. The problem is then characterized by five dimensionless parameters, namely, the Rayleigh number, the Coriolis parameter, the Prandtl number, the magnetic Prandtl number P_m, and the non-dimensional current density given by

$$\hat{R} = \frac{\alpha g \beta r_0^6}{\eta \kappa}, \quad \tau = \frac{2\Omega r_0^2}{\eta}, \quad P = \frac{\nu}{\kappa}, \quad P_m = \frac{\nu}{\eta}, \quad j, \tag{2}$$

respectively. In fact, in the results obtained below the two Prandtl numbers enter only as their ratio $S = P/P_m = \eta/\kappa$. To signify that in our definition of the Rayleigh number the magnetic diffusivity replaces the kinematic viscosity we have attached a hat to $\hat{R}$.

3. Perturbation Analysis Results

Without loss of generality we assume that the velocity, the magnetic field, and the temperature perturbations have an exponential dependence on time t and on the azimuthal angle ϕ. Further, as both the velocity field and the magnetic field are solenoidal we use the poloidal-toroidal decomposition

$$\tilde{\mathbf{u}} = \mathbf{u} \exp\big(i(\omega\tau t + m\phi)\big) = \big(\nabla \times (\nabla v \times \mathbf{r}) + \nabla w \times \mathbf{r}\big) \exp\big(i(\omega\tau t + m\phi)\big), \tag{3a}$$

$$\tilde{\mathbf{b}} = \mathbf{b} \exp\big(i(\omega\tau t + m\phi)\big) = \big(\nabla \times (\nabla h \times \mathbf{r}) + \nabla g \times \mathbf{r}\big) \exp\big(i(\omega\tau t + m\phi)\big), \tag{3b}$$

$$\tilde{\Theta} = \Theta \exp\big(i(\omega\tau t + m\phi)\big). \tag{3c}$$

Equation (1b) can now be written in the form

$$\mathbf{b} = \frac{m\gamma}{\omega} \mathbf{u} - \frac{i}{\omega\tau} \nabla^2 \mathbf{b}, \tag{4}$$

where the parameter γ is defined as $\gamma = j/\tau$, and in the ∇-operator the ϕ-derivative is replaced by its eigenfactor im. This allows us to transform Equation (1a) in the form

$$i\omega\left(1 - \frac{m^2}{\omega^2}\gamma^2\right)\mathbf{u} + \left(1 - \frac{m}{\omega}\gamma^2\right)\mathbf{k} \times \mathbf{u} - \nabla\check{\pi} = \frac{1}{\tau}\Theta\mathbf{r} + \frac{P_m}{\tau}\nabla^2\mathbf{u} + \frac{m^2\gamma^2}{\omega^2\tau}\nabla^2\mathbf{u}$$
$$+ \frac{2m\gamma^2}{i\omega^2\tau}\mathbf{k} \times \nabla^2\mathbf{u} + \frac{m\gamma}{\omega\tau}\nabla^2\mathbf{b}_b + \frac{2\gamma}{i\omega\tau}\mathbf{k} \times \nabla^2\mathbf{b}_b + \frac{P_m}{\tau}\nabla^2\mathbf{u}_b, \tag{5}$$

where $\check{\pi}$ is the effective pressure. In Equation (5), the magnetic field $\mathbf{b}$ appears only in the form of the boundary layer correction $\mathbf{b}_b$, which is required as the basic dissipationless solution does not satisfy all boundary conditions [25]. For the same reason, the Ekman layer correction $\mathbf{u}_b$ must be introduced [22].

Following the procedure of earlier papers [22,26], we use a perturbation approach and solve Equation (1a) in the limit of large τ, using the ansatz

$$\mathbf{u} = \mathbf{u}_0 + \tau^{-1}\mathbf{u}_1 + ..., \quad \omega = \omega_0 + \tau^{-1}\omega_1 + ..., \quad \mathbf{b} = \mathbf{b}_0 + \tau^{-1}\mathbf{b}_1 + ..., \tag{6}$$

The heat equation is solved unperturbed.

3.1. Zeroth-Order Approximation

In the following we shall assume the limit of large τ such that in zeroth order of approximation the right hand side of Equation (5) can be neglected. The left hand side together with the condition $\nabla \cdot \mathbf{u} = 0$ is of the same form as the equation for inertial modes [22,26]. In the nonmagnetic case, the inertial modes corresponding to the sectorial spherical harmonics yield the lowest critical Rayleigh numbers for the onset of convection [26]. We shall assume that this property continues to hold as long as the parameter γ is sufficiently small so that the nonmagnetic limit is approached in the left-hand side of Equation (5). The sectorial inertial modes are given by

$$v_0 = P_m^m(\cos\theta) f(r), \quad w_0 = P_{m+1}^m(\cos\theta)\psi(r), \tag{7a}$$

with

$$f(r) = r^m - r^{m+2}, \qquad \psi(r) = r^{m+1}\frac{2im(m+2)}{(2m+1)(\lambda_0(m^2+3m+2)-m)}, \tag{7b}$$

where λ_0

$$\lambda_0 = \frac{1}{m+2}\left(1 \pm \sqrt{\frac{m^2+4m+3}{2m+3}}\right), \tag{7c}$$

is the frequency of the inertial modes. The sectorial magneto-inertial modes are then described by the same velocity field (7a) and by a magnetic field $\mathbf{b}_0 = m\gamma\mathbf{u}_0/\omega_0$. In the above expressions, the subscript 0 refers to the dissipationless solution of Equation (1). The frequency ω_0 of the magneto-inertial waves is determined by

$$\lambda_0 = \frac{\omega_0^2 - m^2\gamma^2}{\omega_0 - m\gamma^2}, \tag{8}$$

which yields

$$\omega_0 = \frac{\lambda_0}{2} \pm \sqrt{\frac{\lambda_0^2}{4} + m\gamma^2(m-\lambda_0)}. \tag{9}$$

With account of (7c), this dispersion relation allows for a total of four different frequencies ω_0. For small values of γ^2, these are given by

$$\omega_{01,2} = \frac{1}{m+2}\left(1 \pm \sqrt{\frac{m^2+4m+3}{2m+3}}\right) + m^2\gamma^2(m+2)\left(1 \pm \sqrt{\frac{m^2+4m+3}{2m+3}}\right)^{-1} - m\gamma^2, \tag{10a}$$

$$\omega_{03,4} = -m^2\gamma^2(m+2)\left(1 \pm \sqrt{\frac{m^2+4m+3}{2m+3}}\right)^{-1} + m\gamma^2. \tag{10b}$$

The upper sign in expression (10a) refers to retrogradely propagating modified inertial waves, while the lower sign corresponds to the progradely traveling variety. The effect of the magnetic field tends to increase the absolute value of the frequency in both cases. Expression (10b) describes the dispersion of the slow magnetic waves. The upper sign refers to the progradely traveling modified Alfven waves and the lower sign corresponds to retrogradely propagating modified Alfven waves.

3.2. First-Order Approximation

The magneto-inertial waves described by expressions (7a) satisfy the condition that the normal component of the velocity field vanishes at the boundary. This property implies that the normal component of the magnetic field vanishes there as well. Additional boundary conditions must be specified when the full dissipative problem described by (5) is considered. We shall assume a stress-free boundary with either a fixed temperature (case A) or a thermally insulating boundary (case B),

$$\mathbf{r}\cdot\mathbf{u} = \mathbf{r}\cdot\nabla(\mathbf{r}\times\mathbf{u})/r^2 = 0 \quad \text{and} \quad \left\{ \begin{array}{ll} \Theta = 0 & \text{(case A)} \\ \partial_r\Theta = 0 & \text{(case B)} \end{array} \right\} \quad \text{at} \quad r = 1. \tag{11}$$

Additionally, we shall assume an electrically insulating exterior of the sphere which requires

$$g = 0 \quad \text{at} \quad r = 1 \tag{12}$$

and the matching of the poloidal magnetic field to a potential field outside the sphere.

After the ansatz (6) has been inserted into Equation (5) such that terms with $\mathbf{u}_1$ appear on the left hand side, while those with $\mathbf{u}_0$ and ω_0 appear on the right hand side, we obtain the solvability condition for the equation for $\mathbf{u}_1$ by multiplying it with $\mathbf{u}_0^*$ and averaging it over the fluid sphere,

$$\begin{aligned} i\omega_1\left\langle|\mathbf{u}_0|^2\right\rangle\left(1+\left(\frac{m^2}{\omega_0^2}-\frac{m(\omega_0^2-m^2\gamma^2)}{\omega_0^2(\omega_0-m\gamma^2)}\right)\gamma^2\right) \\ = \langle\Theta\mathbf{r}\cdot\mathbf{u}_0^*\rangle+\left(\langle\mathbf{u}_0^*\cdot\nabla^2\mathbf{u}_0\rangle\frac{m\gamma}{\omega_0}+\langle\mathbf{u}_0^*\cdot\nabla^2\mathbf{b}_{0b}\rangle\right)\left(\frac{m}{\omega}-\frac{\omega_0^2-m^2\gamma^2}{\omega_0^2-m\omega_0\gamma^2}\right)\frac{\gamma}{\tau}, \end{aligned} \tag{13}$$

where the brackets $\langle ... \rangle$ indicate the average over the fluid sphere and the $*$ indicates the complex conjugate. We have neglected all terms connected with viscous dissipation, i.e., we have assumed the vanishing of P_m, as we wish to focus on the effect of ohmic dissipation. The effects of viscous dissipation have been dealt with in the earlier paper [26]. Because $\langle\mathbf{u}_0^*\cdot\nabla^2\mathbf{u}_0\rangle$ vanishes, as demonstrated in [27], we must consider only the influence of the boundary layer magnetic field $\mathbf{b}_{0b}$. It is determined by the equation

$$i\omega_0\tau\mathbf{b}_{0b} = \nabla^2\mathbf{b}_{0b}. \tag{14}$$

As the solutions of this equation are characterized by gradients of the order $\sqrt{\tau}$, the boundary layer correction needed for the poloidal component is of the order $\sqrt{\tau}$ smaller than the correction needed for the toroidal component. For large τ we need to take into account only the contribution g_{0b} given by

$$\begin{aligned} g_{0b} &= -g_0(r=1)\exp\left(-(1+is)(1-r)\sqrt{|\omega_0|\tau/2}\right) \\ &= -\frac{m\gamma}{\omega_0}w_0(r=1)\exp\left(-(1+is)(1-r)\sqrt{|\omega_0|\tau/2}\right), \end{aligned} \tag{15}$$

where s denotes the sign of ω_0. The solvability condition thus becomes reduced to

$$i\omega_1\left\langle|\mathbf{u}_0|^2\right\rangle\left(1+\left(\frac{m\gamma^2(m-\omega_0)}{\omega_0(\omega_0-m\gamma^2)}\right)\right) \tag{16}$$
$$=\frac{1}{\tau}\langle\Theta\mathbf{r}\cdot\mathbf{u}_0^*\rangle-\frac{3}{2}\frac{m\gamma^2(m-\omega_0)(s+i)}{(\omega_0-m\gamma^2)\sqrt{2|\omega_0|\tau}}\int_{-1}^{1}|P_m^{m+1}|^2\mathrm{d}(\cos\theta)$$
$$\times(m+1)(m+2)\left|\frac{2m(m+2)}{(2m+1)\left(\frac{\omega_0^2-m^2\gamma^2}{\omega_0-m\gamma^2}(m+1)(m+2)-m\right)}\right|^2.$$

3.2.1. Explicit Expressions in the Limit $\tau S \ll 1$

Equation (1c) for Θ can most easily be solved in the limit of vanishing $\omega_0\tau S$. In this limit, we obtain for Θ,

$$\Theta = P_m^m(\cos\theta)\exp(im\varphi+i\omega\tau t)q(r), \tag{17}$$

with

$$q(r)=\hat{R}\left(\frac{m(m+1)r^{m+4}}{(m+5)(m+4)-(m+1)m}-\frac{m(m+1)r^{m+2}}{(m+3)(m+2)-(m+1)m}-cr^m\right), \tag{18}$$

where the coefficient c is given by

$$c=\begin{cases}\dfrac{m(m+1)}{(m+5)(m+4)-(m+1)m}-\dfrac{m(m+1)}{(m+3)(m+2)-(m+1)m}, & \text{case A,}\\[2ex] \dfrac{(m+4)(m+1)}{(m+5)(m+4)-(m+1)m}-\dfrac{(m+2)(m+1)}{(m+3)(m+2)-(m+1)m}, & \text{case B.}\end{cases} \tag{19}$$

As Θ and the left hand side of Equation (16) is imaginary, the real parts of the two terms on the right hand side must balance. We thus obtain for $\hat{R}$ the result

$$\hat{R}=s\sqrt{\frac{\tau}{2|\omega_0|}}\frac{\gamma^2(m-\omega_0)}{(\omega_0-m\gamma^2)}\left|\frac{m(m+2)}{\frac{\omega_0^2-m^2\gamma^2}{\omega_0-m\gamma^2}(m+1)(m+2)-m}\right|^2 \tag{20}$$
$$\times(2m+9)(2m+7)(2m+5)^2(2m+3)\frac{m+2}{m+1}\frac{1}{b},$$

where the coefficient b assumes the values

$$b=\begin{cases}m(10m+27) & \text{case A,}\\ 14m^2+59m+63 & \text{case B.}\end{cases} \tag{21}$$

Obviously the lowest value of $\hat{R}$ is usually reached for $m=1$, but the fact that there are four different possible values of the frequency ω_0 complicates the determination of the critical value $\hat{R}_c$. Expression (20) is also of interest, however, in the case of spherical fluid shells when the $(m=1)$-mode is affected most strongly by the presence of the inner boundary. Convection modes corresponding to higher values of m may then become preferred at onset as their r-dependence decays more rapidly with distance from the outer boundary according to relationships (7b).

3.2.2. Solution of the Heat Equation in the General Case

For the solution of Equation (1c), in the general case it is convenient to use the Green's function method. The Green's function $G(r,a)$ is obtained as solution of the equation

$$\Big[\partial_r r^2 \partial_r + \big(-i\omega_0 \tau S\, r^2 - m(m+1)\big)\Big] G(r,a) = \delta(r-a), \tag{22}$$

which can be solved in terms of the spherical Bessel functions $j_m(\mu r)$ and $y_m(\mu r)$,

$$G(r,a) = \begin{cases} G_1(r,a) = A_1 j_m(\mu r) & \text{for } 0 \le r < a, \\ G_2(r,a) = A j_m(\mu r) + B y_m(\mu r) & \text{for } a < r \le 1, \end{cases} \tag{23}$$

where

$$\mu \equiv \sqrt{-i\omega_0 \tau S}, \quad A_1 = \mu\left(y_m(\mu a) - j_m(\mu a)\frac{y_m(\mu)}{j_m(\mu)}\right), \tag{24a}$$

$$A = -\mu j_m(\mu a)\frac{y_m(\mu)}{j_m(\mu)}, \quad B = \mu j_m(\mu a). \tag{24b}$$

A solution of Equation (1c) can be obtained in the form

$$q(r) = -m(m+1)\hat{\mathrm{R}}\left(\int_0^r G_2(r,a)\left(a^m - a^{m+2}\right)a^2 \mathrm{d}a + \int_r^1 G_1(r,a)\left(a^m - a^{m+2}\right)a^2 \mathrm{d}a\right). \tag{25}$$

Evaluations of these integrals for $m = 1$ yield the expressions

$$q(r) = \begin{cases} \dfrac{2\hat{\mathrm{R}}}{(\omega_0 \tau S)^2}\left(r(\mu^2+10) - \mu^2 r^3 - \dfrac{10(\mu r\cos(\mu r) - \sin(\mu r))}{r^2(\mu\cos\mu - \sin\mu)}\right) & \text{case A}, \\ \dfrac{2\hat{\mathrm{R}}}{(\omega_0 \tau S)^2}\left(r(2\mu^2+10) - \mu^2 r^3 - \dfrac{(\mu^2-10)(\mu r\cos(\mu r) - \sin(\mu r))}{r^2(2\mu\cos\mu - (2-\mu^2)\sin\mu)}\right) & \text{case B}. \end{cases} \tag{26}$$

Lengthier expressions are obtained for $m > 1$. This first-order approximation of the temperature perturbation is illustrated in Figure 2 for the preferred modes of inertial magnetoconvection. The preferred modes of convection at onset are determined by minimizing the values of the critical Rayleigh number $\hat{R}$ at given values of the other parameters. The critical Rayleigh number $\hat{R}$ and frequency ω_1 are calculated on the basis of Equation (16) using expressions (26). In the case $m = 1$ we obtain

$$\begin{aligned} \hat{\mathrm{R}} = \frac{189}{20}\, & \frac{s\sqrt{2\tau}\gamma^2(\omega_0 - 1)}{\sqrt{|\omega_0|}(\omega_0 - \gamma^2)(6\lambda_0 - 1)^2} \\ & \times \begin{cases} \left(\mu^{-4} - 525\mu^{-8} - 175\,\mathrm{Re}\left\{\dfrac{\sin\mu}{\mu^6(\mu\cos\mu - \sin\mu)}\right\}\right)^{-1} & \text{case A}, \\ \left(\mu^{-4} + 231\mu^{-8} + 7\,\mathrm{Re}\left\{\dfrac{(\mu^5 - 8\mu^3 + 9\mu)\cos\mu - 9\sin\mu}{\mu^8((\mu^2-2)\sin\mu + 2\mu\cos\mu)}\right\}\right)^{-1} & \text{case B}, \end{cases} \end{aligned} \tag{27}$$

where Re{} indicates the real part of the term enclosed by {}.

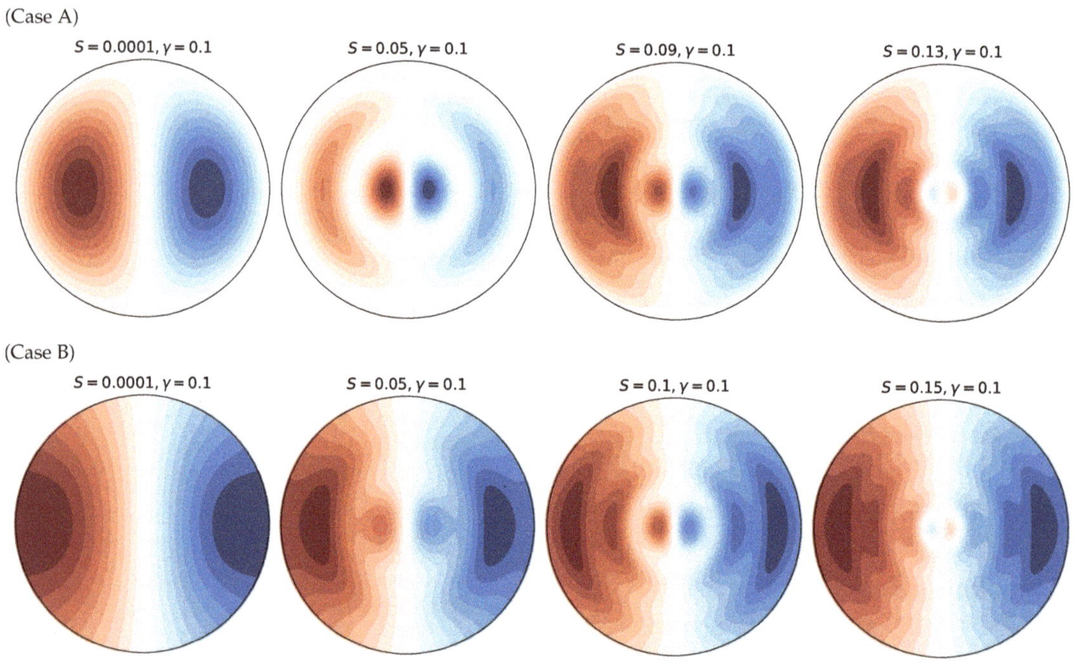

Figure 2. Contour plots of the (normalized) temperature perturbation $\Theta(r)$ of the preferred mode given by Equations (17) and (26) in case A (top row) and case B (bottom row) with values of S and γ as specified in the panels and $\tau = 10^4$, $m = 1$ and frequency ω_{01}. Expressions (17) and (18) for the limit $\tau S \ll 1$ appear identical to the plots in the first column.

4. Discussion

Expressions (27) have been plotted as functions of S in Figures 3c and 4c for cases A and B, respectively. Four distinct curves appear as there are four possible values of ω_0 for each m. For values S of the order 10^{-2} or less, expressions (20) are well approached. The retrograde mode corresponding to the positive sign in (7c) always yields the lower value of $\hat{R}$, but it loses its preference to the progradely traveling modified Alfven mode corresponding to the upper sign in (10b) as S becomes of the order 10^{-1} or larger. This transition can be understood on the basis of the increasing difference in phase between Θ and u_r with increasing S. While the mode with the largest absolute value of ω is preferred as long as Θ and u_r are in phase, the mode with the minimum absolute value of ω becomes preferred as the phase difference increases as the latter is detrimental to the work done by the buoyancy force. The frequency perturbation ω_1 usually makes only a small contribution to ω which tends to decrease the absolute value of ω. This transition shifts towards smaller values of S and γ as τ is increased as illustrated in Figure 5. The magneto-inertial convective modes corresponding to higher values of $m = 1 \ldots 8$ exhibit similar behavior as Figures 3d and 4d demonstrate for the cases A and B, respectively. The value $m = 1$ is always the preferred value of the wavenumber, except possibly in a very narrow range near $\gamma = 0.03$, as indicated by Figure 3a,b in the case A, and possibly near $\gamma = 0.02$ in the case B and Figure 4a,b. The axisymmetric mode $m = 0$, given for comparison in Figures 3c,d and 4c,d, is never preferred in contrast to the purely non-magnetic case where it becomes the critical one near the transition from retrograde to prograde inertial convection modes as seen in Figure 6.

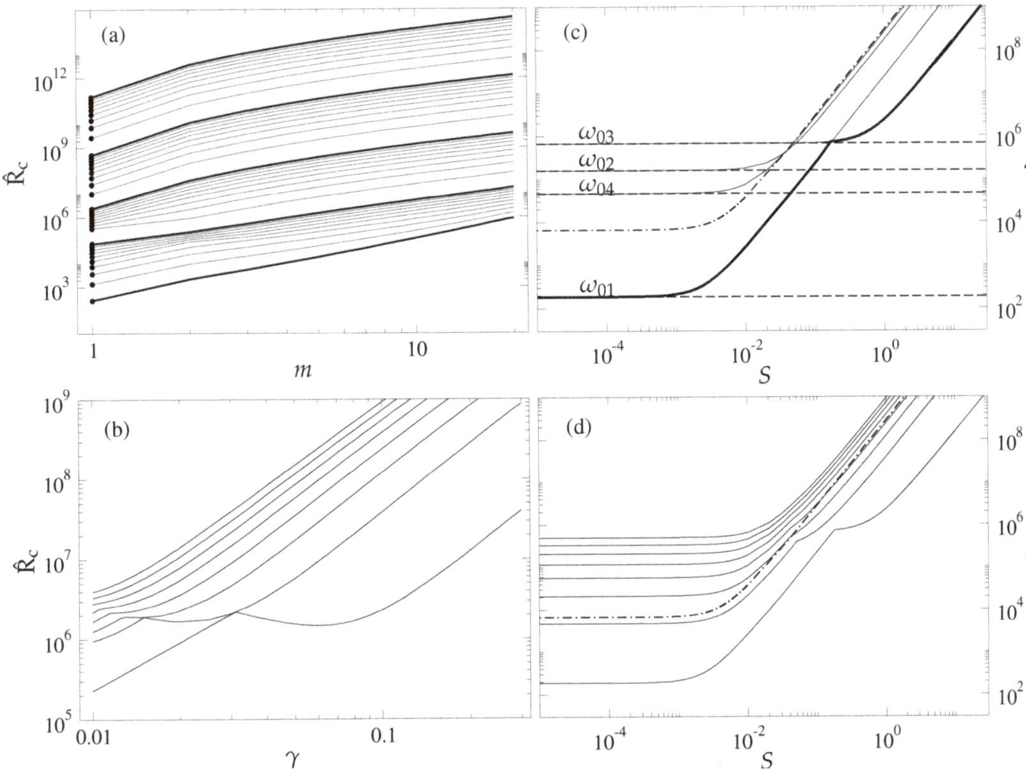

Figure 3. Case A. (**a**) The critical Rayleigh number $\hat{R}_c$ as a function of the wave number m for $\gamma = 0.1$ and $\tau = 10^2 \ldots 10^6$ increasing from bottom with log-scale decades given by the five thick lines. (**b**) The critical Rayleigh number $\hat{R}_c$ as a function of γ for $S = 1$ and $m = 1 \ldots 8$ increasing from bottom. (**c**) Competition of modes with increasing S for $\gamma = 0.1$ and $m = 1$. Explicit expressions (20) in the limit $\tau S \ll 1$ are shown by broken lines. (**d**) The critical Rayleigh number $\hat{R}_c$ as a function of S for $\gamma = 0.1$ and $m = 1 \ldots 8$ increasing from bottom. The axisymmetric mode $m = 0$ is given for comparison in panels (**c**,**d**) by a dot-dashed line. In panels (**b**–**d**) $\tau = 10^4$.

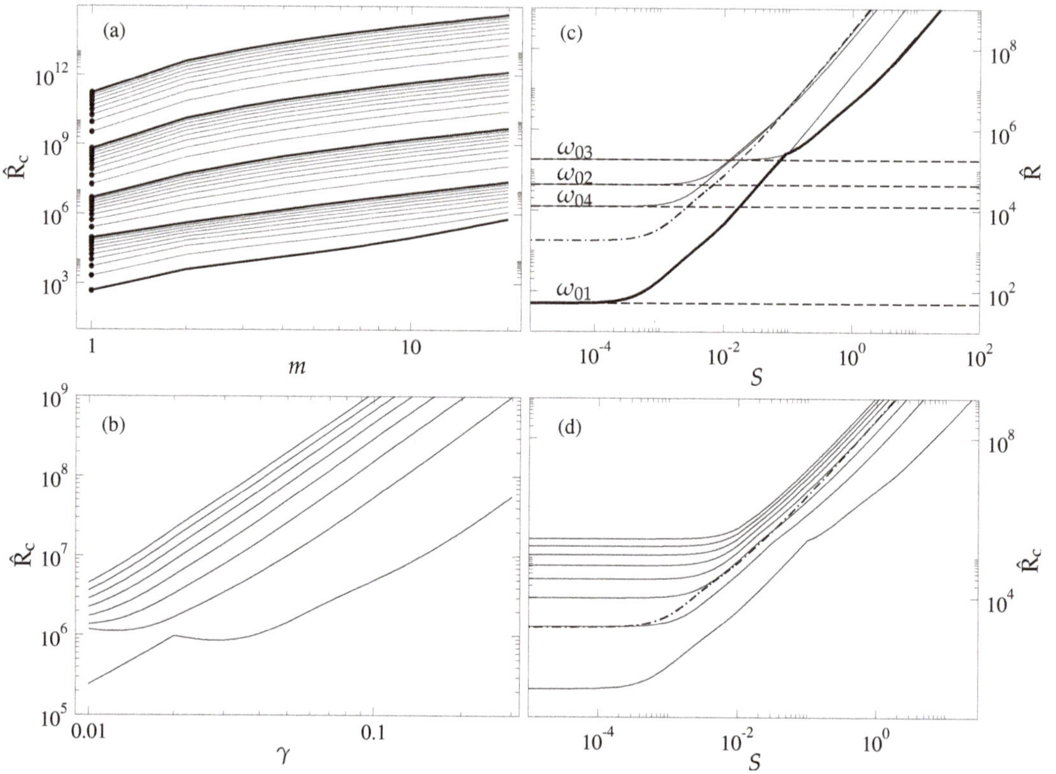

Figure 4. Case B. (**a**) The critical Rayleigh number $\hat{R}_c$ as a function of the wave number m for $\gamma = 0.1$ and $\tau = 10^2 \dots 10^6$ increasing from bottom with log-scale decades given by the five thick lines. (**b**) The critical Rayleigh number $\hat{R}_c$ as a function of γ for $S = 1$ and $m = 1 \dots 8$ increasing from bottom. (**c**) Competition of modes with increasing S for $\gamma = 0.1$ and $m = 1$. Explicit expressions (20) in the limit $\tau S \ll 1$ are shown by broken lines. (**d**) The critical Rayleigh number $\hat{R}_c$ as a function of S for $\gamma = 0.1$ and $m = 1 \dots 8$ increasing from bottom. The axisymmetric mode $m = 0$ is given for comparison in panels (**c**,**d**) by a dot-dashed line. In panels (**b**–**d**) $\tau = 10^4$.

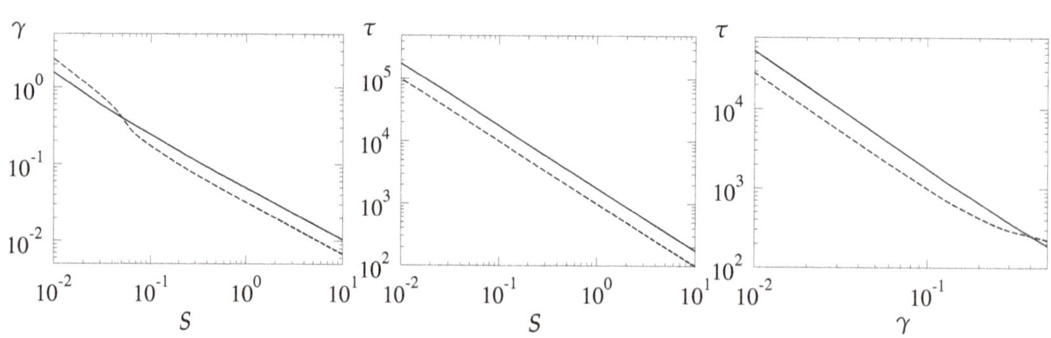

Figure 5. The boundary where the transition from modes characterised by ω_{01} to modes characterised by ω_{03} occurs in various cross-sections of the parameter space. The value of the parameters are $m = 1$, $S = 1$, $\gamma = 0.1$, and $\tau = 5000$ where they are not varied on the axes. Case A is denoted by a solid lines and Case B by broken lines.

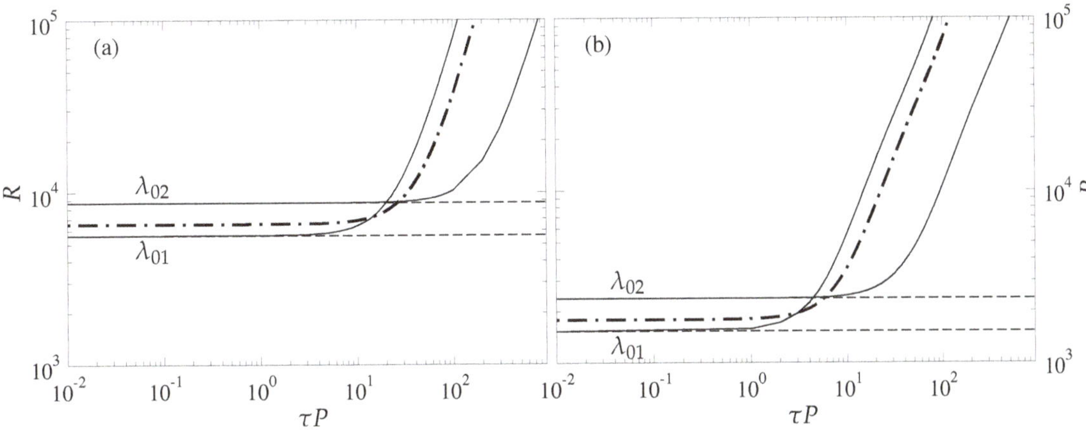

Figure 6. Competition of modes with increasing τP in the non-magnetic case discussed in [26]. The Rayleigh number R as a function of τP for $m = 0$ (thick dash-dotted lines) and $m = 1$ (thin lines). Results based on the explicit expressions (4.6) and (3.4) from [26] are shown in solid lines and broken lines, respectively, in the case $m = 1$. (**a**) Case A, fixed temperature boundary conditions. (**b**) Case B, insulating thermal boundary conditions.

For very large values of τ and S, the Rayleigh number $\hat{R}$ increases in proportion to $\sqrt{\tau}(\tau S)^2$ for fixed m. In spite of this strong increase, Θ remains of the order $\tau^{3/2}S$ on the right hand side of Equation (1a). The perturbation approach thus continues to be valid for $\tau \longrightarrow \infty$ as long as $S \ll 1$ can be assumed. For any fixed low value of S, however, the onset of convection in the form of prograde inertial modes will be replaced with increasing τ at some point by the onset in the form of columnar magneto-convection because the latter obeys an approximate asymptotic relationship for R of the form $\tau^{4/3}$ (see, for example, Eltayeb et al. [28]). This second transition depends on the value of S and will occur at higher values of τ and R for lower values of S. There is little chance that magneto-inertial convection occurs in the Earth's core, for instance, as S is of the order 30,000 while the usual estimate for τ is 10^{15}, but it might be relevant for understanding of rapidly rotating stars with strong magnetic fields.

5. Conclusions

A main result of the analysis of this paper is that for small values of the magnetic Prandtl number P_m and γ an azimuthal magnetic field exerts a stabilizing influence on the onset of convection in the form of sectorial magneto-inertial modes. As a consequence, magneto-convection with azimuthal wave number $m = 1$ is generally preferred at onset for both thermally-infinitely conducting and thermally-insulating boundaries. In contrast, in the absence of a magnetic field, inertial modes with azimuthal wave number $m = 1$ are preferred, but only in the case of thermally-insulating boundaries, while in the case with infinitely conducting thermal boundaries large azimuthal wave numbers are preferred soon after moderately large rotation is reached [26] and magnetic field is absent. Axisymmetric magneto-convection is never the preferred mode at onset while in the non-magnetic case it appears to be realized in a minute region of the parameter space only. These results are also in contrast to previous magnetoconvection results obtained for larger values of P_m where a destabilizing role of the azimuthal magnetic field has been found.

The region of the parameter space investigated in the present paper differs considerably from those analyzed in previous work. Most authors have emphasized regimes of high magnetic flux density where the magnetic field exerts a destabilizing influence and strongly decreases the critical Rayleigh number for onset of convection (see, for example, in [28,29]). Unfortunately, no explicitly analytical results are possible in that region of the parameter space. Moreover, the choice of parameter values has often been motivated by applications to the problem of the geodynamo in which case the parameter S is large, perhaps as large as 10^5, when molecular diffusivities are used. On the other hand, small values of S may be relevant for magneto-convection in stars where a high thermal diffusivity is generated by radiation.

Author Contributions: Conceptualization, F.H.B. and R.D.S.; formal analysis, F.H.B. and R.D.S.; data curation, R.D.S.; writing—original draft preparation, F.H.B.; writing—review and editing, F.H.B. and R.D.S.; visualization, R.D.S. funding acquisition, R.D.S. All authors have read and agreed to the published version of the manuscript.

Funding: The research of R.S. was funded by the Leverhulme Trust grant number RPG-2012-600.

Institutional Review Board Statement: Not applicable.

Informed Consent Statement: Not applicable.

Data Availability Statement: The data presented in this study are plotted from the analytical expressions printed here.

Conflicts of Interest: The authors declare no conflict of interest.

References

1. Jones, C.A. Planetary Magnetic Fields and Fluid Dynamos. *Annu. Rev. Fluid Mech.* **2011**, *43*, 583–614. [CrossRef]
2. Roberts, P.H.; King, E.M. On the genesis of the Earth's magnetism. *Rep. Prog. Phys.* **2013**, *76*, 096801. [CrossRef] [PubMed]
3. Charbonneau, P. Solar Dynamo Theory. *Annu. Rev. Astron. Astrophys.* **2014**, *52*, 251–290. [CrossRef]
4. Ogilvie, G.I. Astrophysical fluid dynamics. *J. Plasma Phys.* **2016**, *82*, 205820301. [CrossRef]
5. Glatzmaier, G.A. Geodynamo Simulations—How Realistic Are They? *Annu. Rev. Earth Planet. Sci.* **2002**, *30*, 237–257. [CrossRef]
6. Miesch, M.; Matthaeus, W.; Brandenburg, A.; Petrosyan, A.; Pouquet, A.; Cambon, C.; Jenko, F.; Uzdensky, D.; Stone, J.; Tobias, S.; et al. Large-Eddy Simulations of Magnetohydrodynamic Turbulence in Heliophysics and Astrophysics. *Space Sci. Rev.* **2015**, *194*, 97–137. [CrossRef]
7. Chandrasekhar, S. *Hydrodynamic and Hydromagnetic Stability*; International Series of Monographs on Physics; Clarendon Press: Oxford, UK, 1961.
8. Zhang, K.; Schubert, G. Magnetohydrodynamics in Rapidly Rotating spherical Systems. *Annu. Rev. Fluid Mech.* **2000**, *32*, 409–443. [CrossRef]
9. Weiss, N.O.; Proctor, M.R.E. *Magnetoconvection*; Cambridge University Press: Cambridge, UK, 2014. [CrossRef]
10. Hide, R. Free Hydromagnetic Oscillations of the Earth's Core and the Theory of the Geomagnetic Secular Variation. *Philos. Trans. R. Soc. Math. Phys. Eng. Sci.* **1966**, *259*, 615–647. [CrossRef]
11. Malkus, W.V.R. Hydromagnetic planetary waves. *J. Fluid Mech.* **1967**, *28*, 793–802. [CrossRef]
12. Finlay, C.C.; Dumberry, M.; Chulliat, A.; Pais, M.A. Short Timescale Core Dynamics: Theory and Observations. *Space Sci. Rev.* **2010**, *155*, 177–218. [CrossRef]
13. Hori, K.; Takehiro, S.; Shimizu, H. Waves and linear stability of magnetoconvection in a rotating cylindrical annulus. *Phys. Earth Planet. Inter.* **2014**, *236*, 16–35. [CrossRef]
14. Hori, K.; Jones, C.A.; Teed, R.J. Slow magnetic Rossby waves in the Earth's core. *Geophys. Res. Lett.* **2015**, *42*, 6622–6629. [CrossRef]
15. Busse, F.H. Thermal instabilities in rapidly rotating systems. *J. Fluid Mech.* **1970**, *44*, 441. [CrossRef]
16. Busse, F.H. Asymptotic theory of convection in a rotating, cylindrical annulus. *J. Fluid Mech.* **1986**, *173*, 545. [CrossRef]
17. Simitev, R.; Busse, F. Prandtl-number dependence of convection-driven dynamos in rotating spherical fluid shells. *J. Fluid Mech.* **2005**, *532*, 365. [CrossRef]
18. Zhang, K.K.; Busse, F.H. On the onset of convection in rotating spherical shells. *Geophys. Astrophys. Fluid Dyn.* **1987**, *39*, 119–147. [CrossRef]
19. Ardes, M.; Busse, F.; Wicht, J. Thermal convection in rotating spherical shells. *Phys. Earth Planet. Int.* **1997**, *99*, 55–67. [CrossRef]
20. Simitev, R.; Busse, F. Patterns of convection in rotating spherical shells. *New J. Phys.* **2003**, *5*, 97. [CrossRef]
21. Plaut, E.; Busse, F.H. Multicellular convection in rotating annuli. *J. Fluid Mech.* **2005**, *528*, 119–133. [CrossRef]
22. Zhang, K. On coupling between the Poincaré equation and the heat equation. *J. Fluid Mech.* **1994**, *268*, 211–229. [CrossRef]
23. Zhang, K. On coupling between the Poincaré equation and the heat equation: non-slip boundary condition. *J. Fluid Mech.* **1995**, *284*, 239–256. [CrossRef]
24. Zhang, K.; Liao, X. *Theory and Modeling of Rotating Fluids: Convection, Inertial Waves and Precession*; Cambridge Monographs on Mechanics, Cambridge University Press: Cambridge, UK, 2017. [CrossRef]
25. Zhang, K.; Busse, F.H. On hydromagnetic instabilities driven by the Hartmann boundary layer in a rapidly rotating sphere. *J. Fluid Mech.* **1995**, *304*, 263–283. [CrossRef]
26. Busse, F.H.; Simitev, R. Inertial convection in rotating fluid spheres. *J. Fluid Mech.* **2004**, *498*, 23–30. [CrossRef]

27. Zhang, K.; Earnshaw, P.; Liao, X.; Busse, F.H. On inertial waves in a rotating fluid sphere. *J. Fluid Mech.* **2001**, *437*, 103–119. [CrossRef]
28. Eltayeb, I.A.; Kumar, S.; Hide, R. Hydromagnetic convective instability of a rotating, self-gravitating fluid sphere containing a uniform distribution of heat sources. *Proc. R. Soc. Lond. A* **1977**, *353*, 145–162. [CrossRef]
29. Fearn, D.R. Thermally driven hydromagnetic convection in a rapidly rotating sphere. *Proc. R. Soc. Lond. A* **1979**, *369*, 227–242. [CrossRef]

fluids

MDPI

Article

Numerical Study of Rotating Thermal Convection on a Hemisphere

Patrick Fischer [1,*], Charles-Henri Bruneau [1] and Hamid Kellay [2]

1 Institut de Mathématiques de Bordeaux (IMB), Université de Bordeaux, CNRS UMR 5251, F-33400 Talence, France; charles-henri.bruneau@u-bordeaux.fr

2 Laboratoire Ondes et Matiere d'Aquitaine (LOMA), Université de Bordeaux, CNRS UMR 5798, F-33400 Talence, France; hamid.kellay@u-bordeaux.fr

* Correspondence: patrick.fischer@math.u-bordeaux.fr

Received: 29 September 2020; Accepted: 12 October 2020; Published: 20 October 2020

check for updates

Abstract: Numerical simulations of rotating two-dimensional turbulent thermal convection on a hemisphere are presented in this paper. Previous experiments on a half soap bubble located on a heated plate have been used for studying thermal convection as well as the effects of rotation on a curved surface. Here, two different methods have been used to produce the rotation of the hemisphere: the classical rotation term added to the velocity equation, and a non-zero azimuthal velocity boundary condition. This latter method is more adapted to the soap bubble experiments. These two methods of forcing the rotation of the hemisphere induce different fluid dynamics. While the first method is classically used for describing rotating Rayleigh–Bénard convection experiments, the second method seems to be more adapted for describing rotating flows where a shear layer may be dominant. This is particularly the case where the fluid is not contained in a closed container and the rotation is imposed on only one side of it. Four different diagnostics have been used to compare the two methods: the Nusselt number, the effective computation of the convective heat flux, the velocity and temperature fluctuations root mean square (RMS) generation of vertically aligned vortex tubes (to evaluate the boundary layers) and the energy/enstrophy/temperature spectra/fluxes. We observe that the dynamics of the convective heat flux is strongly inhibited by high rotations for the two different forcing methods. Also, and contrary to classical three-dimensional rotating Rayleigh–Bénard convection experiments, almost no significant improvement of the convective heat flux has been observed when adding a rotation term in the velocity equation. However, moderate rotations induced by non-zero velocity boundary conditions induce a significant enhancement of the convective heat flux. This enhancement is closely related to the presence of a shear layer and to the thermal boundary layer just above the equator.

Keywords: computational fluid dynamics; turbulence; rotating thermal convection; Rayleigh–Bénard

1. Introduction

Many atmospheric and geophysical flows are actually fluid flows on curved surfaces and can be modeled by two-dimensional thermal convection. In nature and technological applications [1], many physical phenomena are actually related to buoyancy-driven flows whether affected by background rotation or not. In particular, buoyancy driven by temperature, and affected by the rotation of the Earth, is one of the leading force in oceans [2–4] and atmosphere. The combined effects of rotation and buoyancy lead to the formation of large structures called cells (such as polar, Ferrel or Hadley [5] cells). Large scale flows generated by rotation and buoyancy are also observed in the outer core of the earth [6–9]. These large scale zonal flows are also observed on the Sun and other planets of our solar system [10,11]. The Rayleigh–Bénard convection model is commonly used for studying such physical phenomena. This model consists, in general, in a cylindrical convection cell on a rotating table [12–19]. Much insight about the role of rotation on such flows has been gained from such an experimental model. Other systems have been proposed to study the role of rotation on thermal convection and notably in a curved geometry which bears some relevance to atmospheric flows. A half soap bubble heated at the equator has been recently proposed in some numerical and physical experiments [20,21] to study thermal convection. Indeed, in these experiments, thermal convection and the movement of large scale structures have been observed on the surface of the bubble. The results obtained in these experiments have shown some strong similarities with atmospheric flows [20–22]. It has been shown in particular in these studies that the dynamics of vortices on the bubble surface and tropical cyclones [20–22] present the same statistical behavior. Heating on one side (the equator) is responsible for the creation of intense thermal convection leading to turbulent velocity and temperature fields whose statistical properties can be described by theories of turbulent thermal convection elaborated by Corrsin, Obukhov and Bolgiano [20,21]. Further and by subjecting the bubble to rotation, strong effects of the rotation have been observed on the nature of the fluctuations. These experiments of hemispherical bubbles heated at the equator allow to study thermal convection without the presence of lateral walls. A variety of physical phenomena of relevance to atmospheric and geophysical flows can be explored with such numerical and experimental setup.

We here consider such a hemispherical system numerically and explore two different ways for forcing the rotation of the bubble. The first method to force a rotation on a system is to add a rotation term in the mathematical model. This method is classically used in numerical simulations of rotating Rayleigh–Bénard convection where the cells (top and bottom plates, lateral walls) are submitted to a rotating force. It has also been used in our first study of rotating half soap bubble where second-order temperature structure functions have been used to compare the numerical simulations to the physical experiments [23]. However, this solid rotation does not exactly correspond to the experimental protocol where the bubble is located on a rotating table (it does not correspond to classical rotating Rayleigh–Bénard convection experiments either since, in this case, it is the boundaries, both upper and lower as well as lateral, of the container which produce the rotation of the fluid). In the experiments on rotating half soap bubbles, only the equator is subjected to the rotation force leading to the presence of a shear layer. The effects of this forcing as well as the differences with a global rotation forcing are the subject of this paper. As we will see below, the two methods affect the flow but their effects are markedly different from each other. The mathematical way to reproduce this phenomenon is to impose a non-zero azimuthal velocity at the equator. We will show in the sequel that these two different ways for forcing the half bubble into rotation, global rotation and azimuthal velocity at the equator, lead to different flow behaviors.

Even if our setup is different from the classical Rayleigh–Bénard setting, we will however compare our results with those obtained with classical rotating Raleigh–Bénard experiments. A stereographic transform, fully described in our previous papers, allows us to project the Navier–Stokes equations onto the equatorial plane. Regular Cartesian grid based methods can then be used to numerically solve the equations. The two rotation forcing models are described in Section 2 and the results are analyzed and discussed in Sections 3–6. In these sections, we will present and discuss the effects of rotation on the Nusselt number, the convective heat flux, the temperature and velocity fluctuations, and the spectral properties.

2. Mathematical Model and Numerical Computations

We first describe the mathematical model corresponding to the non-rotating case [24]. The equations for two-dimensional thermal convection under the Boussinesq approximation [25] can be used to describe the fluid behavior which does not depend on the coordinate system:

$$\left\{ \begin{array}{rcl} \dfrac{D\mathbf{U}}{Dt} & = & -\dfrac{1}{\rho}\nabla p + \nu\nabla^2\mathbf{U} - \beta T\mathbf{g} - F\mathbf{U} \\ \nabla\cdot\mathbf{U} & = & 0 \\ \dfrac{DT}{Dt} & = & \alpha\nabla^2 T - ST \end{array} \right. \tag{1}$$

where $\frac{D}{Dt} \equiv \frac{\partial}{\partial t} + (\mathbf{U}\cdot\nabla)$ is the total time derivative, $\mathbf{U}$ denotes the velocity, p the pressure, ρ the mass density, ν the kinematic viscosity, β the coefficient of thermal expansion, $\mathbf{g}$ the gravity field, T the temperature of the fluid, α the coefficient of thermal diffusion, F is the friction factor and S is the thermal dissipation coefficient. Unlike Rayleigh–Bénard convection, two additional terms, $F\mathbf{U}$ and ST, have been added in order to compensate respectively the injection of kinetic and thermal energies. As explained in [24], if the parameters F and S are kept equal to zero, an accumulation of kinetic and thermal energies leading to a numerical burst of the bubble can be observed. Indeed, in usual Rayleigh–Bénard experiments, the top plate is cooled allowing to keep a constant global averaged temperature. In the experiments on soap bubbles, we do not have a specific cooling process. Only the exchange of heat between the bubble and the surrounding air and the friction with the air stabilize experimental conditions. We have to take into consideration these interactions with the air in the numerical simulations to stabilize the computations. This is why we have to artificially remove a small part of thermal and kinetic energies in the equations. This stabilization process can be interpreted as a way to take into consideration the interaction between the bubble and the surrounding air. See [24] for more details about this stabilization process.

As mentioned in the introduction, the purpose of this paper is the study of two different methods for mathematically rotating a hemispherical cap subjected to a thermal gradient:

- Method 1: adding a rotation term $2\mathbf{\Omega}\times\mathbf{U}$ in the velocity equation, where $\mathbf{\Omega}$ denotes the rotation vector
- Method 2: adding a non-zero azimuthal velocity boundary condition U_f at the equator.

The equations with the rotation forcing term has to be written in a non-dimensional form. Using L, δT and the characteristic velocity $U = \sqrt{\beta g\delta TL}$ (where δT denotes the difference of temperature between the pole and the equator, and L the characteristic length equal to the radius of the bubble ($L = r$)), the equations can be made non-dimensional. As in many papers about thermal convection, the control parameters in our numerical simulations are the Rayleigh number $Ra = \frac{g\beta L^3\delta T}{\nu\alpha}$ and the Prandtl number $Pr = \frac{\nu}{\alpha}$. Various values for the Rayleigh and Prandtl numbers have been used, but the only numerical results presented in this article are for $Ra = 3.10^8$ and $Pr = 7$. We also have to define the corresponding dimensionless friction factor $\tilde{F} = \frac{FL}{U}$ and the dimensionless thermal dissipation coefficient $\tilde{S} = \frac{SL}{U}$. We remind here that the goal of these dimensionless factors is to take into consideration the interaction between the bubble and the air. The empirical evaluation of their respective values has been described in [24] and we chose here $\tilde{F} = \tilde{S} = 0.06$, which leads to a stationary state in the non-rotating case. Using the same notation for the dimensionless pressure, velocity and temperature, the non-dimensional equations, with a rotation forcing term (Method 1), can thus be written as:

$$\begin{cases} \frac{D\mathbf{U}}{Dt} &= -\nabla p + \sqrt{\frac{Pr}{Ra}}\nabla^2\mathbf{U} + T\mathbf{k_0} - \frac{1}{Ro}\mathbf{k_0}\times\mathbf{U} - \tilde{F}\mathbf{U} \\ \nabla\cdot\mathbf{U} &= 0 \\ \frac{DT}{Dt} &= \frac{1}{\sqrt{Pr.Ra}}\nabla^2 T - \tilde{S}T \end{cases} \tag{2}$$

where $Ro = \frac{\sqrt{\beta g \delta T L}}{2\Omega L}$ denotes the Rossby number, $\Omega = |\mathbf{\Omega}|$ the rotation rate and $\mathbf{k_0}$ the vertical unit vector. Adding a rotation term in the equation is classically used for modeling rotating Rayleigh–Bénard experiments. However, method 2 best describes our physical experiment where the bubble lays on top of a rotating table. Only the part of the bubble in contact with the table is submitted to the rotation forcing thus creating a shear stress and possibly shear induced instabilities. Both methods induce a rotation of the bubble but the corresponding flows present different physical behaviors in terms of the capacity of convecting the heat flux. Four analysis tools will be used in the next parts to compare both methods:

- the Nusselt number criterion
- the convective heat flux (defined in the sequel)
- the velocity and temperature fluctuations RMS (to evaluate the shear and thermal boundary layers)
- the energy/enstrophy/temperature spectra and fluxes

Two kinds of coordinate systems are considered here: $\mathbf{C_0}(x_0, y_0, z_0)$ the usual 3D orthogonal Cartesian coordinates and $\mathbf{C}(x, y)$ a 2D orthogonal Cartesian coordinate system obtained by a stereographic projection onto the equatorial plane. The stereographic projection is a well known mapping allowing to project a sphere onto a plane. Usually, the projection is defined on the entire sphere, but in the present study we do not need to described the whole sphere. Thus, we use in this study a version of the stereographic projection restricted to the northern hemisphere. It can be shown that this mapping is bijective and preserves the angles. The consequences being that a Cartesian grid on the plane corresponds to a Cartesian grid on the surface of the sphere. The two kind of coordinates are related to each other by the following equalities:

$$\begin{cases} x_0 &= \frac{2xr^2}{r^2 + x^2 + y^2} \\ y_0 &= \frac{2yr^2}{r^2 + x^2 + y^2} \\ z_0 &= r\frac{r^2 - (x^2 + y^2)}{r^2 + x^2 + y^2} \end{cases} \tag{3}$$

and

$$\begin{cases} x &= \frac{rx_0}{r + \sqrt{r^2 - (x_0^2 + y_0^2)}} \\ y &= \frac{ry_0}{r + \sqrt{r^2 - (x_0^2 + y_0^2)}} \end{cases} \tag{4}$$

where r denotes the radius of the sphere. Figure 1 illustrates the stereographic projection where a point P on the sphere is projected on P' on the equatorial plane. Using the stereographic coordinates and the calculations detailed in [24], the Navier–Stokes Equation (2) for a hemisphere of radius $r = 1$ can thus be written using stereographic coordinates:

$$\begin{aligned}
&\frac{\partial u}{\partial t}+u\frac{\partial u}{\partial x}+v\frac{\partial u}{\partial y}+\frac{2}{1+x^2+y^2}\left(-xu^2-2yuv+xv^2\right)= \\
&\quad\sqrt{\frac{Pr}{Ra}}\left[\frac{(1+x^2+y^2)^2}{4}\left(\frac{\partial^2 u}{\partial x^2}+\frac{\partial^2 u}{\partial y^2}\right)-u-\left(1+x^2+y^2\right)\left(x\frac{\partial u}{\partial x}+y\frac{\partial u}{\partial y}+y\frac{\partial v}{\partial x}-x\frac{\partial v}{\partial y}\right)\right] \\
&\quad-\left(\frac{1+x^2+y^2}{2}\right)^2\frac{\partial p}{\partial x}-x\beta gT+\frac{1}{Ro}v-\tilde{F}u. \\
&\frac{\partial v}{\partial t}+u\frac{\partial v}{\partial x}+v\frac{\partial v}{\partial y}+\frac{2}{1+x^2+y^2}\left(yu^2-2xuv-yv^2\right)= \\
&\quad\sqrt{\frac{Pr}{Ra}}\left[\frac{(1+x^2+y^2)^2}{4}\left(\frac{\partial^2 v}{\partial x^2}+\frac{\partial^2 v}{\partial y^2}\right)-v-\left(1+x^2+y^2\right)\left(x\frac{\partial u}{\partial y}-y\frac{\partial u}{\partial x}+x\frac{\partial v}{\partial x}+y\frac{\partial v}{\partial y}\right)\right] \\
&\quad-\left(\frac{1+x^2+y^2}{2}\right)^2\frac{\partial p}{\partial y}-y\beta gT-\frac{1}{Ro}u-\tilde{F}v.
\end{aligned} \tag{5}$$

$$\frac{\partial u}{\partial x}+\frac{\partial v}{\partial y}=\frac{4\,(xu+yv)}{1+x^2+y^2} \tag{6}$$

$$\frac{\partial T}{\partial t}+u\frac{\partial T}{\partial x}+v\frac{\partial T}{\partial y}=\frac{1}{\sqrt{Pr.Ra}}\frac{(1+x^2+y^2)^2}{4}\left(\frac{\partial^2 T}{\partial x^2}+\frac{\partial^2 T}{\partial y^2}\right)-\tilde{S}T \tag{7}$$

where (u,v) are the two components of the velocity field in the (x,y) stereographic coordinates. As can be noticed, the stereographic projection of the Navier–Stokes equations induced much more complicated expressions, with a couple of extra terms, compared to the usual cartesian coordinates formulation. But since the stereographic coordinates form a two-dimensional orthogonal basis, traditional numerical methods to solve the equations on cartesian grids can be used. The circular domain in the equatorial plane, a no-slip boundary condition on the equator, as well as a Dirichlet boundary condition for T are defined by using a L^2-penalization method. This kind of method consists in adding two penalization terms: $\frac{T-T_b}{C}$, in the temperature equation and $\frac{\mathbf{U}-\mathbf{U_b}}{C}$, in the momentum equation and to consider the set of equations on the whole square domain. The boundaries are then considered like a porous medium of very low permeability [26]. In our numerical simulations, C is set equal to 10^{16} in the bubble, and equal to 10^{-10} outside. The boundary conditions are then imposed by taking $\mathbf{U_b} = 0$ and $T_b = 1$ for method (1) and $\mathbf{U_b} = U_f$ and $T_b = 1$ for method (2). When $C = 10^{16}$ in the bubble, these extra terms vanish and we solve the regular Equation (1), and when $C = 10^{-10}$ outside the bubble, all the other terms are numerically very small compared to the extra terms except the pressure gradient. There is a coupling between these terms and the pressure gradient to yield Darcy equations and we get $\mathbf{U} = U_f$ and $T = 1$ at the boundary. Finite differences schemes have been chosen for solving the equations described in the previous part. A staggered uniform Cartesian grid with a mesh size $l = \delta x = \delta y$, where δx and δy are the discretization steps in each direction on the projection plane, is used to discretize the spatial domain. The corresponding uniform time steps is denoted δt. The unsteady term in the governing equations is solved by using a second-order Gear scheme. The convection terms are solved using an explicit scheme whereas the linear terms are solved using an implicit one. If one denotes by $\mathbf{U}^n$ the approximation of $\mathbf{U}$ at time $t_n = n\delta t$, then the approximation of the total time derivative of a general variable ϕ using the second-order Gear scheme can be written as:

$$\frac{D\phi}{Dt}\approx\frac{3\phi^n}{2\delta t}-\frac{4\phi^{n-1}-\phi^{n-2}}{2\delta t}+2(\mathbf{U}^{n-1}\cdot\nabla)\phi^{n-1}-(\mathbf{U}^{n-2}\cdot\nabla)\phi^{n-2} \tag{8}$$

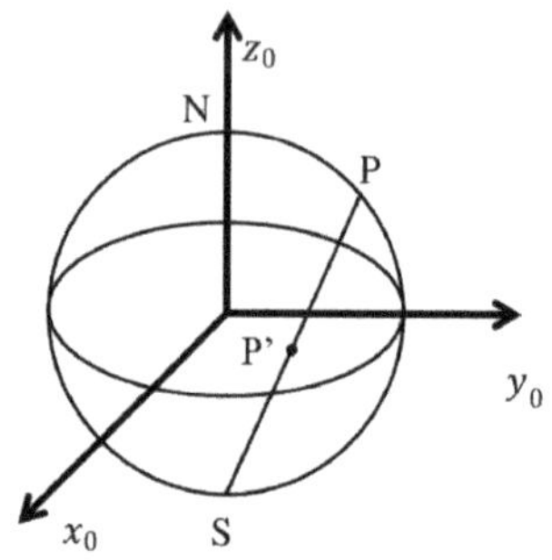

Figure 1. Stereographic projection: the point P of coordinates (x_0, y_0, z_0) on the sphere is projected on P′ of coordinates (x, y) on the plane. N and S denote respectively the north and south poles. (Reproduced with permission from [24]).

As shown in Figure 2, the discrete values of the velocity field are located at the middle of the cell sides and the discrete values of the pressure p and the temperature T are located at the center of each cell. A Murman-like scheme is used for the approximation of the convection terms. This kind of scheme has been fully described in [27]. A time splitting has been used to first solve the temperature equation and then the coupled pressure and velocity equations since the temperature equation can be solved separately. A parallel process by Message Passing Interface (MPI) is used to solve both of them. A linear system of equations $AT = E$, where A is a pentadiagonal matrix, T is the temperature vector and E the explicit part of the discretized temperature equation is obtained by the discretization of the temperature equation. The matrix A being not a self-adjoint matrix anymore in our present curved surface problem, we have to solve this linear problem using a biconjugated gradient method. The discretization of the coupled velocity-pressure system of equations leads to solving a discrete linear system $L_l V_l^n = B_l^{n-1}$ where L_l represents the discrete operator, B_l^{n-1} is the discrete equivalent of the right hand side of (8) and $V_l^n = (\mathbf{U}_l^n, p_l^{n+1/2})$ is the approximate solution we are looking for. All the numerical results presented in the following have been obtained with $Ra = 3 \times 10^8$, $Pr = 7$ on a 1024×1024 grid. A grid convergence analysis in order to verify that the grid is dense enough to describe a turbulent flow had been carried out in [24]. We had computed the global temperature on the bubble for different sizes of the grid and for a given Rayleigh number. The results obtained with the grids 1024×1024 and 2048×2048 for various values of the Rayleigh number, and various values of the rotation rates, were almost exactly the same, whereas the results obtained with coarser grids (256×256 and 512×512) were slightly different. As shown in [24], a fine grid (2048×2048) would be better to describe the boundary layer at the equator, but would not modify the global behavior of the fluid in the bubble. The grid 1024×1024 has thus been chosen in the present study for performing the various numerical simulations. Different values for the parameters $\tilde{F}$ and $\tilde{S}$ have been tested in [24] to stabilize the computations. The bubble can be stabilized with $\tilde{F} = 0.06$ and $\tilde{S} = 0.06$ which are the values used in the numerical simulations. Examples, in the non-rotating case, for the temperature and the vorticity fields are given in Figure 3 for two different times: at the beginning of the simulation (top) and in the stationary state (bottom). We can observe in Figure 3 thermal plumes emerging at the base of the bubble. The experiment is similar to a long two-dimensional toric Bénard cell in the beginning of our simulation: the cold fluid is heated from below, and the plumes are convected upward. In the beginning, there is no interaction between the plumes and the center of the domain remains empty. We can also observe these small thermal structures emerging from the heated equator on the vorticity field. The plumes continue to grow over time and start to interact with each other. As can be observed in Figure 3, the small structures along the equator are now giving birth to vortices that move along the bubble surface. A stationary state can be achieved and the results presented in the sequel have been obtained in this regime of the numerical experiments.

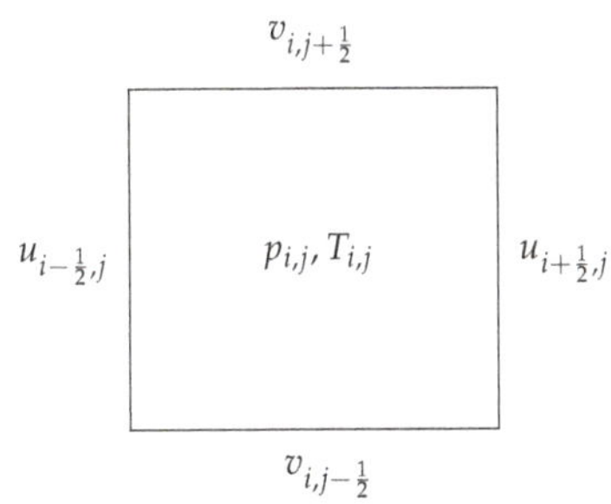

Figure 2. Description of numerical spatial cell: the velocity components are computed in the middle of the cell borders whereas the pressure and the temperature are computed at the center of the cell. (Reproduced with permission from [24]).

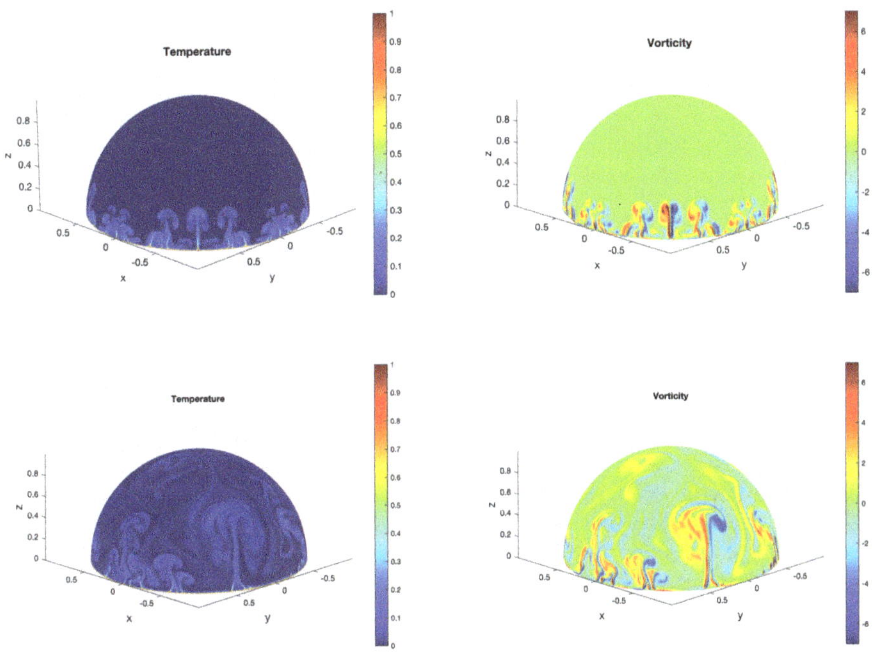

Figure 3. Top: Beginning of the simulations; mushroom shape structures are created at the equator. Bottom: Stationary state; plumes move up on the bubble and the structures start to interact with each other. Left: Temperature fields. Right: Vorticity fields.

3. Nusselt Number Criterion

The Nusselt number is usually defined as the ratio of the total vertical heat flux and the conductive heat flux, and can be evaluated on any horizontal plane in a Rayleigh–Bénard cell. But since our experiment is not a Rayleigh–Bénard cell, we have to first define a local Nusselt number and then compute the average of all the local Nusselt numbers on the bubble. For a given latitude l, we can define the local Nusselt number $Nu(l)$ as:

$$Nu(l) = \frac{< U_{k_0} T >_l - \alpha \partial_{k_0} < T >_l}{\alpha \delta T L^{-1}} \tag{9}$$

where U_{k_0} denotes the vertical component of the velocity, along the vertical unit vector $\mathbf{k_0}$, T is the temperature, and $< . >_l$ stand for a temporal average in time over a latitude l. We can then obtain a global Nusselt number Nu by averaging the local Nusselt numbers over all latitudes. In classical Rayleigh–Bénard experiments a constant heat transfer is established between the two plates and the difference between "local" vs. "global" Nusselt numbers is meaningless. In these cases, it is possible to calculate a Nusselt number by averaging over any horizontal plane. However, local and global Nusselt numbers are important physical values to evaluate when the heat transfer depends on the height, as in an open channel or in our bubble experiment (see Equations (4) and (5) in [28] for instance). The global Nusselt number has to be understood as a global indicator of the convection vs conduction balance in the cell.

Physical and numerical experiments found in the literature show that the rotation introduces three different regimes [12,14–19] in three-dimensional Rayleigh–Bénard experiments. The three regimes can be characterized using the Nusselt number. For low rotation (small Ω and large Ro), the flow remains almost unaffected by the rotation. The heat transfer measured by the Nusselt number does not change in this regime and a large scale circulation is observed in the cell. For moderate rotations, the flow is affected by the rotation leading to an enhancement of the heat flux. In this regime, in 3D, the large scale circulation is replaced by vortical structures (also called plumes in the literature). The enhanced convective heat flux is directly related to the vortical structures created by the moderate rotation of the system. For high rotation, the fluid is totally dominated by the rotation forcing and the convective heat flux is decreased. The transition between the moderate and the high rotation regimes is defined in [29] as the rotation for which the Nusselt number is maximum.

In our results, we also observe three different regimes. The ratio $Nu(Ro)/Nu(+\infty)$ for various values of Ro and the ratio $Nu(U_f)/Nu(0)$ for various values of U_f are reported in Figure 4. The relative Nusselt number is hardly affected for high values of Ro or small azimuthal velocities. For low values of Ro or high values of the azimuthal velocity, the relative Nusselt number decreases, showing that the flow is dominated by rotation. For intermediate values of both control parameters, the relative Nusselt number is greater than 1. As can be observed, the rotation created with the rotation forcing term (Method 1) does not significantly increase the heat flux. The maximum increase being less than 5%.

However, an increase of more than 15% is observed when the rotation is created by a moderate azimuthal velocity boundary condition (Method 2). This increase of the convective heat flux is of the same order as the one observed in the 3D rotating Rayleigh–Bénard case [12,14–19]. Note here that the Nusselt number ratio does not present a narrow maximum as in 3D experiments and simulations but a wide and plateau-like region, making it difficult to define an exact transition limit as in [29] between the regime where the Nusselt number is enhanced by rotation and the regime where the Nusselt number is decreased at high rotation rates. So we use a slightly modified definition for this transition in our study and define the three regimes by comparison to the non-rotational case as:

- Regime I where the Nusselt number is unchanged: low rotation,
- Regime II where the Nusselt number is increased: moderate rotation,
- Regime III where the Nusselt number is decreased: high rotation.

Considering the results shown in Figure 4, this definition for the transition from Regime II to Regime III seems to be more natural and better adapted for the functional shape of the variation of the relative Nusselt number versus Ro or U_f. The structures responsible for the Nusselt enhancement in regime II will be described in the next part.

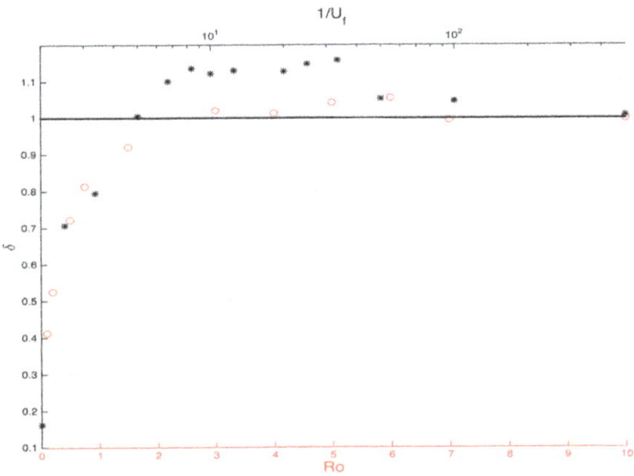

Figure 4. Red circles: Method 1, ratio δ of the Nusselt number $Nu(Ro)$ in the presence of global rotations to $Nu(\infty)$ (no rotation). Black stars: Method 2, ratio δ of the Nusselt number $Nu(U_f)$ in the presence of azimuthal velocity boundary conditions to $Nu(0)$ (no rotation).

4. Convective Heat Flux

The convective heat flux, created by the buoyant forces, corresponds to the transport of heat by the movement of the fluid from hot areas (at the equator) to cold ones (north pole). In classical Rayleigh–Bénard experiments, the structures responsible for the convective heat flux are called "plumes" [13,30] and are usually characterized by a strong correlation between the temperature fluctuations $\theta = T - \langle T \rangle$ (where $< . >$ denotes the overall time and space average) and the vertical velocity. In our numerical simulations, the vertical velocity is replaced by the radial component (pointing to the north pole) U_R of the velocity, and the convective heat flux is then defined by $\Phi(l) = \langle \theta . U_R \rangle_l$ where l denotes the latitude in radians. This mathematical object precisely measures the heat that is moved upward from the equator to the north pole. We obtain with our convective heat flux $\Phi(l)$ the same kind of information as the plumes counting described by Pieri et al. (see Figures 5 and 6 in [13]). Our results for various values of the Rossby number (Method 1) and of the velocity boundary condition U_f (Method 2) are summarized in Figure 5.

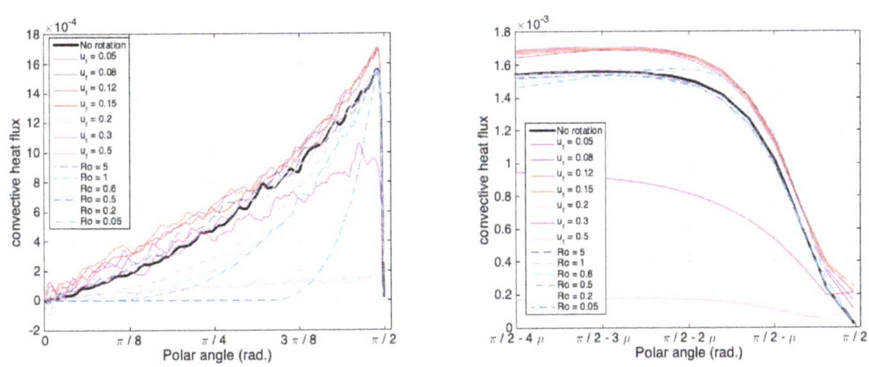

Figure 5. Left: Convective heat flux $\Phi(l)$ for various values of the Rossby number and of velocity boundary condition U_f. Right: Zoom close to the equator with $\mu = 0.01$.

The first striking observation is that moderate velocity rotations induced by the non-zero velocity boundary condition (Method 2) create increases of the convective heat flux at the equator, and this initial bonus of heat flux is then maintained in the bulk of the bubble. This increase is observed for all the values of U_f for which the Nusselt number was increased ($Nu(U_f)/Nu(+\infty)$ ratio larger than 1). It can be observed that the profile of the heat flux for $U_f = 0.2$ is the same as the profile of the heat flux without rotation confirming the result obtained with the Nusselt number criterion. Relatively strong velocity rotations in Method 2 decrease the convective heat flux at the equator leading to a decrease of the Nusselt number. This observation can be linked to our analysis about the instabilities and thermal boundary layers of the next part. Thermal plumes are created at the edge of the thermal boundary layer. When the shear layer is thinner than the thermal boundary layer then the "new born" plumes are boosted by a non-zero average velocity field. But when the shear layer limit reaches the thermal boundary layer limit, the "new born" plumes are swept by the rotations.

No increase is observed with the rotations induced by Method 1, and we even observe that the convective heat flux is strongly decreased in the bulk. Again, this result will be discussed in light of the analysis of the RMS of velocity fluctuations in the next section: high velocity rotations prevent the structures from reaching the high latitude areas. However, we can observe in the equator area (Figure 5 right) that very high velocity rotations ($Ro = 0.05$ for instance) induce the same kind of increase as those obtained with Method 2. But, since the global rotations also completely inhibit the dynamics in the bulk, the benefit of the increase is immediately erased as we get farther from the equator. The extra convective heat flux in the equator area is not sufficient to compensate the lack of heat flux in the bulk. Thus, the overall balance in this situation is negative and leads to a Nusselt number lower than the one obtained in the no-rotation case.

We can also visualize the plumes responsible for the convective heat flux in the convective heat fields $\theta.U_R$ for various values of the parameters Ro and U_f. In Figure 6, mushroom-shaped plumes can be observed in the non-rotating case. Short and medium sized plumes are present all around the bubble. The effects of rotation through Method 1 for $Ro = 0.2$ can be observed in Figure 6. The long plumes have completely been suppressed and only the short plumes are present very close to the equator. Most of the surface of the bubble is now completely free of structures. In the case of Method 2, moderate velocity rotations increase the length of the plumes as can be observed in Figure 6. However, for high velocity rotations (top Figure 7), long plume structures are still present but are very localized in a small part of the bubble. For very high rotations (bottom Figure 7), very few and very localized structures can be observed almost reaching the north pole. This is a completely different phenomenon from what we can observe with a very small Rossby number in Method 1. With Method 1, the long plumes are suppressed but short plumes are still present all over the bubble. With Method 2, however, short plumes are absent and a few, localized, and weak heat structures succeed in detaching from the equator to escape to the north pole. This is a clear observation of the intermittent character of the dynamics induced by the rotations created by Method 2.

Even if our setup is completely different we can make a parallel with two new studies about 3D rotating Rayleigh–Bénard Convection. Rajaei et al. [31] recently published new results about the rotation dominated regime in rapidly rotating Rayleigh–Bénard convection. They found that the transition to the rotation dominated regime coincides with the suppression of vertical motions, the strong penetration of vortical plumes into the bulk and a reduced interaction of vortical plumes with their surroundings. In Alards et al. [32], the authors study in detail, with a Lagrangian approach, the transition from the rotation-unaffected regime, where the heat transfer is constant, to the rotation-affected regime, where the heat is enhanced. They found a sharp transition in the horizontal acceleration statistics near the top plane at a Rossby number lower ($Ro \approx 2.25$) than the one typically found for the transition between rotation-unaffected/affected regimes ($Ro \approx 2.7$). According to their study, the steepening of the velocity gradients at the boundary and the generation of swirling convective flows in the boundary layers are two partially separate processes: heat transfer enhancement at $Ro \approx 2.7$ and generation of vertically aligned vortex tubes at $Ro \approx 2.25$. They also observe a crossing between the thermal and the Ekman layers for a Rossby number equal to 1.4 whereas the maximum of the Nusselt enhancement is located for Ro between 0.1 and 0.2.

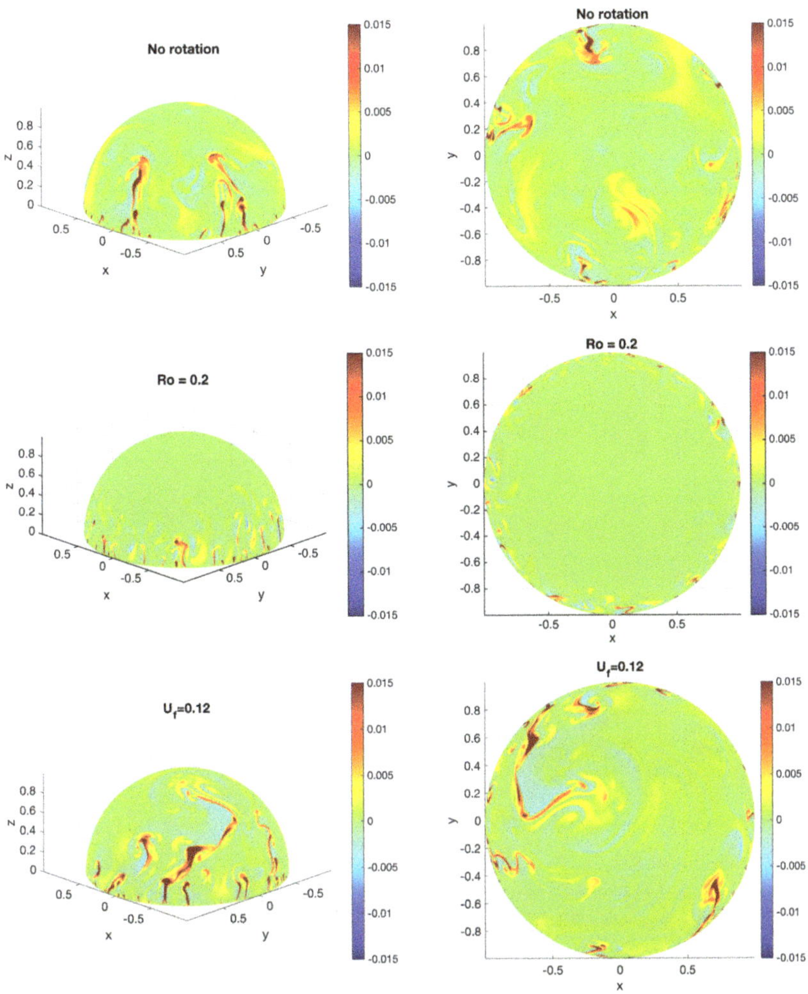

Figure 6. Convective heat flux fields. Top: No-rotation case. Middle: Method 1 with $Ro = 0.2$. Bottom: Method 2 with $U_f = 0.12$. Left: Side view. Right: North pole view.

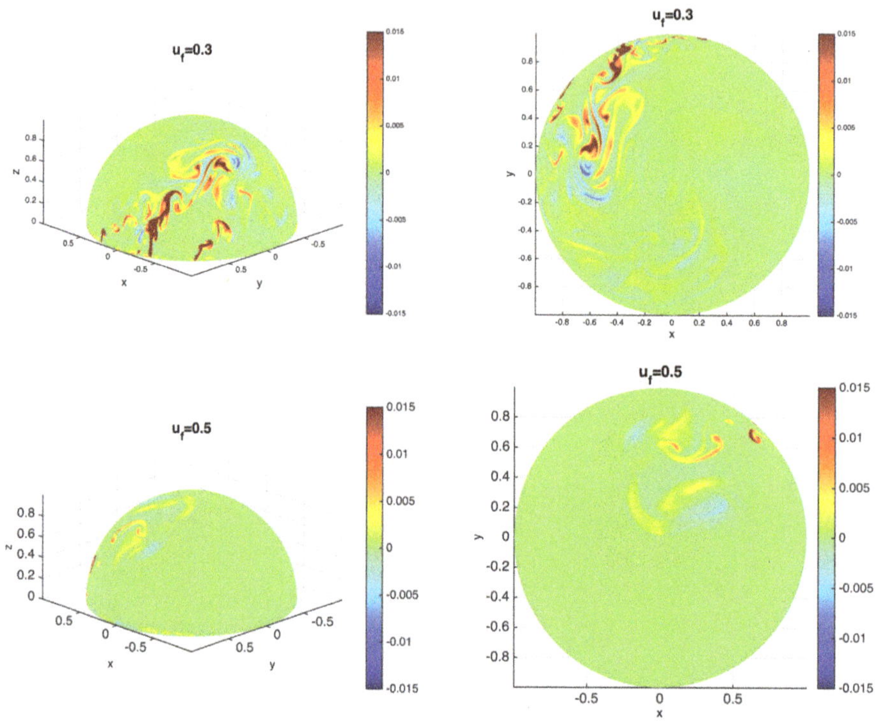

Figure 7. Convective heat flux fields with method 2. Top: $U_f = 0.3$. Middle: $U_f = 0.5$. Left: Side view. Right: North pole view.

5. Velocity and Temperature Fluctuations RMS

It has been shown in Section 3 that the Nusselt number is greatly influenced by the way the rotation is forced on the bubble. The difference between the two methods is that in one case the rotation is imposed on the whole bubble (Method 1 with the rotation term in the equation) whereas in the other method the rotation is imposed at the equator only (Method 2). The latter method creates a shear layer that might be responsible for the Nusselt number enhancement. So in this part, the RMS of the fluctuations of azimuthal (parallel to the equator) and radial (pointing to the north pole) velocity components are studied and compared to the thermal boundary layer (evaluated with the temperature fluctuations RMS as in [24]).

It can be noticed in Figure 8 that the rotations created by Method 1 significantly change the temperature fluctuations in the bulk of the bubble, but not so much at the equator. Indeed, in the bulk, the temperature fluctuations are suppressed by high rotation whereas close to the equator the location and the height of the maximum of the RMS is not significantly modified by the rotation forcing. This can be also observed in Figure 9 (top), where no difference can be observed in the temperature profiles at the equator, except a small increase beyond the thermal boundary layer for very high rotations.

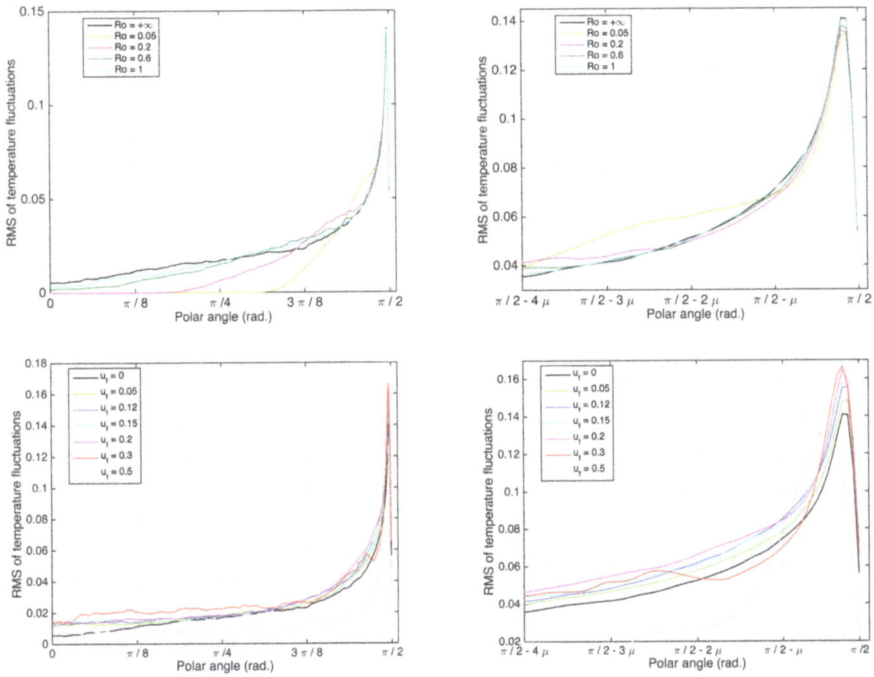

Figure 8. Temperature fluctuations RMS. Top: Method 1 for various values of the Rossby number. Bottom: Method 2 for various values of the forcing U_f. Right: Zoom around the equator; $\mu = 0.05$.

From this first observation, we can conclude that the rotations induced by Method 1 do not modify the thermal boundary layer, but suppresse the temperature fluctuations in the bulk of the bubble. Let us now turn to Method 2. Moderate rotations forced by the velocity boundary condition U_f (Method 2) slightly increase the temperature fluctuations in the bulk but significantly modify them close to the equator. Indeed, a shift of the maximum location of the RMS away from the equator can be observed. This shift is accompanied by a small increase of the maximum for moderate rotations, and a strong decrease for high rotations. The shift is also more important for large rotation velocity U_f (bottom of Figure 8). The temperature increases at the equator can be also verified in Figure 9 (bottom) where the temperature is directly related to the rotation velocity values U_f. So Method 1 influences the flow essentially in the bulk of the bubble while Method 2 essentially influences the thermal boundary layer which grows with the forcing velocity and which eventually also influences the bulk.

The study of the radial (pointing to the north pole) U_R and azimuthal (parallel to the equator) U_A velocity components fluctuations can complete this analysis.

As can be observed in Figure 10, rotations created by Method 1 significantly decrease the velocity fluctuations (both radial U_R and azimuthal U_A). The size of the area where the fluctuations are the largest decreases with the increase of the rotation speed. It can also be observed that the azimuthal velocity component fluctuations present a "double bump" shape. This particular shape corresponds in fact to clockwise and counterclockwise fluctuations. This aspect can be verified by visualizing the averaged azimuthal velocity component Figure 11 (Left). We can observe three bands: close to the equator the average motion is in the same direction as the forcing (counterclockwise), then in the other direction and finally again in the same direction as the forcing. The bandwidth is decreasing as the rotation is increasing. This phenomenon, also highlighted in the soap bubble experiments, might be related to the circulation cells in the atmosphere where easterly and westerly winds are observed.

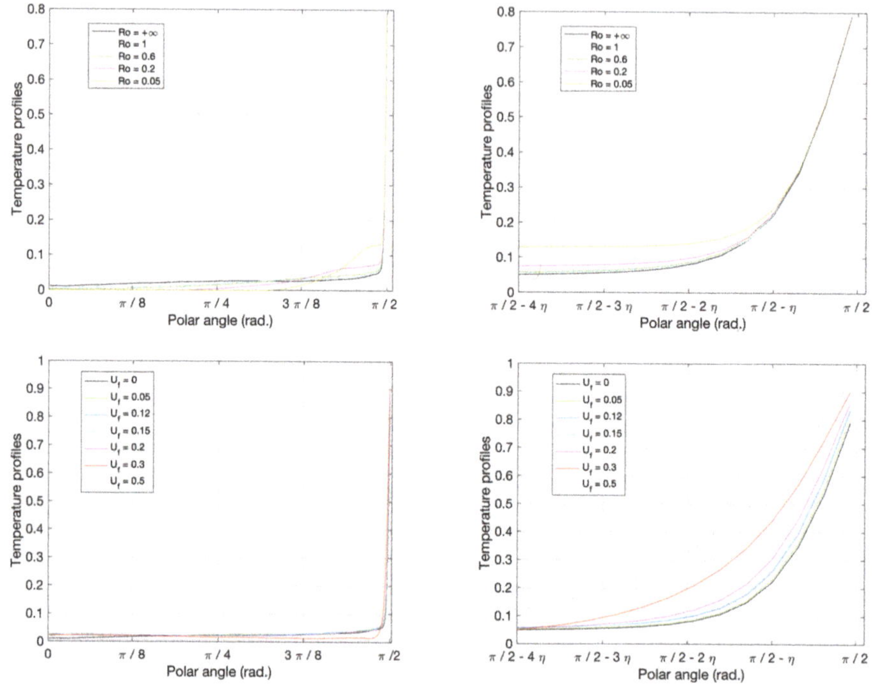

Figure 9. Temperature profiles. Top: Method 1 for various values of the Rossby number. Bottom: Method 2 for various values of the forcing U_f. Right: Zoom around the equator; $\eta = 0.01$.

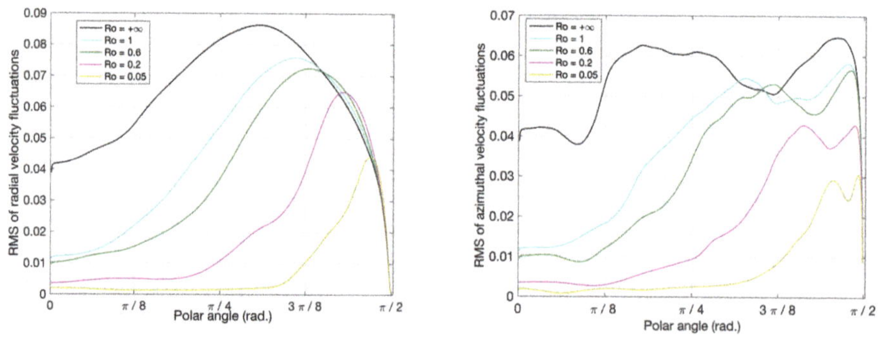

Figure 10. Method 1. Left: Radial velocity component fluctuations RMS for various values of the Rossby number Ro. Right: Azimuthal velocity component fluctuations RMS for various values of the Rossby number Ro.

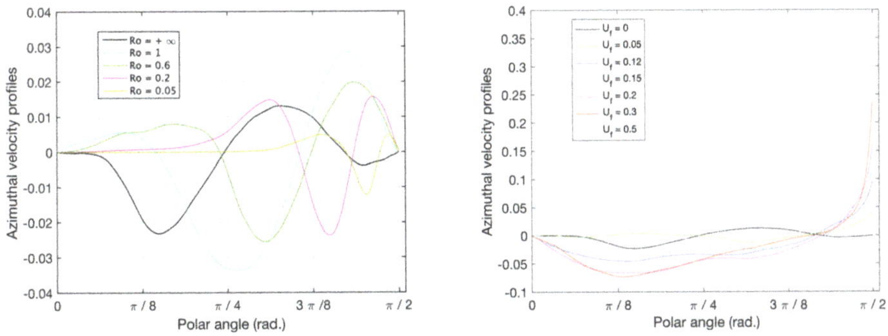

Figure 11. Azimuthal velocity profiles. Left: Method 1 for various values of the Rossby number Ro; Right: Method 2 for various values of U_f.

With Method 2, the influence of the rotation forcing U_f on the fluid can be observed in Figure 12. Except for very large forcings ($U_f = 0.5$ for instance), the radial velocity component fluctuations are increased around the pole. Moderate rotations with Method 2 seem to push more fluid to the pole than the convection alone without rotation. A close look at the area above the equator (Figure 12 right) allows us to remark that the RMS profiles present a minimum when $U_f \neq 0$. We conjecture that this minimum corresponds in fact to the limit of a shear layer, only present with this way of rotating the bubble, and has to be compared to the limit of the thermal boundary layer (BL). The size of these two layers for various values of the rotation forcing U_f are summarized in Figure 13. Two linear fits obtained from the first six values of U_f, from $U_f = 0$ to $U_f = 0.3$, are also materialized in this figure. We note a linear growth of this shear layer thickness with the velocity forcing. The slope of the linear fit for the shear layer thickness is estimated at 3.5×10^{-2}, whereas the slope for the thermal boundary layer one is about 3.7×10^{-3}. The shear layer thickness grows faster than the thermal layer, but we could not find any theoretical explanations for the values of these linear growth.

The thermal boundary layer size increases slowly while the shear layer size increases faster. When the size of this shear layer reaches the size of the thermal boundary layer near $U_f \approx 0.25$ both keep the same size and continue to grow with increasing forcing velocity U_f. This observation might be related to the phenomenon discussed in the previous part: the Nusselt number increases for moderate velocity rotations up to $U_f \approx 0.2$. As long as the shear layer size is smaller than the thermal boundary layer size the convective heat flux is enhanced by the rotation forcing, leading to an enhancement of the Nusselt number. But when the shear layer is the same size as the thermal boundary layer then the enhancement is stopped, and the Nusselt number starts to decrease.

This kind of competition between different characteristic scales has been also observed in three-dimensional Rayleigh–Bénard numerical experiments in [29,33,34] in the context of the transition between a regime where the heat transfer is not significantly affected by rotation and a rotationally controlled regime. In the non-rotationally affected regime, Rajaei observed that the kinematic boundary layer thickness remains constant and then starts to decrease when the rotation dominates the dynamics (see Figure 3.4 in [29]). This transition corresponds to a transition from a Prandtl–Blasius type layer to an Ekman type layer. The boundary layer related to the low velocity rotation regime (where they observe a Large Scale Circulation) is of Prandtl–Blasius type and the one related to the moderate velocity rotations regime is of Ekman type. Other observations were provided by King et al. [33] (see in particular Figure 3b) who found that the transition from non-rotating to rotationally controlled heat transfer occurs when the thermal boundary layer and the Ekman layer thicknesses are crossing each other.

Our observations show that for the transition between our Regime II and our Regime III, the thickness of this shear layer (found with Method 2 only) becomes comparable to the thickness of the thermal boundary layer. In fact, this transition corresponds to the progressive disappearance of the thermal plumes described in the previous part. Also, compared to Method 1, we do not observe the "double bumps" in the azimuthal velocity component RMS which was related to circulation cells. This can be also verified in the azimuthal velocity profiles in Figure 11 (right) where we do not find the caracteristic three bands. This phenomenon seems thus to be a consequence of the global rotation of the system like in the Earth's system, and cannot be reproduced with Method 2. Also in Figure 11 (right), we can clearly observe that the azimuthal velocity at the equator is directly linked to the forcing value U_f. In short, in the context of our two-dimensional numerical experiments, the convective heat enhancement (measured by the Nusselt number) can be produced by Method 2. Method 1 can produce a very small enhancement of the Nusselt number (less than 5%). Large circulations, detected in the azimuthal velocity component, can be produced with Method 1 but not with Method 2. As it will be shown in the next part, more insight into the dynamics of the flow can be obtained using global physical properties like spectra and fluxes.

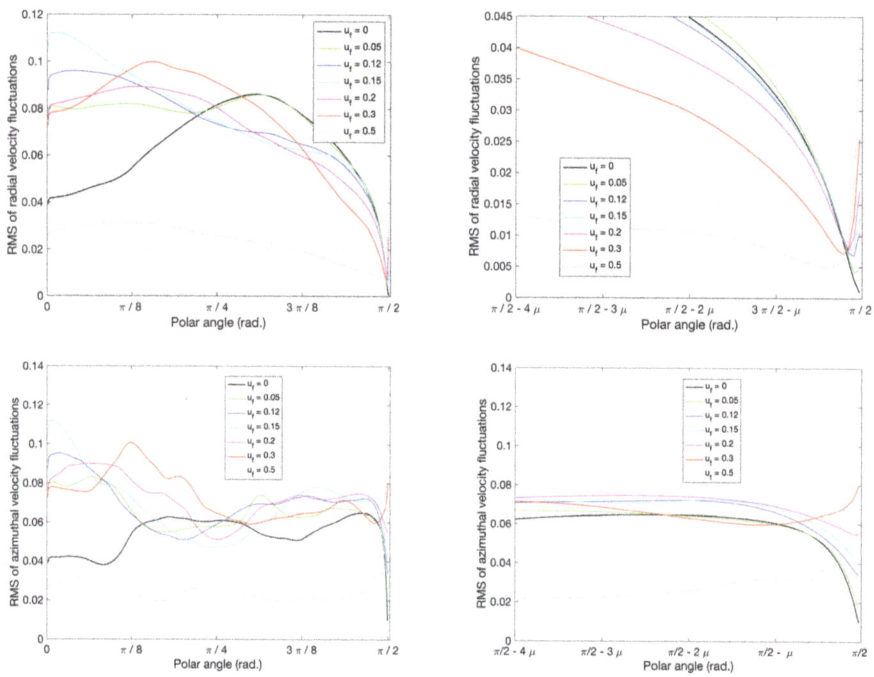

Figure 12. Velocity fluctuations RMS for various values of U_f (method 2). Top: Radial velocity component. Bottom: Azimuthal velocity component. Right: Zoom close to the equator; $\mu = 0.05$.

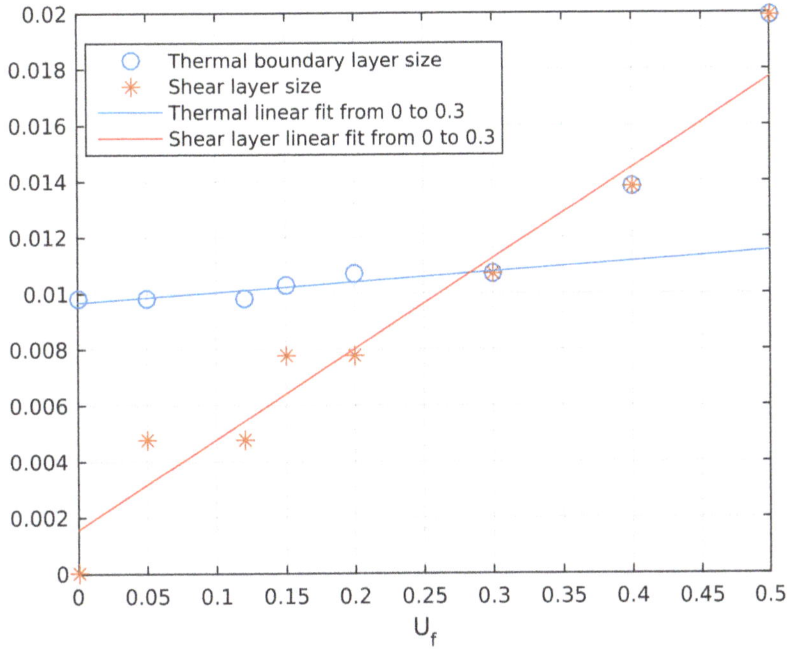

Figure 13. Thermal boundary and shear layers sizes (polar angle in radians) for various values of the velocity forcing U_f.

6. Spectra and Fluxes

Energy, enstrophy and temperature spectra/fluxes are commonly used to study two or three-dimensional turbulence as well as Rayleigh–Bénard convection cells. Indeed, according to the corresponding theories, it is well known that inverse or direct cascades of energy/enstrophy in 2D or 3D turbulence or the presence of a Bolgiano–Obukhov regime in Rayleigh–Bénard convection can be detected using energy/enstrophy/temperature spectra and fluxes [24]. Very recently, Sharma et al. [35,36] published new results on rapidly rotating forced and decaying turbulence. Their studies are based on numerical simulations in a cube of size $(2\pi)^3$ with periodic boundary conditions on all the sides. In the decaying case, they observed that the turbulent flow evolves in time with a real Rossby number decreasing to $\sim 10^{-3}$ and the flow becoming quasi two-dimensional with strong coherent columnar structures arising due to the inverse cascade of energy (as in classical two-dimensional turbulence where strong vortices are observed). They propose a new scaling for the energy spectrum of three-dimensional rapidly decaying turbulence: $E(k) = C\epsilon_\omega^{2/3}k^{-3}\exp\left(-C(k/k_d)^2\right)$, where ϵ_ω is the enstrophy dissipation rate, k_d the enstrophy dissipation number, and C a positive real constant. For the rapidly rotating forced turbulence, they propose a scaling split into two components related to the horizontal plane and the vertical rotation axis in the range of wavenumbers smaller than the forcing scale: $E(k_\perp, k_\parallel) \approx k_\perp^{-5/2}k_\parallel^{-1/2}$ where $\perp$ and $\parallel$ denote the directions perpendicular and parallel to the vertical rotation axis. However,our numerical setup being different, and our rotation speed being weaker, we cannot expect to obtain the same results.

Mathematical tools specifically designed for spherical problems has to be used to obtain global properties, such as spectra or fluxes, for our numerical simulations. The definitions of the spherical harmonics decomposition are first recalled and then used to compute these various physical indicators (spectra and fluxes) of interest.

Any function f with spherical symmetry can be easily written as a linear combination of special functions $\{Y_k^m\}$,

$$f(\theta,\varphi)=\sum_{k=0}^{+\infty}\sum_{m=-k}^{+k} f_k^m Y_k^m(\theta,\varphi), \tag{10}$$

the so-called spherical harmonics defined by:

$$Y_k^m(\theta,\varphi)=\sqrt{\frac{2k+1}{4\pi}\frac{(k-m)!}{(k+m)!}}P_k^m(\cos(\theta))\,e^{im\varphi}. \tag{11}$$

Here, $P_k^m(.)$ denotes the associated Legendre polynomials. The expansion coefficients f_k^m are then obtained by projection of f onto the spherical harmonics:

$$\begin{aligned} f_k^m &= \int_0^{2\pi}\int_0^{\pi/2} f(\theta,\varphi)\overline{Y_k^m(\theta,\varphi)}\sin(\theta)d\theta d\varphi \\ &= \sqrt{\frac{2k+1}{4\pi}\frac{(k-m)!}{(k+m)!}}\int_0^{2\pi}\int_0^{\pi/2} f(\theta,\varphi)P_k^m(\cos(\theta))\,e^{-im\varphi}\sin(\theta)d\theta d\varphi \end{aligned} \tag{12}$$

The integration over φ is actually a Fourier transform and is numerically computed with a traditional routine. The power spectrum can then be computed from the coefficients f_k^m:

$$||f||^2=\sum_{k=0}^{+\infty} f_k^2=\sum_{k=0}^{+\infty}\sum_{m=-k}^{k}|f_k^m|^2. \tag{13}$$

The coefficients f_k^m are thus equivalent to the Fourier coefficients and the coefficients f_k are equivalent to the Fourier wavenumbers in a traditional Fourier power spectrum. For instance, the thermal energy is defined as:

$$||T||^2=\sum_{k=0}^{+\infty} T(k)^2=\sum_{k=0}^{+\infty}\sum_{m=-k}^{k}|T_k^m|^2, \tag{14}$$

where

$$T_k^m=\int_0^{2\pi}\int_0^{\pi/2} T(\theta,\varphi)\overline{Y_k^m(\theta,\varphi)}\sin(\theta)d\theta d\varphi \tag{15}$$

Energy, enstrophy and thermal fluxes can be also computed using the spherical harmonics coefficients. The energy flux is usually computed in the Fourier domain from the nonlinear term in the Navier–Stokes equation:

$$\Pi_E(k)=\int_k^{+\infty} S_E(k')dk' \tag{16}$$

where $S_E(k)$ is the nonlinear energy transfer function and is obtained by angular integration of $\widehat{\boldsymbol{U}^*(\boldsymbol{k})}.\widehat{(\boldsymbol{U}.\nabla)\boldsymbol{U}(\boldsymbol{k})}$. Here, the symbol $\widehat{.}$ denotes the usual Fourier transform. The enstrophy and thermal fluxes are obtained in the same way:

$$\begin{aligned} \Pi_Z(k) &= \int_k^{+\infty} S_Z(k')dk' \\ \Pi_T(k) &= \int_k^{+\infty} S_T(k')dk' \end{aligned} \tag{17}$$

where $S_Z(k)$ and $S_T(k)$ are the enstrophy and thermal transfer functions. They are obtained by angular integration of $\widehat{w^*(\boldsymbol{k})}.\widehat{(\boldsymbol{U}.\nabla)w(\boldsymbol{k})}$ and $\widehat{T^*(\boldsymbol{k})}.\widehat{(\boldsymbol{U}.\nabla)T(\boldsymbol{k})}$.

So, in our study, we replace the usual Fourier decompositions by spherical harmonics decompositions, and the angular integration of the Fourier coefficients by a summation of the spherical coefficients f_k^m over the degree $m = -k$ to $m = k$. The results are strictly equivalent to classical energy, enstrophy and thermal fluxes and can be considered as such.

Energy, enstrophy and temperature spectra and fluxes are given in three separate Figures 14–16 for relatively strong ($Ro = 0.75$ and $U_f = 0.3$), moderate ($Ro = 6$ and $U_f = 0.15$) and no rotations. The values for relatively strong rotations are those leading to 80% of the initial (without rotation) Nusselt number, and the values for moderate rotations are those leading to the highest enhancement of the Nusselt number.

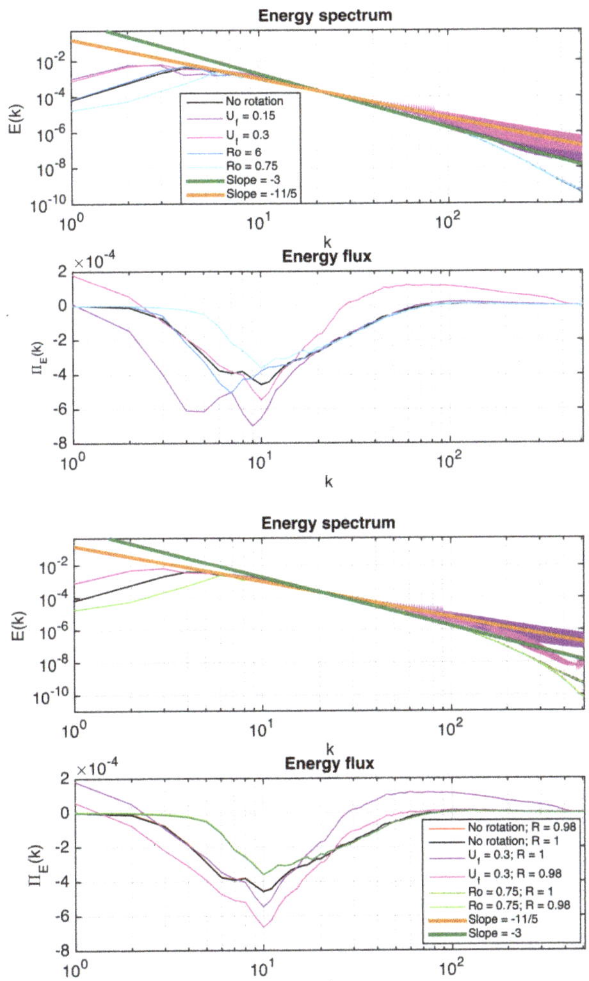

Figure 14. First and second rows: Energy spectra and fluxes for strong, moderate and no rotation for methods 1 and 2. Third and fourth rows: Tukey windowed energy spectra and fluxes for strong rotations. Method 1 with $Ro = 0.75$. Method 2 with $U_f = 0.3$. Radius of the window $R = 0.98$.

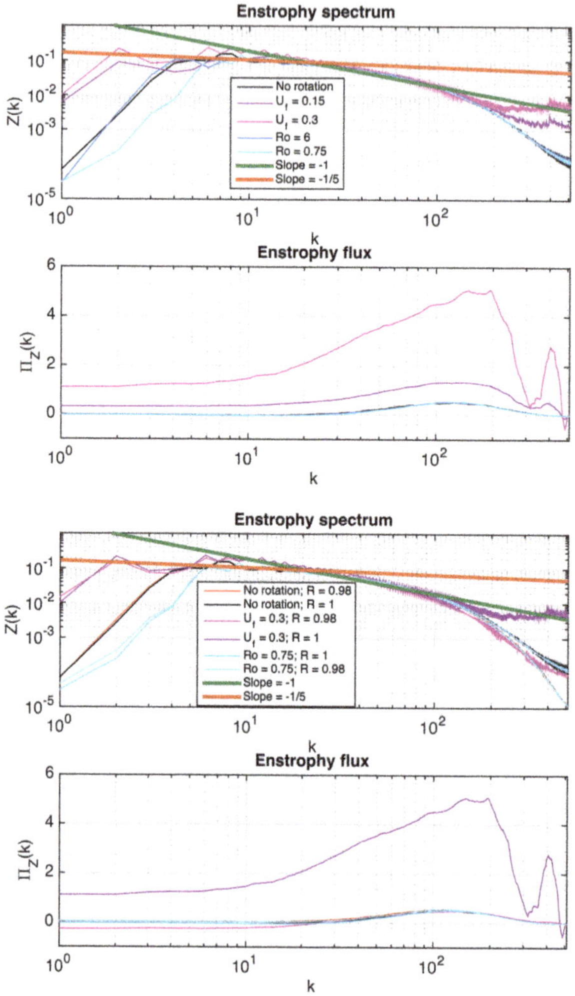

Figure 15. First and second rows: Enstrophy spectra and fluxes for strong, moderate and no rotation for methods 1 and 2. Third and fourth rows: Tukey windowed enstrophy spectra and fluxes for strong rotations. Method 1 $Ro = 0.75$. Method 2 $U_f = 0.3$. Radius of the window $R = 0.98$.

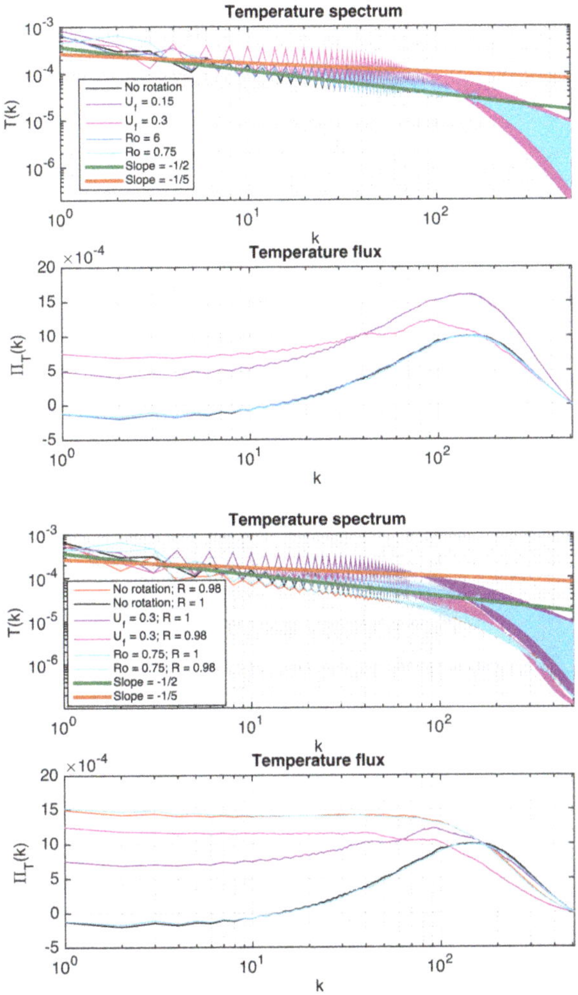

Figure 16. First and second rows: Temperature spectra and fluxes for strong, moderate and no rotation for methods 1 and 2. Third and fourth rows: Tukey windowed temperature spectra and fluxes for strong rotations. Method 1 $Ro = 0.75$. Method 2 $U_f = 0.3$. Radius of the window $R = 0.98$.

6.1. Energy Spectra and Fluxes

The first observation is that the energy spectra corresponding to the rotating bubbles are above the spectrum of the non-rotating one at large and small scales (except at large scales for the case $Ro = 0.75$. As previously noticed, strong rotations with Method 1 prevent large structures to develop towards the pole). One can also notice that the slope of the energy spectrum (observed for moderate wave numbers) is not modified by the rotations. Rotations created by Method 2 induce an energy increase at large scales. The third observation concerns the small scales. Rotations created by Method 1 do not modify the energy spectrum at small scales whereas strong oscillations are detected for rotations created by Method 2. This pattern had been previously observed in temperature spectra and it had been proven that it is due to the boundary conditions at the equator [24]. These oscillations in the energy spectra are due to the non-zero velocity boundary condition. What we observe here is that Method 2 intensifies the shear layer and now the energy spectrum is also affected by these instabilities. These oscillations can be removed by excluding the shear layer above the equator from the computations. This can be done by using a circular version of a Tukey window [37] to smoothly remove the shear layer without introducing spurious artifacts. We recall here that a Tukey window allows to smooth a step from 1 to 0 with a cosine function shape decrease. We present in bottom of Figure 14 the effects of this windowing process on the energy spectra and fluxes for the strong rotations cases ($Ro = 0.75$ and $U_f = 0.3$). The windowed and non-windowed non-rotating cases are also shown in this figure. The radius from which the window decreases has been chosen equal to $R = 0.98$. Even if it is a very short window, it still allows to significantly reduce the oscillations. It had been shown in [37] that a larger window would allow to completely remove them. As expected, the windowing process affects particularly the results obtained with Method 2 where a strong shear layer is present whereas it does not significantly modify the results obtained with Method 1 where there are kinematic and thermal layers but almost no shear layer.

In our previous paper, we have shown that, in absence of rotation, as the Rayleigh number increases, the energy (resp. enstrophy) spectra tend to follow a $k^{-11/5}$ (resp. $k^{-1/5}$) scaling corresponding to the Bolgiano regime [38–41]. Our results were consistent with Verma et al. [40] and Kumar et al. [41] confirming the two-dimensional nature of our experiment with a strong kinetic energy inverse cascade. In our present numerical simulations with an intermediate Rayleigh number ($Ra = 3 \times 10^8$) we observe a k^{-3} slope at intermediate scale, and the rotations created with method 2 tend to strengthen the buoyancy and a $k^{-11/5}$ slope is observed. Particularly at large scales. The shape of the energy fluxes remains essentially the same, and we can observe that the strongest inverse energy transfer correspond to the highest Nusselt number enhancement. Strong rotations created by Method 1 lead to a weaker energy flux.

6.2. Enstrophy Spectra and Fluxes

We can observe in the enstrophy spectra that strong rotations created by Method 1 decrease the quantity of enstrophy at large scales whereas strong or moderate rotations created by Method 2 increase the quantity of enstrophy in this range of scales (first and second rows of Figure 15). This can be explained by the fact that in Method 2 very long structures stretching from the equator to the pole appear in the bubble, while Method 1 rotations pack the structures in a band above the equator. The size of this band depends on the rotation velocity. As already observed in the energy spectra, the rotations created with Method 2 strengthen the buoyancy and enable a Bolgiano regime at large scales where a $k^{-1/5}$ can be observed. At intermediate scales, a Kraichnan-like regime with a k^{-1} slope takes place. As mentioned above, the rotations created with Method 1 weaken the buoyancy at large scales.

We can also observe in Figure 15 that Method 2 substantially modifies the enstrophy flux in the bubble. Method 1 does not really change the enstrophy flux, but Method 2 strongly increases it. In particular, it can be noticed the appearance of a bump at small scales around $k = 400$. The higher the azimuthal velocity, the higher the bump. We believe that this bump is created by instabilities in the shear layer, and can be removed by excluding the shear layer above the equator from the computations. Again, this can be done by using the circular version of a Tukey window [37] to smoothly remove the shear layer without introducing spurious artifacts. We present in Figure 15 (third and fourth rows) the effects of this windowing process on the enstrophy spectra and fluxes for the strong rotations cases ($Ro = 0.75$ and $U_f = 0.3$). The non-rotating non-windowed case is also shown in this figure. The size of the window has been chosen as $R = 0.98$. The windowing process does not significantly change the enstrophy spectra at large scales. But we can verify that the instabilities in the shear layer are responsible for the shape of the spectrum at very small scales (in particular for Method 2). The difference is even more striking in the enstrophy flux. We can observe that the small shear layer is influencing the whole flux, and is in particular responsible for the bump located at $k = 400$. If we remove from the figure the flux obtained with $U_f = 0.3, R = 1$, we can zoom and observe that the remaining fluxes are very similar at small scales (Figure 17). It can be noticed that concerning Method 1 the windowed enstrophy flux is very similar to the windowed non-rotating enstrophy flux. Even in the non-rotating case the boundary layer plays an important role in the transfer of enstrophy through scales. Thus we can conclude that the shear layer is really dominant in our experiments, for both methods of rotation, but in particular when rotating the bubble with Method 2.

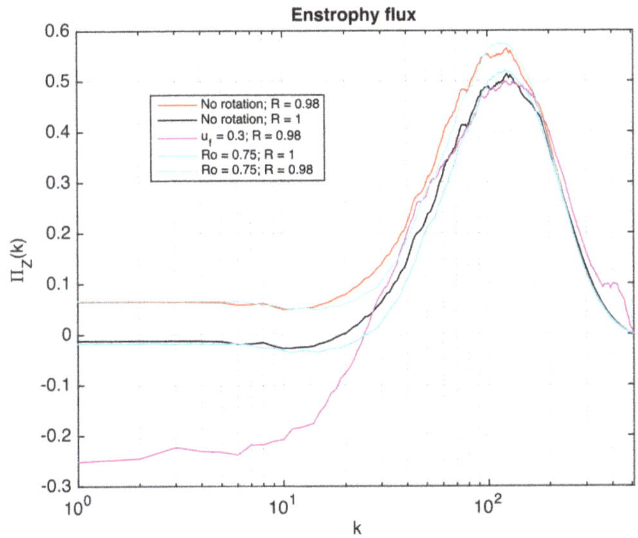

Figure 17. Tukey windowed enstrophy fluxes for strong rotations. Method 1 $Ro = 0.75$. Method 2 $U_f = 0.3$ with radius of the window $R = 0.98$ only.

6.3. Temperature Spectra and Fluxes

The temperature spectra are composed of two branches leading to an oscillatory behavior. This characteristic feature has been studied and explained in detail in [24]. It was shown that the temperature spectrum presents dual branches because the mean temperature profile has a steep variation near the equator. These oscillations can be removed by subtracting the average temperature profile from the temperature fields.

Here, we observe that the rotations created by Method 1 do not significantly modify the temperature spectrum, whereas the rotations created by Method 2 largely increase the oscillations, in particular at small scales. This is consistent with our previous results since rotations induced by Method 2 strengthen the instabilities in the shear layer and thus enhance its influence in the fluid. Without rotation, we can observe a $k^{-1/2}$ scaling in the temperature spectrum. This scaling changes slightly with rotation values and a $k^{-1/5}$ for strong rotation with Method 2 ($U_f = 0.3$) is observed. We do not have any explanation for theses results. We can observe in the temperature fluxes that the rotation leading to the highest enhancement of the Nusselt number (for $U_f = 0.15$) has the highest temperature flux. Once again the influence of the instabilities in the shear layer can be studied by removing them with the Tukey windowing process. The results are summarized in Figure 16 (third and fourth rows). Again, removing the shear layer allows to reduce the oscillations as observed in [24]. However, here we just remove a very small area above the equator whereas in [24] the average profile of temperature was removed. We can also observe that the windowing significantly modifies the temperature fluxes in the non-rotating case and also in the case of the rotation induced by Method 1. The effect is actually the same for both cases.

In conclusion, we observe a mix of the Bolgiano and Kraichnan regimes. The rotations created by Method 2 strengthen the buoyancy at large scales and the Bolgiano regime seems to be dominant in this range of scales. This is particularly observed in the enstrophy spectra where a $k^{-1/5}$ slope is detected in the results for Method 2. Also, two characteristics seem to play a very important role: the boundary layers that clearly influence the whole flow, and the intermittency induced by the shear layer in Method 2. Different physical mechanisms can be responsible for these phenomena and can be different in two and three-dimensional thermal turbulence: for instance, tubes of vorticity can clearly not be produced in two-dimensional experiments. However, plumes of convective heat can still be produced in two-dimensional experiments.

7. Conclusions

We have presented numerical simulations of a rotating thermal convection experiment. The results for the non-rotating version of this experiment have been reported in [20–22,24]. The purpose of the present paper was to analyze the reaction of the fluid when two different kinds of rotation are applied to the bubble: a solid rotation forced by a rotation term in the equations (Method 1), and a global rotation forced by a non-zero azimuthal velocity boundary condition at the equator (Method 2). The second-order temperature structure functions have already been analyzed in [23] for a rotating bubble with Method 1 for various values of the Rossby number. Three different regimes were observed in the temperature structure functions depending on the Rossby number. We have also found in the present study three different regimes when analyzing the convective heat transfer: regime I when the Nusselt number is unchanged, regime II when the Nusselt number is increased, and regime III when the Nusselt number is decreased. Three-dimensional Rayleigh–Bénard experiments also present three different regimes depending on the Rossby number [12,14–19]. In our numerical simulations, when the velocity is low (Regime I), the fluid remains unaffected and we do not observe any increase of the heat transfer. For moderate velocity rotations (Regime II), the convective heat transfer from the equator to the north pole is improved by about 5% with Method 1 and more than 15% with Method 2. For relatively high rotations (Regime III), we observe a large drop in the heat transfer efficiency with both methods. These numerical simulations have been analyzed with four different tools: the Nusselt number, the convective heat flux, the RMS of velocity and temperature fluctuations, and the energy/enstrophy/temperature spectra/fluxes. The enhancement of the heat transfer, measured by the Nusselt number, is created by the elongation of the plumes that stretch from the equator to the north pole for moderate rotations. These plumes are completely suppressed by high rotations induced by Method 1 and strongly inhibited by high rotations induced by Method 2. Rotations induced by Method 2 create a shear layer with strong velocity fluctuations near the equator.

We have shown that the Nusselt number enhancement is controlled by the relative thicknesses of the thermal and shear layers, in the same way as in [33] where they compare the thermal and the viscous layers. When the thermal boundary layer is larger than the shear layer, the heat transfer is enhanced by the rotation and we observe an increase of 15–18% of the Nusselt number compared to the non-rotating case (Regime II). When the thickness of the shear layer reaches that of the thermal boundary layer, the creation of the plumes is affected by the rotation and we observe shorter plumes in the bulk of the bubble. With Method 2, when the velocity rotation is relatively high, most of the plumes are suppressed but intermittent and rare very long structures that stretch from the equator to the pole are present. In this case, the dynamics of the fluid is completely modified and the global Nusselt number is decreased compared to the non-rotating one. The analysis of the energy/enstrophy/temperature spectra and fluxes also confirms that the small shear layer, strengthened by high azimuthal velocity rotations, plays a very important role in the statistical properties of the bubble. This method for producing the rotations strengthens the buoyancy leading to a Bolgiano-like scaling at large scales. At intermediate scales, a Kraichnan-like scaling is observed in the energy and enstrophy spectra. We also observe a dominant inverse cascade of energy and a dominant direct cascade of enstrophy.

Author Contributions: Conceptualization, methodology, validation, writing–original draft preparation, writing–review and editing, project administration, funding acquisition: P.F., C.-H.B., H.K. Software: P.F., C.-H.B. All authors have read and agreed to the published version of the manuscript.

Funding: Numerical experiments presented in this paper were carried out using the PlaFRIM experimental testbed, supported by Inria, CNRS (LABRI and IMB), Université de Bordeaux, Bordeaux INP and Conseil Régional d'Aquitaine (see https://www.plafrim.fr/). H.K. thanks the Institut Universitaire de France for partial support.

Acknowledgments: The authors wish to thank the editor and the referee for their valuable comments and suggestions.

Conflicts of Interest: The authors declare no conflict of interest.

References

1. Verma, M. *Physics of Buoyant Flows: From Instabilities to Turbulence*; World Scientific: Singapore, 2018.
2. Gascard, J.; Watson, A.; Messias, M.; Olsson, K.; Johannessen, T.; Simonsen, K. Long-lived vortices as a mode of deep ventilation in the Greenland Sea. *Nature* **2002**, *416*, 525–527.
3. Marshall, J.; Schott, F. Open-ocean convection : Observation, theory, and models. *Rev. Geophys.* **1999**, *37*, 1–64.
4. Wadhams, P.; Holfort, J.; Hansen, E.; Wilkinson, J. A deep convective chimney in the winter Greenland Sea. *Geophys. Res. Lett.* **2002**, *29*, 76-1–76-4.
5. Hadley, G. Concerning the cause of the general trade-winds. *Philos. Trans. R. Soc. Lond.* **1735**, *39*, 58–62.
6. Cardin, P.; Olson, P. Chaotic thermal convection in a rapidly rotating spherical shell: consequences for flow in the outer core. *Phys. Earth Planet. Inter.* **1994**, *82*, 235–259.
7. Glatzmaier, G.; Coe, R.; Hongre, L.; Roberts, P. The role of the Earth mantle in controlling the frequency of geomagnetic reversals. *Phys. Earth Planet. Inter.* **1999**, *401*, 885–890.
8. Jones, C. Convection-driven geodynamo models. *Philos. Trans. R. Soc. Lond. Ser. A Math. Phys. Eng. Sci.* **2000**, *358*, 873–897.
9. Sarson, G. Reversal models from dynamo calculations. *Phil. Trans. R. Soc. Lond. A* **2000**, *358*, 921–942.
10. Heimpel, M.; Aurnou, J. Turbulent convection in rapidly rotating spherical shells: A model for equatorial and high latitude jets on Jupiter and Saturn. *Icarus* **2007**, *187*, 540–557.
11. Ingersoll, A. Atmospheric dynamics of the outer planets. *Science* **1990**, *248*, 308–316.
12. Kooij, G.; Botchev, M.; Geurts, B. Direct numerical simulation of Nusselt number scaling in rotating Rayleigh-Bénard convection. *Int. J. Heat Fluid Flow* **2015**, *55*, 1363–1367.
13. Pieri, A.; Falasca, F.; von Hardenberg, J.; Provenzale, A. Plume dynamics in rotating Rayleigh-Bénard convection. *Phys. Lett. A* **2016**, *380*, 26–33.
14. Zhong, J.Q.; Stevens, J.; Clercx, H.; Verzicco, R.; Lohse, D.; Ahlers, G. Prandtl-, Rayleigh-, and Rossby-number dependance of heat transport in turbulent rotatting Rayleigh-Bénard convection. *Phys. Rev. Lett.* **2009**, *102*, 044502.
15. Kunnen, R.; Clercx, H.; Geurts, B. Breakdown of large-scale circulation in turbulent rotating convection. *Europhys. Lett.* **2008**, *84*, 24001.
16. Stevens, J.; Clercx, H.; Lohse, D. Boundary layers in rotating weakly turbulent Rayleigh-Bénard convection. *Phys. Fluids* **2010**, *22*, 085103.
17. Rajaei, H.; Joshi, P.; Alards, K.; Kunnen, R.; Toschi, F.; Clercx, H. Transitions in turbulent rotating convection: A Lagrangian perspective. *Phys. Rev. E* **2016**, *93*, 043129.
18. Rajaei, H.; Joshi, P.; Kunnen, R.; Clercx, H. Flow anisotropy in rotating buoyancy-driven turbulence. *Phys. Rev. Fluids* **2016**, *1*, 044403.
19. Rajaei, H.; Kunnen, R.; Clercx, H. Exploring the geostrophic regime of rapidly rotating convection with experiments. *Phys. Fluids* **2017**, *29*, 045105.
20. Seychelles, F.; Amarouchene, Y.; Bessafi, M.; Kellay, H. Thermal Convection and Emergence of Isolated Vortices in Soap Bubbles. *Phys. Rev. Lett.* **2008**, *100*, 144501.
21. Seychelles, F.; Ingremeau, F.; Pradere, C.; Kellay, H. From Intermittent to Nonintermittent Behavior in Two Dimensional Thermal Convection in a Soap Bubble. *Phys. Rev. Lett.* **2010**, *105*, 264502.
22. Meuel, T.; Xiong, Y.L.; Fischer, P.; Bruneau, C.H.; Bessafi, M.; Kellay, H. Intensity of vortices: From soap bubbles to hurricanes. *Sci. Rep.* **2013**, *3*, 3455.

23. Meuel, T.; Coudert, M.; Bruneau, C.H.; Fischer, P.; Kellay, H. Effects of rotation on temperature fluctuations in turbulent thermal convection on a hemisphere. *Sci. Rep.* **2018**, *8*, 16513.
24. Bruneau, C.H.; Fischer, P.; Xiong, Y.L.; Kellay, H. Numerical simulations of thermal convection on a hemisphere. *Phys. Rev. Fluids* **2018**, *3*, 043502.
25. Boussinesq, J. *Théorie Analytique de la Chaleur Mise en Harmonie avec la Thermodynamique et avec La théorie Mécanique de la Lumière*; Gauthier-Villars: Paris, France, 1903.
26. Angot, P.; Bruneau, C.H.; Fabrie, P. A penalization method to take into account obstacles in incompressible viscous flow. *Numer. Math.* **1999**, *81*, 497–520.
27. Bruneau, C.; Saad, M. The 2D lid-driven cavity problem revisited. *Comput. Fluids* **2006**, *35*, 326–348.
28. Sanvicente, E.; Giroux-Julien, S.; Menezo, C.; Bouia, H. Transitional natural convection flow and heat transfer in an open channel. *Int. J. Therm. Sci.* **2013**, *63*, 87–104.
29. Rajaei, H. Rotating Rayleigh-Bénard Convection. Ph.D. Thesis, Technische Universiteit, Eindhoven, The Netherlands, 2017.
30. Vincent, A.; Yuen, D. Plumes and waves in two-dimensional turbulent thermal convection. *Phys. Rev. E* **1999**, *60*, 2957–2963.
31. Rajaei, H.; Alards, K.; Kunnen, R.; Clercx, H. Velocity and acceleration statistics in rapidly rotating Rayleigh-Bénard convection. *J. Fluid Mech.* **2018**, *857*, 374–397.
32. Alards, K.M.J.; Kunnen, R.P.J.; Stevens, R.J.A.M.; Lohse, D.; Toschi, F.; Clercx, H.J.H. Sharp transitions in rotating turbulent convection: Lagrangian acceleration statistics reveal a second critical Rossby number. *Phys. Rev. Fluids* **2019**, *4*, 074601.
33. King, E.; Stellmach, S.; Noir, J.; Hansen, U.; Aurnou, J. Boundary layer control of rotating convection systems. *Nature* **2009**, *457*, 301–304.
34. Ecke, R.; Niemela, J. Heat Transport in the Geostrophic Regime of Rotating Rayleigh-Bénard Convection. *Phys. Rev. Lett.* **2014**, *113*, 114301.
35. Sharma, M.; Kumar, A.; Verma, M.; Chakraborty, S. Statistical features of rapidly rotating decaying turbulence: Enstrophy and energy spectra and coherent structures. *Phys. Fluids* **2018**, *30*, 045103.
36. Sharma, M.; Verma, M.; Chakraborty, S. On the energy spectrum of rapidly rotating forced turbulence. *Phys. Fluids* **2018**, *30*, 115102.
37. Bruneau, C.; Fischer, P. Spectra and filtering: A clarification. *Int. J. Wavelets Multiresolut. Inf. Process.* **2007**, *5*, 465–483.
38. Bolgiano, R. Turbulent spectra in a stably stratified atmosphere. *J. Geophys. Res.* **1959**, *71*, 2226–2229.
39. Bolgiano, R. Structure of turbulence in stratified media. *J. Geophys. Res.* **1962**, *67*, 3015–3023.
40. Verma, M.; Kumar, A.; Pandey, A. Phenomenology of buoyancy-driven turbulence: Recent results. *New J. Phys.* **2017**, *19*, 025012.
41. Kumar, A.; Chatterjee, A.; Verma, M. Energy spectrum of buoyancy-driven turbulence. *Phys. Rev. E* **2014**, *90*, 023016.

Publisher's Note: MDPI stays neutral with regard to jurisdictional claims in published maps and institutional affiliations.

© 2020 by the authors. Licensee MDPI, Basel, Switzerland. This article is an open access article distributed under the terms and conditions of the Creative Commons Attribution (CC BY) license (http://creativecommons.org/licenses/by/4.0/).

fluids

MDPI

Article

Non-Modal Three-Dimensional Optimal Perturbation Growth in Thermally Stratified Mixing Layers

Helena Vitoshkin [1,*] and Alexander Gelfgat [2]

1 Agricultural Research Organization, The Volcani Center, P.O. Box 6, 68 Dereh Hamacabim, Rishon Lezion 5025001, Israel
2 School of Mechanical Engineering, Faculty of Engineering, Tel-Aviv University, Ramat Aviv, Tel-Aviv 6997801, Israel; gelfgat@tau.ac.il
* Correspondence: elenav@volcani.agri.gov.il

Abstract: A non-modal transient disturbances growth in a stably stratified mixing layer flow is studied numerically. The model accounts for a density gradient within a shear region, implying a heavier layer at the bottom. Numerical analysis of non-modal stability is followed by a full three-dimensional direct numerical simulation (DNS) with the optimally perturbed base flow. It is found that the transient growth of two-dimensional disturbances diminishes with the strengthening of stratification, while three-dimensional disturbances cause significant non-modal growth, even for a strong, stable stratification. This non-modal growth is governed mainly by the Holmboe modes and does not necessarily weaken with the increase of the Richardson number. The optimal perturbation consists of two waves traveling in opposite directions. Compared to the two-dimensional transient growth, the three-dimensional growth is found to be larger, taking place at shorter times. The non-modal growth is observed in linearly stable regimes and, in slightly linearly supercritical regimes, is steeper than that defined by the most unstable eigenmode. The DNS analysis confirms the presence of the structures determined by the transient growth analysis.

Keywords: stratified mixing layer; non-modal instability; Kelvin-Helmholtz instability; Holmboe instability

check for updates

Citation: Vitoshkin, H.; Gelfgat, A. Non-Modal Three-Dimensional Optimal Perturbation Growth in Thermally Stratified Mixing Layers. *Fluids* **2021**, *6*, 37. https://doi.org/10.3390/fluids6010037

Received: 11 December 2020
Accepted: 4 January 2021
Published: 11 January 2021

Publisher's Note: MDPI stays neutral with regard to jurisdictional claims in published maps and institutional affiliations.

Copyright: © 2021 by the authors. Licensee MDPI, Basel, Switzerland. This article is an open access article distributed under the terms and conditions of the Creative Commons Attribution (CC BY) license (https://creativecommons.org/licenses/by/4.0/).

1. Introduction

The mixing layer flow is the simplest configuration allowing for the well-known Kelvin-Helmholtz (or KH) instability that takes place when two parallel flows having different velocities meet. The instability develops as a wave (or KH mode) in the shear layer separating two uniform flows. Studies of this phenomenon started in the early works of Lord Kelvin [1] and Strutt and Lord Rayleigh [2]. The phenomenon appears to be so complicated and to pose so many questions that in spite of hundreds of studies published, it remains the subject of many current researches [3–6]. For the details, the reader is referred to review papers [7–9].

The isothermal mixing layer is known to be linearly unstable either in the inviscid limit or starting from relatively low Reynolds numbers within the Newtonian viscous model [10]. It is also known that the flow remains linearly stable for the streamwise dimensionless wavenumber larger than unity [10,11]. At the same time, studies [12,13] showed that temporal non-modal disturbances growth can take place in the isothermal inviscid and viscous mixing layer flows, respectively. A similar mechanism was discovered recently in round jets [5].

The classical result on the linear stability of an inviscid mixing layer flow stably stratified by density (or temperature) stays that the flow becomes linearly stable for the gradient Richardson number exceeding 0.25 [10–15]. Later studies showed that along with the monotonic Kelvin-Helmholtz instability, an oscillatory instability is also possible [16]. The latter is the Holmboe instability that develops as two waves traveling in opposite directions [10,17]. Furthermore, numerical calculations [18–20] supported by experiments of [21] showed that a three-dimensional mode traveling at an angle to the base velocity may attain the largest growth rate. For example, a shear layer characterized by a relatively thin region of stably stratified fluid can exhibit Holmboe instability [16,21,22], whose growth rate increases with increasing stratification. Despite the valid Squire transformation proved in [23], the flow is predicted to be linearly unstable to a three-dimensional disturbance.

It was shown in [10] that the instability at small Richardson numbers is governed by two monotonic KH modes, one of which can become unstable. With further increase of the Richardson number, the pair of monotonic modes are replaced by a pair corresponding to complex conjugated eigenvalues related to the Holmboe instability. It was shown recently that KH instability plays an important role in the formation of interfaces between coating and substrate materials and metallic materials [24,25]. The non-linear evolution of KH and Holmboe instabilities was studied in quite a large number of articles (e.g., [17,26–33]). A difference in the nonlinear disturbances growth compared with the predictions of linear stability theory was reported in [31]. Surprisingly, the issue of non-modal growth of disturbances at short times was addressed only by [14] for non-stratified mixing layers and by [32] for a stratified inviscid model. The study [32] described the non-modal amplification of the stratified mixing layer in terms of the interaction of gravity wave and vortex energy of the optimal perturbation for large wavenumbers. It was shown that the amplified transient growth can be three-dimensional and is associated with wave generation due to a vertical motion at large wavenumbers.

This study extends the previous results of [12] and [29] to viscous thermally stratified mixing layers. We show that the non-modal growth in this flow is governed mainly by the Holmboe-type modes. In the following, we examine their dynamics and interaction via consideration of a fully nonlinear three-dimensional model.

The paper is organized in the following way. In Section 2, we outline the fundamental details of the instability analysis and give a formulation of the optimal growth methodology, including the derivation of an adjoint operator. Section 3 discusses the implementation of different techniques for the calculation of the largest possible non-modal growth. Section 4 presents non-modal instability properties as a function of the Richardson number and evolution of the optimal disturbances within three-dimensional models. The summary and conclusions are derived in Section 5.

2. Formulation of the Problem and Numerical Techniques

We consider a flow of an incompressible Newtonian fluid in a thermally stratified mixing layer. After the Boussinesq approximation is applied, the flow is governed by the momentum, continuity, and energy equations

$$\frac{\partial \mathbf{u}}{\partial t} + (\mathbf{u}\cdot\nabla)\mathbf{u} = -\frac{1}{\rho}\nabla p + \nu\Delta\mathbf{u} + \mathcal{g}\gamma(T-\overline{T})e_z \quad (1a)$$

$$\nabla\cdot\mathbf{u} = 0 \quad (1b)$$

$$\frac{\partial T}{\partial t} + (\mathbf{u}\cdot\nabla)T = \kappa\Delta T \quad (1c)$$

Here, $\mathbf{u} = (u_x, u_y, u_z)$ is the velocity with components in the streamwise (x), spanwise (y) and vertical (z) directions; p and T are the pressure and the temperature, ρ is the density, ν is the kinematic viscosity, $\mathcal{g}$ is gravitational acceleration, γ is the thermal expansion acceleration, e_z is the unit vector in the z-direction (vertical direction), $\overline{T}$ is the mean temperature, which is defined below, κ is the thermal diffusivity, and Δ denotes the Laplacian operator.

The stratified mixing layer flow is modeled as two horizontal fluid layers moving in opposite directions with velocities $\pm U_{max}$ when the colder fluid layer is located under the warmer one. It is assumed that the base flow and temperature profiles can be described by hyperbolic tangent profiles

$$U(z) = U_{max} tanh\left(\frac{z}{\delta_v}\right), T(z) = T_1 + \overline{T}\left[1 + tanh\left(\frac{z}{\delta_T}\right)\right] \tag{2}$$

where T_1 and T_2 are temperatures of lower and upper fluid layers, respectively, $(T_2 > T_1)$, $\overline{T} = 0.5(T_2 - T_1)$, and δ_v and δ_T are the thicknesses of the velocity and temperature layers (note $\delta_v > \delta_T$ for a Prandtl number larger than unity). The thickness of the velocity layer is defined following assumptions derived in [30] where the authors showed that presentation of base flow based on the "*tanh*" profile provides a better fit to the experimental data and is proportional to momentum thickness.

Seeking a solution as a linear combination of the steady base flow (U, T) and an infinitesimal disturbance $\overline{¶} = (\overline{\mathbf{v}}, \overline{\theta})$, we arrive at the dimensionless linearized equations for the velocity, $\overline{\mathbf{v}}$, and temperature, $\overline{\theta}$, perturbations:

$$\frac{\partial \overline{\mathbf{v}}}{\partial t} + \left(\overline{\mathbf{v}} \cdot \nabla\right) U + (U \cdot \nabla)\overline{\mathbf{v}} = -\nabla p + Re^{-1}\Delta\overline{\mathbf{v}} + Ri\overline{\theta}\mathbf{e}_z \tag{3a}$$

$$\nabla \cdot \overline{\mathbf{v}} = 0 \tag{3b}$$

$$\frac{\partial \overline{\theta}}{\partial t} + \left(\overline{\mathbf{v}} \cdot \nabla\right) T + (U \cdot \nabla)\overline{\theta} = Pe^{-1}\Delta\overline{\theta} \tag{3c}$$

Equations (3a–3c) are rendered dimensionless using the scales δ_v, U_{max}, δ_v/U_{max}, ρU_{max}^2, and $(T_2 - T_1)$ for length, velocity, time, pressure, and temperature, respectively. The Reynolds number is defined by $Re = U_{max}\delta_v/\nu$, $Pr = \nu/\kappa$ is the Prandtl number, $Pe = RePr = U_{max}\delta_v/\kappa$ is the Péclet number, $Ri = \gamma g(T_2 - T_1)\delta_v/U_{max}^2$ is the bulk Richardson number, and $\delta = \delta_v/\delta_T$ is the velocity and temperature thicknesses ratio.

For the modal analysis, we assume an exponential dependence of the perturbations on time and apply the Fourier integral expansion in the infinite x- and y-direction, i.e., $\tilde{¶} = \check{¶}(z)\exp[i(\alpha x + \beta y) + \lambda t]$. Subsequently, the linearized continuity and momentum equations can be rearranged into a system of Orr-Sommerfeld and linearized energy equations (see, e.g., [28]) that read

$$\lambda\tilde{¶}(z) = \mathcal{L}\tilde{¶}(z), \tag{4a}$$

$$\mathcal{L} = \begin{pmatrix} \mathcal{L}_{OS} & \mathcal{L}_{w\theta} \\ \mathcal{L}_{Tw} & \mathcal{L}_{Sq_\theta} \end{pmatrix} \tag{4b}$$

$$\mathcal{L}_{OS} = \Delta^{-1}\left(i\alpha(-U\Delta + U_{zz}) + Re^{-1}\Delta^2\right), \tag{4c}$$

$$\mathcal{L}_{w\theta} = -Ri\left(\alpha^2 + \beta^2\right)\Delta^{-1} \tag{4d}$$

$$\mathcal{L}_{Tw} = -T_z, \tag{4e}$$

$$\mathcal{L}_{Sq_\theta} = Pe^{-1}\Delta - i\alpha U. \tag{4f}$$

Here, λ is a complex time increment, α and β are real wavenumbers in the $x-$ and $y-$ directions respectively, $\mathcal{L}$ is the linearized generalized operator, $\mathcal{L}_{OS}$ is the Orr-Sommerfeld operator that determines the evolution of perturbation of the vertical velocity component w; $\mathcal{L}_{Tw}$ and $\mathcal{L}_{w\theta}$ are operators coupling the velocity perturbation w with temperature perturbations θ; $\mathcal{L}_{Sq_\theta}$ is the Squire operator that determines the evolution of perturbation of temperature θ; T_z and U_{zz} are the first and second derivatives of temperature and velocity profiles (Equation (2)), respectively. Note that the Laplacian operator reduces to $\Delta = \frac{\partial^2}{\partial z^2} - (\alpha^2 + \beta^2)$, and Δ^{-1} denotes the inverse Laplacian operator.

Equations (4a–4f) define an eigenproblem for eigenvalues λ and eigenvectors $\tilde{\mathbb{q}}$. The existence of an eigenvalue λ having a positive real part is a sufficient condition for linear instability. Therefore, the linear stability problem reduces to finding a Reynolds number at which the real part of the leading eigenvalue (i.e., the eigenvalue with the largest real part) changes its sign from negative to positive and then minimizing Re over all possible values of Ri, α, and β. An extended analysis of Squire's theorem [27,30] demonstrated that a three-dimensional unstable mode can become most unstable in strongly stratified viscous flow.

2.1. Non-Modal Analysis

Considering the non-modal disturbances growth, we are interested in a short-time behavior of small finite-amplitude perturbation governed by the linearized equations. We exploit the homogeneity of the x- and y-direction by assuming solutions in the form $\hat{\mathbb{q}} = \mathbb{q}(z,t)\exp(i\alpha x + i\beta y)$. It can be shown that by applying the continuity equation and excluding the pressure, an arbitrary three-dimensional infinitesimal perturbation is described by $\mathbb{q} = (u_x, u_y, u_z, \theta)$. Furthermore, the two velocity components u_x and u_y can be replaced by the vertical vorticity component $\eta = \partial u_y/\partial x - \partial u_x/\partial y$. Thus, consideration of the perturbation vector $(w, \eta, \theta)^T$ where $u_z = w$ is the vertical velocity component yields an equivalent formulation. The resulting linearized dynamical matrix operator $\mathcal{L}$ is expressed as

$$\frac{\partial \mathbb{q}}{\partial t} = \mathcal{L}\mathbb{q}, \tag{5a}$$

$$\mathcal{L} = \begin{pmatrix} \mathcal{L}_{OS} & 0 & \mathcal{L}_{w\theta} \\ \mathcal{L}_{\eta w} & \mathcal{L}_{Sq} & 0 \\ \mathcal{L}_{Tw} & 0 & \mathcal{L}_{Sq\theta} \end{pmatrix} \tag{5b}$$

$$\mathcal{L}_{\eta w} = -i\beta U_z, \tag{5c}$$

$$\mathcal{L}_{Sq} = Re^{-1}\Delta - i\alpha U \tag{5d}$$

In this case, the coupling operator $\mathcal{L}_{\eta w}$ and the Squire operator governing the vorticity perturbation $\mathcal{L}_{Sq}$ are added to the equations system (4a–4f). The non-modal analysis examines the maximum energy growth at a particular time period of the flow disturbance, $\mathbb{q}(0)$, over all possible initial conditions in terms of energy norm, produced by a sum of kinetic and potential energies, and assuming that the temperature perturbation alters the latter. The squared energy norm is represented by the inner product (to be defined below):

$$E(t) = \left|\left|\mathbb{q}\right|\right|_E^2 = \langle \mathbb{q}, \mathbb{q} \rangle_E \tag{6}$$

For fixed wavenumbers, the integrals in x- and y- coordinates scale out of the problem, and the energy norm in terms of w, η, and θ becomes

$$E(t,\alpha,\beta) = \int \left[\frac{1}{2(\alpha^2+\beta^2)} \left(|w|^2 + \left|\frac{dw}{dz}\right|^2 + |\eta|^2 + C|\theta|^2 \right) \right] dz \tag{7}$$

where the coefficient C is a positive definite scalar. The choice of C will not change the norm qualitatively [34,35]. It was shown in [36] that the perturbation energy is associated with the Brunt-Väisäla frequency and can be expressed in terms of the local Richardson number $Ri_l = \frac{Ri}{T_z}$ (see also [37]). Following this result, we define $C = Ri_l(z=0) = \frac{Ri}{T_z(z=0)}$. It is easy to see that the above energy norm is produced by the inner product

$$\langle \mathbb{q}_1, \mathbb{q}_2 \rangle = \int \left[\frac{1}{2(\alpha^2+\beta^2)} \left(w_1 w_2^* + \frac{dw_1}{dz}\frac{dw_2^*}{dz} + \eta_1\eta_2^* + \frac{Ri}{T_z(0)}\theta_1\theta_2^* \right) \right] dz \tag{8}$$

where * denotes the complex conjugate. In the following, the inner product of two complex vectors $\mathbb{q}_i$ and $\mathbb{q}_j$ associates with the energy norm (7), (8) and is given by

$$\langle \mathbb{q}_i, \mathbb{q}_j \rangle_E = \langle \mathbb{q}, \mathrm{M}\mathbb{q} \rangle = \int_\Omega \left[w_j^*, \eta_j^*, \theta_j^* \right]_j \mathrm{M} \begin{bmatrix} w_i \\ \eta_i \\ \theta_i \end{bmatrix}_i d\Omega \text{ with } \mathrm{M} = \begin{bmatrix} -\Delta & 0 & 0 \\ 0 & 1 & 0 \\ 0 & 0 & Ri/T_z(0) \end{bmatrix} \tag{9}$$

So that $E(t, \alpha, \beta) = ||\mathbb{q}||_E^2$ is the energy amplification at time $t = \tau$, or growth function, is defined as ([38–40])

$$G(\tau, \alpha, \beta) = \max_{\mathbb{q}(0) \neq 0} \frac{E(\tau)}{E(0)} \tag{10}$$

Substitution for $E(\tau)$ in (10), with assuming $E(0) = 1$, and recalling that $\mathbb{q}(\tau) = exp(\mathcal{L}\tau)\mathbb{q}(0) = \Phi(\tau)\mathbb{q}(0)$, yields

$$G(\tau, \alpha, \beta) = \max_{\mathbb{q}(0)\neq 0} \langle \Phi(\tau)\mathbb{q}(0), \Phi(\tau)\mathbb{q}(0) \rangle = \max_{\mathbb{q}(0)\neq 0} \langle \mathbb{q}(0), \Phi^*(\tau)\Phi(\tau)\mathbb{q}(0) \rangle = \|\Phi^*(\tau)\Phi(\tau)\| \tag{11}$$

where $\Phi^*(\tau)$ is the adjoint of the operator $\Phi(\tau)$ (see Section 2.2). The operator $\Phi^*(\tau)\Phi(\tau)$ is self-adjoint and normal, its eigenvalues are real and non-negative, and $||\Phi^*(\tau)\Phi(\tau)|| = \max_i \sigma_i = \sigma_m$ is the largest eigenvalue that corresponds to the maximum possible relative growth attainable at the time τ. The eigenvector $\mathbb{q}_m$ corresponding to the eigenvalue σ_m is the disturbance which yields this maximal growth value.

To examine possible non-modal disturbance growth, we applied two independent approaches. The first one is the Gram matrix factorization, followed by the singular value decomposition (SVD) method [41,42]. The energy norm (7) of an arbitrary perturbation $q(t)$ can be connected with the Euclidean norm of the vector $k(t)$ defined by the eigenvector decomposition of the matrix $\mathcal{L}$: $||q||_2^2 = \sum_{i,j} k_i^* \langle p_j^*, p_i \rangle k_j$, as follows

$$||q||_E^2 = q^* \mathcal{M} q = \sum_{i,j} k_i^* p_j^* \mathcal{M}_{ji} p_i k_j = \sum_{i,j} k_i^* S k_j \tag{12}$$

where $\mathcal{M}$ is the mass matrix, $\boldsymbol{p}$ is the eigenvector, and $S = \langle p, \mathcal{M}p \rangle$ is the Gram matrix, which is the complex-valued Hermitian matrix. Then, using factorization of the positive definite Gram matrix $S = F^H F$, the maximum energy growth at time τ can be calculated as the value of the 2-norm of the matrix $Fexp(\Lambda\tau)F^{-1}$ and is equal to the square of its maximal singular value. This method allows for the calculation of non-modal growth based on a separated part of the spectrum. Following the arguments of [13], only a discrete part of the spectrum should be included in the analysis. To separate the discrete part of the spectrum, we tried the approach of [13] and extracted the eigenvectors that have finite amplitude inside and are near the mixing zone and vanish far from it. Alternatively, we included in the analysis only those modes whose pseudospectrum is located at a distance smaller than a given ε from the eigenvalues of the linearized problem.

The second approach is the iterative forward/backward time integration applying a random initial perturbation. The evolution forward is governed by the operator $\mathcal{L}$ and is followed by the integration backward governed by the adjoint operator, $\mathcal{L}^*$. The latter is derived in the next section. This approach involves both discrete and continuous parts of the spectrum and results in the growth function value at a given time and the corresponding optimal initial vector [43].

To validate the results, we apply the full three-dimensional DNS with the optimally perturbed base flow as the initial condition. This validation shows that the second approach that involves forward/backward time integrations yields the correct results.

2.2. Adjoint form of the Orr-Sommerfeld, Squire, and Energy Equations

We derive the adjoint operator considering the derivation of the adjoint form of the first- and second-order spatial derivatives and the first-order time derivatives. As noted by [44], the direct and adjoint parabolic problems have opposite directions of stable time-like evolution. From the definition (5a,b) and using (9) the adjoint operator $\mathcal{L}^*$ can be obtained as follows:

$$\langle \mathbb{q}, \mathcal{L}\mathrm{M}\mathbb{q}\rangle = \int_\Omega \begin{bmatrix} -\Delta\left((\mathcal{L}_{OS}w)^* + (\mathcal{L}_{\eta w}\eta)^* + (\mathcal{L}_{Tw}\theta)^*\right) \\ (\mathcal{L}_{Sq}\eta)^* \\ \frac{Ri}{T_z(0)}\left((\mathcal{L}_{w\theta}w)^* + (\mathcal{L}_{Sq_\theta}\theta)^*\right) \end{bmatrix}^{\mathrm{T}} \begin{bmatrix} w \\ \eta \\ \theta \end{bmatrix} d\Omega = \langle (\mathcal{L}\mathrm{M}\mathbb{q})^*, \mathbb{q}\rangle \tag{13}$$

Then, the adjoint system may be obtained by the integration by parts and reads

$$\left[\Delta\frac{\partial}{\partial t} + 2U_z\frac{\partial^2}{\partial x\partial z} + \left(U\frac{\partial}{\partial x} + \frac{1}{Re}\Delta\right)\Delta\right]w^* = -U_z\frac{\partial}{\partial y}\eta^* - T'\theta^* \tag{14a}$$

$$\left[\frac{\partial}{\partial t} + \frac{1}{Re}\Delta + U\frac{\partial}{\partial x}\right]\eta^* = 0 \tag{14b}$$

$$\left[\frac{\partial}{\partial t} + \frac{1}{Pe}\Delta + U\frac{\partial}{\partial x}\right]\theta^* = -\Delta_{2D}Riw^* \tag{14c}$$

where $\Delta_{2D} = \frac{\partial^2}{\partial x^2} + \frac{\partial^2}{\partial y^2}$ is the 2D Laplacian and U_z is the first derivative of the velocity profile (Equation (2)). Evidently, the adjoint operator $\mathcal{L}^*$ of the adjoint system $\partial\mathbb{q}^*/\partial t = -(\mathcal{L}\mathbb{q})^*$ in terms of a periodical in the stream- and span-wise direction perturbation will be expressed as

$$\mathcal{L}^* = \begin{bmatrix} \mathcal{L}^*_{OS} & \mathcal{L}^*_{\eta w} & \mathcal{L}^*_{Tw} \\ 0 & \mathcal{L}^*_{Sq} & 0 \\ \mathcal{L}^*_{w\theta} & 0 & \mathcal{L}^*_{Sq\theta} \end{bmatrix} \tag{15}$$

where the entries of the matrices are

$$\mathcal{L}^*_{OS} = \Delta^{-1}\left(-2i\alpha U_z\frac{\partial}{\partial z} - \left(i\alpha U + Re^{-1}\Delta\right)\Delta\right), \quad \mathcal{L}^*_{w\theta} = -Ri\left(\alpha^2 + \beta^2\right) \tag{16a}$$

$$\mathcal{L}^*_{\eta w} = -\Delta^{-1}i\beta U_z, \qquad \mathcal{L}^*_{Sq} = -Re^{-1}\Delta - i\alpha U \tag{16b}$$

$$\mathcal{L}^*_{Tw} = -\Delta^{-1}T_z, \qquad \mathcal{L}^*_{Sq\theta} = -Pe^{-1}\Delta - i\alpha U \tag{16c}$$

Finally, in accordance with Equation (11), the time integration forward/backward iterative procedure will result in the optimal initial vector, which maximizes growth at the specified time and a given set of parameters (α, β, Re, Ri, Pe) [43,45].

3. Calculations Techniques

For the non-modal analysis, the linear system (Equation (5)) was discretized via a second-order central finite difference method. The eigenvalues and eigenvectors of matrix $\mathcal{L}$ were calculated using the QR algorithm. The verification of the method and test calculation for the isothermal mixing layer can be found in [13]. Table 1 presents an example of the convergence of four leading eigenvalues belonging to the discrete part of the spectrum for a non-isothermal case. It is shown that the use of 1000 grid points yields four converged decimal digits for the first mode; however, the convergence slows down for the next modes. Despite this, it is shown that the growth function calculated by the iterative integration method converges to the value of 21.71 already for 700 grid points.

Table 1. Convergence of four least stable eigenvalues belonging to the discrete spectrum for stratified flow $\alpha = 0.5$, $\beta = 0$, $Re = 1000$, $Ri = 0.8$.

N	1st Mode		2nd Mode		3rd Mode		4th Mode		Growth Function, G (t = 10)
	λr	λi	λr	λi	λr	λi	λr	λi	
500	0.0260	±0.3653	−0.0460	±0.2653	−0.1379	±0.2710	−0.1466	±0.0778	21.70
600	0.0334	±0.3986	−0.0460	±0.2653	−0.1361	±0.3004	−0.1466	±0.0684	21.70
700	0.0324	±0.4591	−0.0460	±0.2653	−0.1352	±0.3017	−0.1466	±0.0650	21.71
800	0.0324	±0.4597	−0.0459	±0.2654	−0.1334	±0.3112	−0.1454	±0.0583	21.71
900	0.0324	±0.4597	−0.0459	±0.2654	−0.1317	±0.3310	−0.1497	±0.0287	21.71
1000	0.0324	±0.4597	−0.0458	±0.2654	−0.1250	±0.3302	−0.1497	±0.0259	21.71
1100	0.0324	±0.4597	−0.0458	±0.2654	−0.1250	±0.3302	−0.1496	±0.0259	21.71
1200	0.0324	±0.4597	−0.0458	±0.2653	−0.1262	±0.3302	−0.1492	±0.0256	21.71
1300	0.0324	±0.4597	−0.0457	±0.2653	−0.1262	±0.3302	−0.1489	±0.0256	21.71
1400	0.0324	±0.4597	−0.0457	±0.2652	−0.1262	±0.3302	−0.1489	±0.0255	21.71
1500	0.0324	±0.4597	−0.0457	±0.2652	−0.1262	±0.3302	−0.1489	±0.0255	21.71

The calculated spectrum was examined via analysis of its pseudospectrum [46]. The ε-pseudospectrum was computed as the minimal singular value of the matrix $(\lambda I - \mathcal{L})$, as was proposed in [47–49]. If u_s is the singular vector of the matrix corresponding to the minimal singular value, then $\varepsilon = ||(\lambda I - \mathcal{L})u_s||_2$ estimates the accuracy of the calculation of the eigenvalue λ. It was found that for the stratified mixing layer configuration, the relatively large part of the eigenvalues corresponds to a relatively large ε-pseudospectrum ($\varepsilon = 10^{-2}$, 10^{-3}), consequently, the eigenvalues calculation accuracy can be even more demanding than in the isothermal case. Results of [13] show that only the eigenmodes with the pseudospectrum $\varepsilon < 10^{-6}$ contributed to the optimal growth in isothermal flow. This is the main limitation of applying the SVD-based method [41] here. A series of tests show that we cannot use the SVD-based method based on a full spectrum for the same reasons as in the case of isothermal flow [10], i.e., we again have to separate part of the spectrum that corresponds to a non-accurate replication of the continuous modes. Figure 1 shows that the growth functions calculated based on extracting the vectors vanishing outside the shear layer (Figure 1a) and based on extracting the vectors corresponding different ε values (Figure 1b) do not yield converged consistent results within the SVD-based method. It is emphasized, however, that despite the obvious scatter in the results shown in Figure 1, they do show that significant non-modal growth can be expected even in cases of a strong, stable stratification. It also supports the results of [32], indicating that optimal perturbation can be located outside the shear layer.

We succeeded in obtaining reliable converged quantitative results using the iterative integration method, which, similarly to the isothermal case [13], converges to within three decimal places already on the 700 nodes grid. This method is used for the computation of the results presented below. It should be noted that this method, compared to the SVD-based one, requires much longer computational runs. To validate the non-modal analysis results, the direct numerical simulations taking the optimal vector as an initial condition were carried out. The fully non-linear time-dependent problem was solved using the approach described in [13].

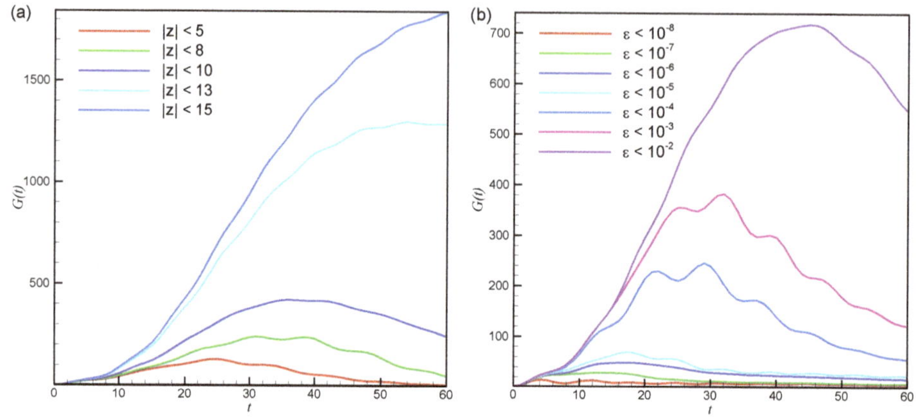

Figure 1. Comparison of growth functions calculated using the singular value decomposition (SVD) method based on extracted vectors (**a**) located in $|z| < L$, i.e., vanishing outside the shear zone and (**b**) corresponding to the values of their pseudospectrum, for $Ri = 0.9$, $Re = 1000$, $\alpha = 0.5$, $\beta = 0$.

4. Results

In the following, we describe results on non-modal growth in the stratified mixing layer for $Pr = 9$, $Re = 1000$, and $R_\delta = \sqrt{Pr} = 3$, assuming the fluid properties close to water, where heat diffuses much slower than momentum. The test calculations (Table 2) show that the least unstable eigenmode (or KH mode) is only slightly affected by the increase of the Prandtl number. The results of [48] also show that for a bounded channel flow, heat diffusivity does not affect linear stability. Conversely, the eigenvalues corresponding to the oscillatory Holmboe modes change noticeably with the increase of the Prandtl number, followed by a narrowing of the layer where the temperature varies.

Table 2. Least unstable eigenvalues at various Prandtl numbers for $\alpha = 0.5$, $\beta = 0$, $Re = 1000$.

Pr	**KH Modes, *Ri* = 0.01**	**Holmboe Modes, *Ri* = 0.8**
0.7	0.140	$(1.45 \times 10^{-3}, \pm 0.5806)$
1	0.139	$(2.69 \times 10^{-3}, \pm 0.5455)$
7	0.131	$(3.84 \times 10^{-3}, \pm 0.4594)$
9	0.129	$(3.24 \times 10^{-3}, \pm 0.4597)$

4.1. Non-Modal Instability as a Function of the Richardson Number

In this section, we present a characteristic case of non-modal energy growth for varying Ri and fixed α, β, Pe, and R_δ. Results for the target time $t = 10$ are summarized in Table 3 (2D case) and Table 4 (3D case). It is seen from Table 2 that the flow is unstable to 2D perturbations ($\alpha = 0.5$, $\beta = 0$, $Re = 1000$) for $Ri \leq 0.8$. At small Richardson numbers ($Ri < 0.1$), the monotonic KH mode grows similarly to the isothermal case. With the increase of Ri, the two KH modes turn into a pair of Holmboe modes [10]. Further increase of the Richardson number, $Ri > 0.8$, makes the flow linearly stable; however, the non-modal growth remains possible. Although all the eigenvalues are negative, the Holmboe modes remain to be the least stable ones. An examination of the growth function, G, at a relatively short time, $t = 10$, shows that the highest amplification occurs at a small Richardson number, and the growth function decreases with the increase of Ri. Note that non-modal amplification reaches its value, for example, 76.4, for $Ri = 0.1$, while amplification of the least unstable mode at $t = 10$ is much smaller, approximately 1.4. This result may be important for the choice of the initial vector for non-linear calculation where the least unstable mode is usually taken as the initial condition. Based on the foregoing results, we argue that at small times the non-modal growth is expected to dominate, i.e., the optimal vector will grow faster than just one least unstable eigenmode. With the steepening stratification, i.e., increase of the Richardson number, a contribution of two-dimensional disturbances into transient non-modal growth diminishes.

Table 3. The variations of the leading eigenvalues and values of amplification function, *G*, (at an arbitrary time $t = 10$) with increasing Richardson number for $\alpha = 0.5$, $\beta = 0$, $R_e = 1000$ (2D case).

Richardson Number	Least Unstable/Leading Stable Mode		G (t = 10)
	λ_{real}	λ_{imag}	
0.01	0.129	0.00	97.2
0.1	3.76×10^{-2}	± 0.16	76.4
0.5	4.42×10^{-2}	± 0.34	22.9
0.8	3.24×10^{-2}	± 0.46	21.7
0.9	-3.64×10^{-2}	± 0.26	21.5
1.0	-2.93×10^{-2}	± 0.26	20.9
2.0	-1.42×10^{-2}	± 0.25	15.3
3.0	-1.29×10^{-2}	± 0.25	12.4

Table 4. The leading eigenvalues, leading monotone and oscillatory (KH-, Holmboe-type) modes, and values of growth function, *G*, at $t = 10$ for the increasing Richardson number at $\alpha = 0.5$, $\beta = 1$, $Re = 1000$ (3D case). The linearly unstable range is denoted by the gray color.

Richardson Number R_i	Least Unstable/Leading Stable Modes		Monotone Modes	Holmboe-Type Modes		G (t = 10)
	λ_{real}	λ_{imag}	λ_{imag}	λ_{real}	λ_{imag}	
0.001	-1.42×10^{-4}	± 0.50	-0.054	-	-	132.6
0.01	-1.42×10^{-4}	± 0.50	-0.11	-4.22×10^{-2}	± 0.16	63.8
0.2	-1.42×10^{-4}	± 0.50	-0.27	-6.06×10^{-2}	± 0.29	43.4
0.3	-1.42×10^{-4}	± 0.50	-0.25	-4.98×10^{-3}	± 0.45	47.4
0.5	-1.43×10^{-4}	± 0.50	-0.17	-4.43×10^{-3}	± 0.29	53.5
0.7	5.35×10^{-3}	± 0.30	-0.17	5.35×10^{-3}	± 0.30	55.0
1.0	9.04×10^{-3}	± 0.33	-0.17	9.04×10^{-3}	± 0.33	73.9
2.0	-1.42×10^{-4}	± 0.50	-0.27	-1.24×10^{-3}	± 0.32	122
3.0	-1.42×10^{-4}	± 0.50	-0.27	-2.74×10^{-3}	± 0.34	158
4.0	-1.42×10^{-4}	± 0.50	-0.27	-3.53×10^{-3}	± 0.36	192
5.0	-1.42×10^{-4}	± 0.50	-0.27	-4.88×10^{-3}	± 0.40	221
10.0	-1.42×10^{-4}	± 0.50	-0.27	-5.60×10^{-3}	± 0.42	378

Based on the results for the isothermal mixing layer [13], where the largest amplification is yielded by 3D optimal disturbance, we examine a particular parameter set, $\alpha = 0.5$, $\beta = 1$, $Re = 1000$, and vary only the Richardson number. These parameters characterize the perturbation that attains a large non-modal growth in the case of isothermal flow. Table 4 shows that the leading mode belongs to the continuous spectrum, Imag $(\lambda) = \pm\alpha$, so that the leading monotone ($\lambda = 0$, KH) and oscillatory ($\lambda \neq 0$, Holmboe) modes are presented in additional columns. It is seen that the oscillating Holmboe modes attain larger growth rates than the monotone ones. Computations of the growth function G (t = 10), using the adjoint forward/backward time integration technique, show large non-modal growth of linearly stable 3D disturbances for $Ri < 0.01$. The growth function decreases with the increase of the Richardson number for $Ri \leq 0.2$. Above the value $Ri = 0.3$, the growth function G ($t = 10$) starts to increase. We observe that the value of G ($t = 10$) continues to increase with increasing Ri.

It should be noted that for a relatively strong stratification, in the interval of $0.7 \leq Ri \leq 1$, we found an additional linear instability. A similar result was reported in [19], where it was shown that stratified shear flow could be unstable to 3D disturbances propagating at an angle to the mean flow. Linear instability at large Richardson numbers for 2D perturbation was also reported in [37,50].

We emphasize that, similarly to the 2D case, the 3D non-modal growth at short times can be significantly larger than the modal one. The least unstable modes in the linearly unstable region are characterized by a minimal growth rate, making them almost neutral. For example, for $Ri = 0.7$ and $Ri = 1$, non-modal amplification reaches values 55 and 73.9, while exponential amplification of the least unstable mode at $t = 10$ is about unity in both cases.

Similar to the isothermal case, the values of the growth function in the 3D case are larger than in the 2D case for stratified flow so that the strongest amplification could be reached via 3D optimal perturbation.

4.2. Transient Dynamics of Optimal Perturbation

We start exploring the optimal vector evolution from the two-dimensional case, $\beta = 0$, and consider $Ri = 1$, for which only non-modal transient growth exists. Figure 2 shows the amplitude of the initial optimal disturbances in terms of the temperature and u- and w- velocity components. We observe that amplitudes of the optimal disturbances are symmetric with respect to the mixing layer midline with the maxima of both of them shifted from the symmetry plane $z = 0$. These shifts indicate that the non-normal growth is governed mainly by the Holmboe modes, whose patterns are discussed in [10]. Contrary to the isothermal case [13], the optimal vector components are not localized only in the shear layer and are noticeably wider. This explains why the selection of vectors lying inside the shear layer is not justified for the stratified model considered here.

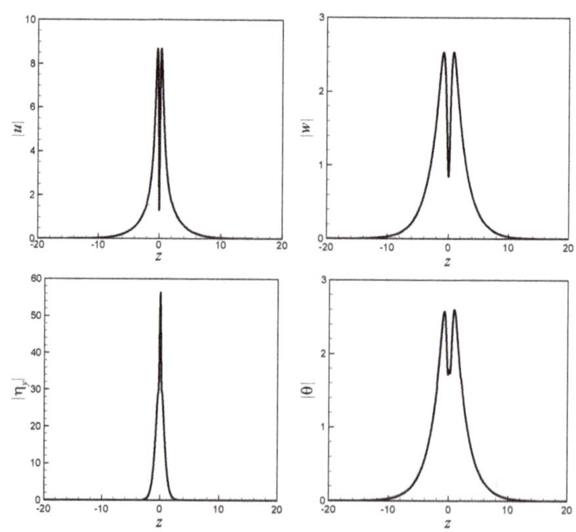

Figure 2. Amplitudes of the initial optimal disturbance for $Ri = 1$, $Re = 1000$, $\alpha = 0.5$, $\beta = 0$ yielding the maximal transient growth at $t = 10$.

To investigate the non-modal growth mechanism, we plot spatial patterns of optimal vector time evolution. Figure 3 shows that the u-velocity component's evolution is represented by two waves traveling in the opposite direction to the mean flow. Such a behavior can be a replication of the well-known Holmboe instability mechanism [16].

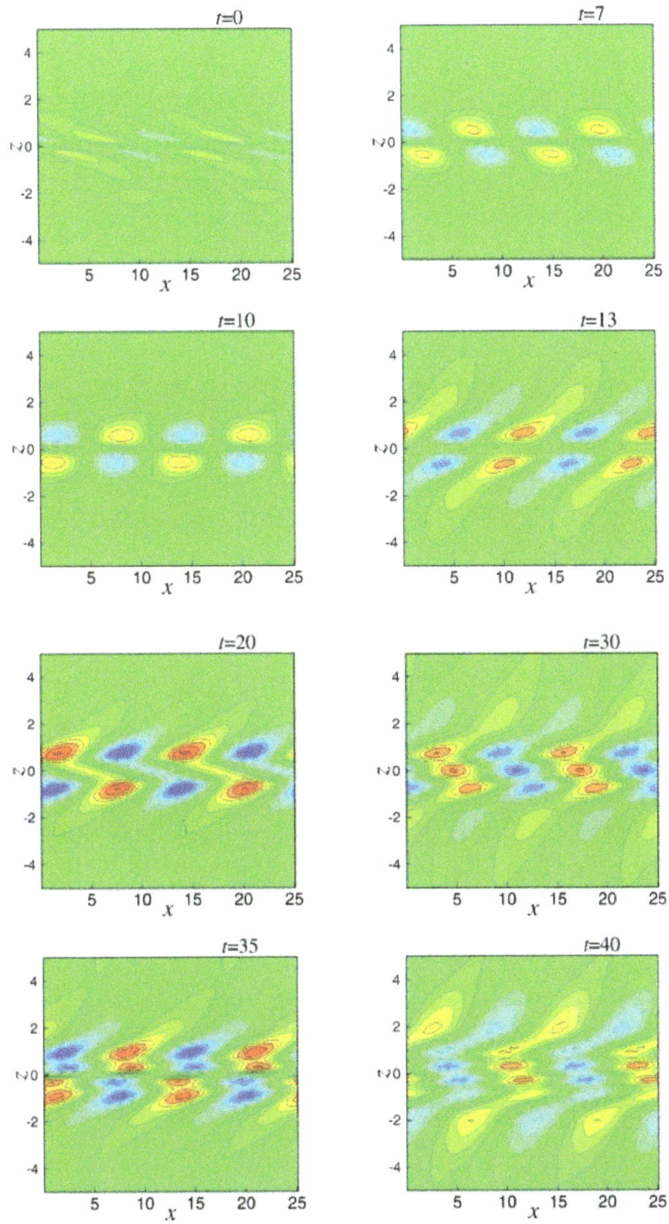

Figure 3. Evolution of the streamwise velocity component, u, of a two-dimensional optimal vector calculated for the maximal growth at $t_{max} = 10$. $Ri = 1$, $Re = 1000$, $\alpha = 0.5$, $\beta = 0$. Solid and dashed lines represent positive and negative values of u. The plotted values are equally spaced between -2 and 2.

Following [51,52] explanations of the linear stability dynamics using the "buoyancy-vorticity wave interaction approach" in an inviscid flow, we presented the η_y- vorticity component at four indicative time moments: initial state, growth, maximum at $t = 10$, and further decaying (Figure 4). Note that similar to the isothermal case [13,14], results for the stratified layer show that the initial optimal vector structures are elongated against the shear in the streamwise direction. This allows us to attribute the initial amplification to the Orr mechanism [14,53]. With the increase of time, the perturbation develops into two series of vortices whose origins are located at the lines $z = \pm 1$, i.e., at the edges of the temperature stratification layer. At the same time, we observe the development of another vortex series with the origins located at the midline. These vortices attain larger amplitude and change their patterns from elongating against the shear to tilting with the shear. Furthermore, the maximal energy growth corresponds to the vertically positioned midline vortices generating vertical velocity. Hence, our calculations confirm the suggestion of [51,52] that vorticity growth occurs due to persistent vertical velocity produced by the propagated waves in the two strong density gradient regions.

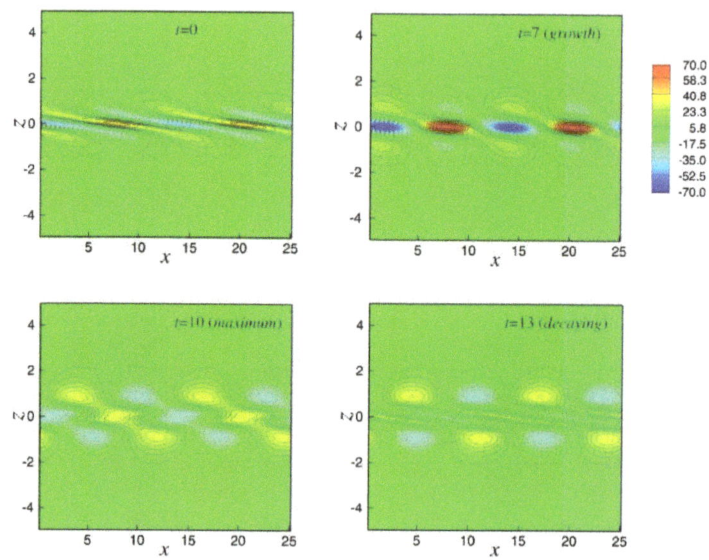

Figure 4. Evolution of the spanwise vorticity component, η_y, of a two-dimensional optimal vector calculated for the maximal growth at $t_{max} = 10$. $Ri = 1$, $Re = 1000$, $\alpha = 0.5$, $\beta = 0$. Solid and dashed lines represent positive and negative values of η_y.

Figure 5 illustrates snapshots of the temperature field where the base temperature profile and the 2D optimal disturbance are superposed. The time evolution of optimal temperature disturbance θ exhibits an extension of the initial temperature layer due to two waves propagating in opposite directions along the x-axis. Such perturbation leads to elevation of the cold fluid and moving down of the warm fluid, which is driven by the vertical velocity component. During the evolution, the perturbation extracts energy from the mean flow, followed by an increase of the growth function. After passing the maximal growth stage, the cold fluid is naturally descending, which corresponds to the growth function decrease. Owing to the small values of the Holmboe modes decay rates, these up-and-down motions of the fluid persist for a relatively long time. The evolution at larger times results in merged heated structures below and above the midplane, corresponding to an increase in the amplitude of the oscillations and an increase of kinetic and potential energy. At later times, the amplitude of the non-modal disturbance decreases, as expected.

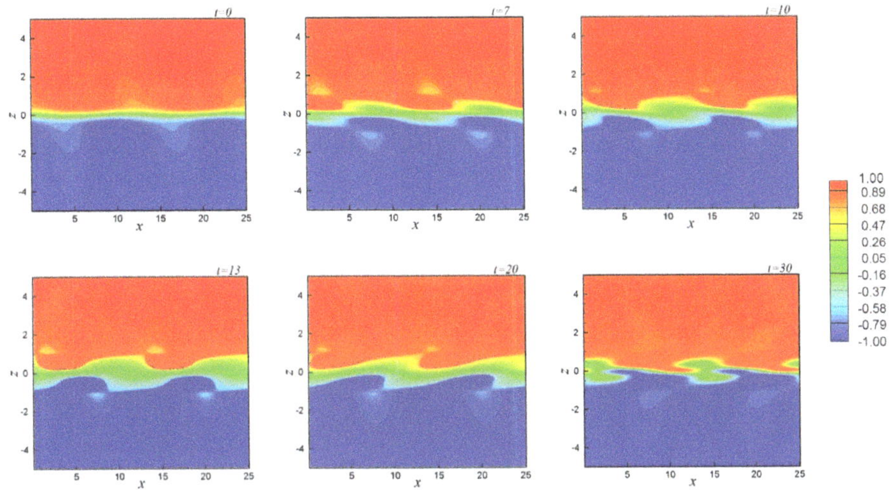

Figure 5. Snapshots of the base flow superposed with 2D optimal temperature perturbation calculated for the maximal growth at $t_{max} = 10$. $Ri = 1$, $Re = 1000$, $\alpha = 0.5$, $\beta = 0$. The blue color corresponds to cold fluid; the red color corresponds to warm fluid.

Figure 6 illustrates the amplitudes of the optimal vector components for $Ri = 2$, $Re = 1000$, $\alpha = 0.5$, $\beta = 1$, the parameters characteristic for 3D transient growth. Similar to the 2D case, the optimal vector amplitude has two symmetric maxima, indicating the dominance of the Holmboe-like modes in the initial optimal disturbance. It is shown that the spanwise vorticity component, η_y, has a maximum also at $z = 0$, supporting the suggestions of [52]. The amplitude of all components of the 3D initial optimal vector decays beyond $z = \pm 10$, making it narrower than the 2D optimal vector.

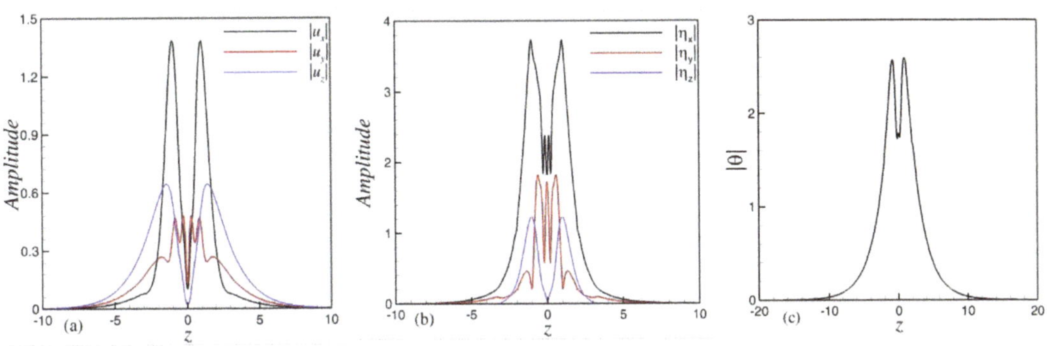

Figure 6. Amplitudes of the initial optimal disturbance for $Ri = 2$, $Re = 1000$, $\alpha = 0.5$, $\beta = 1$ yielding the maximal growth at $t_{max} = 10$; (**a**) three components of velocity magnitude; (**b**) three vorticity components; (**c**) amplitude of the temperature disturbance.

Growth function of the three-dimensional optimal temperature perturbation and patterns of its time evolution are illustrated in Figures 7 and 8, respectively, for the case $\alpha = 0.5$, $\beta = 1$, $Re = 1000$, $Ri = 2$. It is shown that the growth function value at $t = 10$ is similar to those obtained using the forward/backward time integration method, $G = 122$ (Table 4), which justifies the results obtained using two approaches. As already mentioned, the growth rates of the Holmboe modes are low so that the optimal perturbation energy slowly decays at longer times. Similar to the 2D case, the three-dimensional structures exhibit two lines of vortices ordered along with the shear layer and tending to be tilted against the shear slope at the initial stage, constantly changing inclination at later times. At each time moment, corresponding to the maximum of the growth function, the temperature layer's width increases; however, it does not exceed the width observed in the 2D case. Figure 9 shows the evolution of the streamwise velocity at the same times as in Figure 7. The optimal perturbation is composed of two waves traveling in the opposite direction. The wave's amplitude grows and decreases, reflecting in the non-monotonic growth of kinetic energy at longer times.

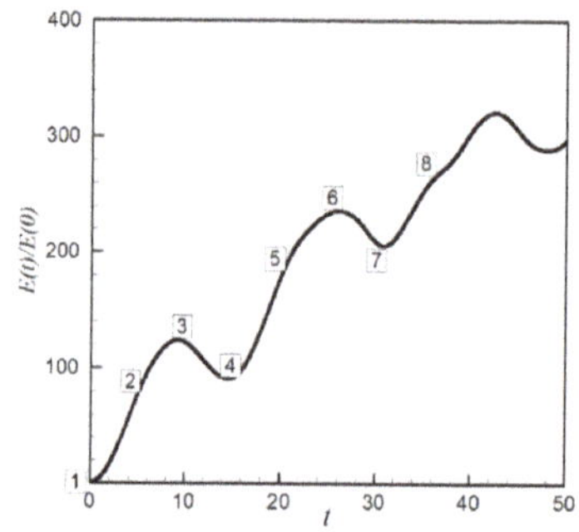

Figure 7. Growth function of the 3D optimal temperature perturbation superposed with base flow profile for $Ri = 2$, $Re = 1000$, $\alpha = 0.5$, $\beta = 1$. The numbers represent different stages of increase and decrease in growth function. The optimal disturbances corresponding to some of these stages are shown in Figure 8.

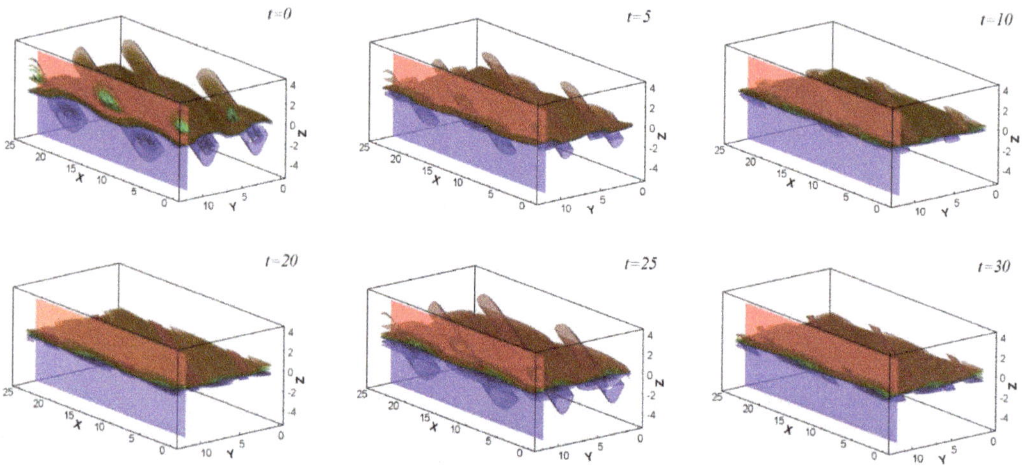

Figure 8. Time evolution of the 3D optimal temperature perturbation superposed with base flow profile for $Ri = 2$, $Re = 1000$, $\alpha = 0.5$, $\beta = 1$, yielding the maximal values at $t_{max} = 10$. The blue color corresponds to cold fluid; the red color corresponds to warm fluid. The growth function increases at $t = 0, 5, 20, 35$, attains maximal values at $t = 10, 25$, and decays at $t = 15, 30$.

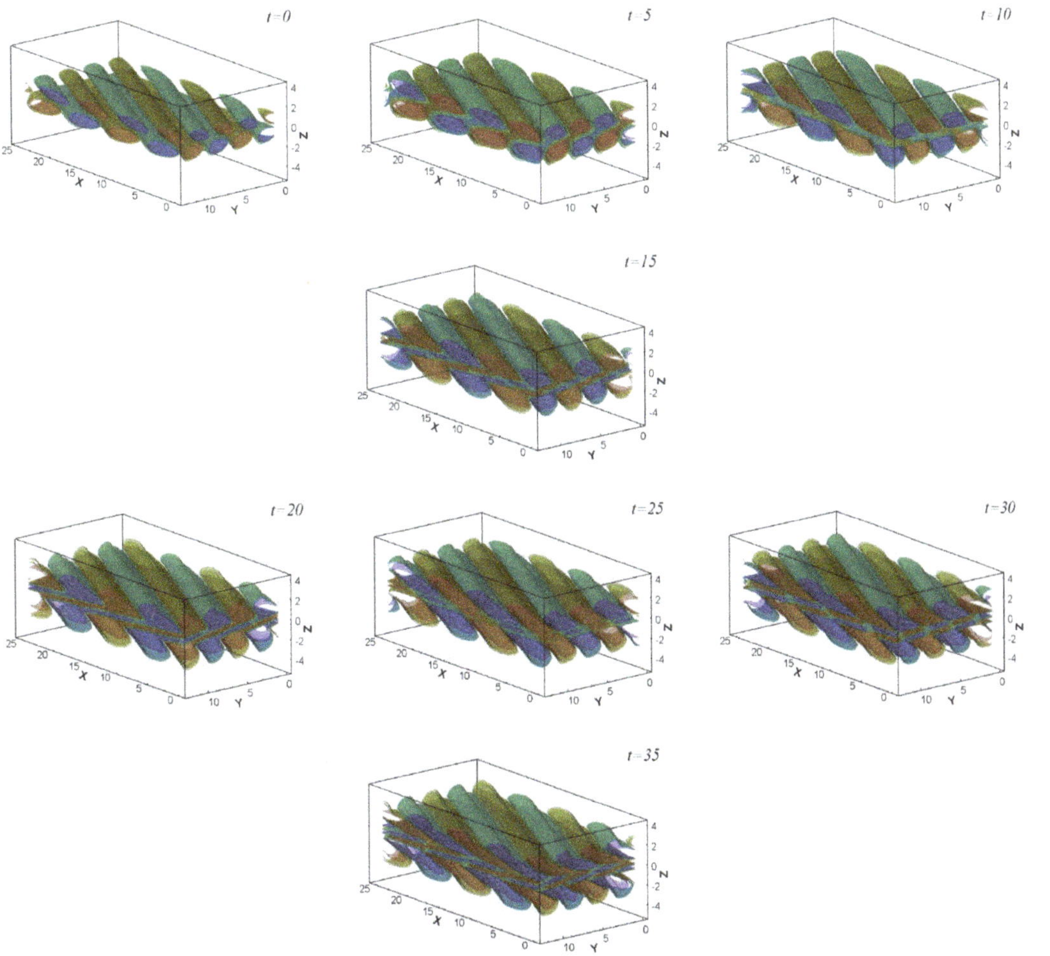

Figure 9. Time evolution of the streamwise velocity component of a three-dimensional optimal vector calculated for the maximal growth at $t = 10$, $Ri = 2$, $Re = 1000$, $\alpha = 0.5$, $\beta = 1$. The growth function increases at $t = 0,\ 5,\ 20,\ 35$, attains maximal values at $t = 10,\ 25$, and decays at $t = 15,\ 30$.

Contrary to the 2D case, the spanwise vorticity component of initial optimal perturbation appears as three lines of vortices with the centers located in the middle plane and the $z = \pm 1$ planes. When the growth function increases, e.g., at times $t = 0, 5, 20, 35$, the vortices are always oriented against the shear. The central vortex splits into two smaller vortices, moves from the midline above and below, and scrolls. At times $t = 10, 25$, at which the growth function attains a maximum, the vortices are strongly tilted with the shear, thus supporting the distinctions between 2D and 3D non-modal growths.

5. Conclusions

The non-modal disturbances growth and transient dynamics of optimal perturbation in stratified viscous mixing layers were investigated. The flow exhibits strong transient growth for streamwise and spanwise wavenumbers, at which the flow is either asymptotically or neutrally stable. Comparing the calculated flow structures with those observed in several previous experimental and numerical studies at the parameters corresponding to linearly unstable regimes, we conclude that the mixing layer flow at early stages of the linear instability development can be strongly affected by the temporal disturbances growth when the optimal perturbation is localized inside the shear zone. The main findings of the performed study are as follows.

- It is known that increasing stratification stabilizes two-dimensional perturbations, which are linearly unstable in the isothermal case. We found that with an increase of the Richardson number, these perturbations exhibit a non-modal growth at relatively short times. This non-modal growth is governed mainly by the Holmboe modes. The non-modal growth weakens and then decays with a further increase of the Richardson number.
- We examined the effect of stratification on linearly stable three-dimensional disturbances, which were found to attain large non-modal amplifications in the stably stratified configuration. It was found that the largest amplification is reached by 3D optimal perturbations, whose growth functions are noticeably larger than those computed for the 2D stability problem.
- It was shown that at short times the non-modal growth of the optimal disturbance gains a larger amplitude than the leading eigenvector growing due to the linear instability. This means that the optimal vector can be a better choice for the initial conditions applied in fully non-linear computations.

In the case of stratified flow, we could not find a clear criterion to define which part of the spectrum should be taken into account to obtain a correct and numerically converged growth function. We could not obtain comparable results using SVD decomposition-based approaches for studying the non-modal growth, trying different extractions of the (seemingly) discrete part of the spectrum. This shows that the stratified mixing layer is a significantly more complicated problem, and the question about which part of the spectrum contributes to the non-modal growth is yet to be investigated.

Author Contributions: Conceptualization, methodology, validation, formal analysis, investigation, resources, data curation, H.V. and A.G.; writing—original draft preparation, H.V.; writing—review and editing, A.G.; visualization, H.V.; supervision, project administration, funding acquisition, A.G. All authors have read and agreed to the published version of the manuscript.

Funding: This research received no external funding.

Conflicts of Interest: The authors declare no conflict of interest. The funders had no role in the design of the study; in the collection, analyses, or interpretation of data; in the writing of the manuscript, or in the decision to publish the results.

References

1. Lord Kelvin, W.T. Hydrokinetic solutions and observations. *Philos. Mag.* **1871**, *42*, 362–377.
2. Strutt, J.W. Lord Rayleigh Investigation of the character of the equilibrium of an incompressible heavy fluid of variable density. *Proc. Lond. Math. Soc.* **1883**, *14*, 170–177.
3. Bestehorn, M. Rayleigh–Taylor and Kelvin–Helmholtz instability studied in the frame of a dimension-reduced model. *Philos. Trans. R. Soc. A* **2020**, *378*, 20190908. [CrossRef]
4. Parker, J.P.; Caulfield, C.P.; Kerswell, R.R. The viscous Holmboe instability for smooth shear and density profiles. *J. Fluid Mech.* **2020**, *896*, A14. [CrossRef]
5. Nastro, G.; Fontane, J.; Joly, L. Optimal perturbations in viscous round jets subject to Kelvin–Helmholtz instability. *J. Fluid Mech.* **2020**, *900*, 1–25. [CrossRef]
6. Penney, J.; Morel, Y.; Haynes, P.; Auclair, F.; Nguen, C. Diapycnal mixing of passive tracers by Kelvin–Helmholtz instabilities. *J. Fluid Mech* **2020**, *900*, A26-1–A26-51. [CrossRef]
7. Caulfield, C.P. Layering, Instabilities, and mixing in turbulent stratified flows. *Ann. Rev. Fluid Mech.* **2021**, *53*, 113–145. [CrossRef]
8. Maslowe, S.A. Critical layers in shear flows. *Ann. Rev. Fluid Mech.* **1986**, *18*, 405–432. [CrossRef]
9. Huerre, P.; Monkewitz, P.A. Local and global instabilities in spatially developing flows. *Ann. Rev. Fluid Mech.* **1990**, *22*, 473–537. [CrossRef]
10. Gelfgat, A.Y.; Kit, E. Spatial versus temporal instabilities in a parametrically forced stratified mixing layer. *J. Fluid Mech.* **2006**, *552*, 189. [CrossRef]
11. Drazin, P.G. *Introduction to Hydrodynamic Stability*; Cambridge University Press: Cambridge, UK, 2002.

12. Arratia, C. Non-Modal Instability Mechanisms in Stratified and Homogeneous Shear Flow. Ph.D. Thesis, Ecole Polytechnique X, Palaiseau, France, 2011.
13. Vitoshkin, H.; Gelfgat, A.Y. Non-modal disturbances growth in a viscous mixing layer flow. *Fluid Dyn. Res.* **2014**, *46*, 041414. [CrossRef]
14. Vitoshkin, H.; Heifetz, E.; Gelfgat, A.Y.; Harnik, N. On the role of vortex stretching in energy optimal growth of three-dimensional perturbations on plane parallel shear flows. *J. Fluid Mech.* **2012**, *707*, 3069–3380. [CrossRef]
15. Taylor, G.I. Effect of variation in density on the stability of superimposed streams of fluids. *Proc. Roy. Soc. Lond. Ser. A* **1931**, *132*, 499–523.
16. Holmboe, J. On the behavior of symmetric waves in stratified shear layers. *Geophys. Publ.* **1962**, *24*, 67–113.
17. Smyth, W.D.; Winters, K.B. Turbulence and Mixing in Holmboe Waves. *J. Phys. Oceanogr.* **2003**, *33*, 694–711. [CrossRef]
18. Smyth, W.D.; Peltier, W.R. Instability and transition in finite-amplitude Kelvin–Helmholtz and Holmboe waves. *J. Fluid Mech.* **1990**, *228*, 387–415. [CrossRef]
19. Caulfield, C.; Peltier, W. Three-dimensionalization of the stratified mixing layer. *Phys. Fluids* **1994**, *6*, 3803–3805. [CrossRef]
20. Deloncle, A.; Chomaz, J.M.; Billant, P. Three-dimensional stability of a horizontally sheared flow in a stably stratified fluid. *J. Fluid Mech.* **2007**, *570*, 297–305. [CrossRef]
21. Browand, F.K.; Winant, C.D. Laboratory observations of shear-layer instability in a stratified fluid. *Bound.-Layer Meteorol.* **1973**, *5*, 67–77. [CrossRef]
22. Smyth, W.D.; Klaassen, G.P.; Peltier, W.R. Finite amplitude Holmboe waves. *Geophys. Astrophys. Fluid Dyn.* **1988**, *43*, 181–222. [CrossRef]
23. Yih, C.S. Stability of two-dimensional parallel flows for three-dimensional disturbances. *Q. Appl. Math.* **1955**, *12*, 434–435. [CrossRef]
24. Nevskii, S.; Sarychev, V.; Konovalov, S.; Granovskii, A.; Gromov, V. Wave instability on the interface coating/substrate material under heterogeneous plasma flows. *J. Mater. Res. Technol.* **2020**, *9*, 539–550. [CrossRef]
25. Konovalov, S.; Chen, X.; Sarychev, V.; Nevskii, S.; Gromov, V.; Trtica, M. Mathematical modeling of the concentrated energy flow effect on metallic materials. *Metals* **2017**, *7*, 4. [CrossRef]
26. Goldstein, S. On the stability of superposed streams of fluids of different densities. *Proc. R. Soc. Lond. Ser. A* **1931**, *31*, 524–548.
27. Chandrasekhar, S. *Hydromagnetic and Hydrodynamic Stability*; Oxford University Press: Oxford, UK, 1961.
28. Lappa, M. *Thermal Convection: Patterns, Evolution and Stability*; Wiley & Sons: Singapore, 2010.
29. Smyth, W.D.; Peltier, W.R. Three-dimensional primary instabilities of a stratified, dissipative, parallel flow. *Geophys. Astrophys. Fluid Dyn.* **1990**, *52*, 249–261. [CrossRef]
30. Smyth, W.D.; Peltier, W.R. The Transition between Kelvin–Helmholtz and Holmboe Instability: An Investigation of the Over reflection Hypothesis. *J. Atmos. Sci.* **1989**, *46*, 3698–3720. [CrossRef]
31. Olson, B.J.; Larsson, J.; Lele, S.K.; Cook, A.W. Nonlinear effects in the combined Rayleigh-Taylor/Kelvin-Helmholtz instability. *Phys. Fluids* **2011**, *23*, 114107. [CrossRef]
32. Arratia, C.; Ortiz, S.; Chomaz, J.M. Inviscid transient growth on horizontal shear layers with strong vertical stratification. In *Nonlinear Dynamics: Materials, Theory and Experiments*; Springer: Cham, Switzerland, 2016; pp. 199–206.
33. Gaster, M.; Kit, E.; Wygnanski, I. Large-Scale Structures in a Forced Turbulent Mixing Layer. *J. Fluid Mech.* **1985**, *150*, 23–39. [CrossRef]
34. Hanifi, A.; Henningson, D.S. The compressible inviscid algebraic instability for streamwise independent disturbances. *Phys. Fluids* **1998**, *10*, 1784–1786. [CrossRef]
35. Biau, D.; Bottaro, A. Transient growth and minimal defects: Two possible initial paths of transition to turbulence in plane shear flows. *Phys. Fluids* **2004**, *16*, 3515–3529. [CrossRef]
36. Farrell, B.F.; Ioannou, P.J. Transient development of perturbations in stratified shear flow. *J. Atmos. Sci.* **1993**, *50*, 2201–2214. [CrossRef]
37. Constantinou, N.C.; Ioannou, P.J. Optimal excitation of two-dimensional Holmboe instabilities. *Phys. Fluids* **2011**, *23*, 074102. [CrossRef]
38. Farrell, B.F. Developing disturbances in shear. *J. Atmos. Sci.* **1987**, *44*, 2191–2199. [CrossRef]
39. Farrell, B.F. Optimal excitation of perturbations in viscous shear flow. *Phys. Fluids* **1988**, *31*, 2093–2102. [CrossRef]
40. Butler, R.M.; Farrell, B.F. Three-dimensional optimal perturbations in viscous shear flows. *J. Phys. Fluids* **1992**, *4*, 1367–1654. [CrossRef]
41. Reddy, S.C.; Henningson, D.S. Energy growth in viscous channel flows. *J. Fluid Mech.* **1993**, *252*, 209–238. [CrossRef]
42. Schmid, P.J.; Henningson, D.S. *Stability and Transition in Shear Flows*; Springer: New York, NY, USA, 2001.
43. Corbett, P.; Bottaro, A. Optimal perturbations for boundary layers subject to stream-wise pressure gradient. *Phys. Fluids* **2000**, *12*, 120–130. [CrossRef]
44. Luchini, P.; Bottaro, A. Görtler vortices: A backward-in-time approach to the receptivity problem. *J. Fluid Mech.* **1998**, *363*, 1–23. [CrossRef]
45. Luchini, P. Reynolds-number-independent instability of the boundary layer over a flat surface: Optimal perturbations. *J. Fluid Mech.* **2000**, *404*, 289–309. [CrossRef]

46. Trefethen, L.N.; Embree, M. *Spectra and Pseudospectra. The Behavior of Nonnormal Matrices and Operators*; Princeton University Press: Princeton, NJ, USA, 2005.
47. Mao, X.; Sherwin, S.J. Continuous spectra of the Batchelor vortex. *J. Fluid Mech.* **2011**, *681*, 1–23. [CrossRef]
48. Schmid, P. Nonmodal stability theory. *Ann. Rev. Fluid Dyn.* **2007**, *39*, 129–162.
49. Sameen, A.; Govindarajan, R. Some indications from instability results about the effectiveness of wall heating as a control option for channel flow. *Sadhana* **2007**, *32*, 65–81. [CrossRef]
50. Alexakis, A. Stratified shear flow instabilities at large Richardson numbers. *Phys. Fluids* **2009**, *21*, 054108. [CrossRef]
51. Harnik, N.; Heifetz, E.; Umurhan, O.M.; Lott, F. A buoyancy–vorticity wave interaction approach to stratified shear flow. *J. Atmos. Sci.* **2008**, *65*, 2615–2630. [CrossRef]
52. Rabinovich, A.; Umurhan, O.M.; Harnik, N.; Lott, F.S.C.; Heifetz, E. Vorticity inversion and action-at-a-distance instability in stably stratified shear flow. *J. Fluid Mech.* **2011**, *670*, 301–325. [CrossRef]
53. Orr, W.M.F. The stability or instability of the steady motions of a perfect liquid and of a viscous liquid. Part II: A viscous liquid. *Proc. R. Ir. Acad.* **1907**, *27*, 9–69.

fluids

MDPI

Article

Experimental Study on Coherent Structures by Particles Suspended in Half-Zone Thermocapillary Liquid Bridges: Review

Ichiro Ueno

Department of Mechanical Engineering, Faculty of Science & Technology, Tokyo University of Science, Chiba 278-8510, Japan; ich@rs.tus.ac.jp

Abstract: Coherent structures by the particles suspended in the half-zone thermocapillary liquid bridges via experimental approaches are introduced. General knowledge on the particle accumulation structures (PAS) is described, and then the spatial–temporal behaviours of the particles forming the PAS are illustrated with the results of the two- and three-dimensional particle tracking. Variations of the coherent structures as functions of the intensity of the thermocapillary effect and the particle size are introduced by focusing on the PAS of the azimuthal wave number $m = 3$. Correlation between the particle behaviour and the ordered flow structures known as the Kolmogorov–Arnold—Moser tori is discussed. Recent works on the PAS of $m = 1$ are briefly introduced.

Keywords: thermocapillary-driven convection; half-zone liquid bridges; particles; coherent structures; particle accumulation structure (PAS); high Prandtl number fluids

check for updates

Citation: Ueno, I. Experimental Study on Coherent Structures by Particles Suspended in Half-Zone Thermocapillary Liquid Bridges: Review. *Fluids* **2021**, *6*, 105. https://doi.org/10.3390/fluids6030105

Academic Editor: Marcello Lappa

Received: 26 January 2021
Accepted: 22 February 2021
Published: 4 March 2021

Publisher's Note: MDPI stays neutral with regard to jurisdictional claims in published maps and institutional affiliations.

Copyright: © 2021 by the author. Licensee MDPI, Basel, Switzerland. This article is an open access article distributed under the terms and conditions of the Creative Commons Attribution (CC BY) license (https://creativecommons.org/licenses/by/4.0/).

1. Introduction

Coherent structures by suspended particles were first found in thermocapillary half-zone liquid bridges by Schwabe et al. [1]. In the geometry of a "half-zone liquid bridge," a liquid is bridged between the coaxial face-to-face disks by its surface tension, and each disk is kept at a constant but different temperature to impose a non-uniform temperature distribution over the free surface of the liquid bridge. One defines the intensity of the thermocapillary effect by the Reynolds number $\mathrm{Re} = |\gamma_T|\Delta TL/(\rho\nu^2)$, where γ_T is the temperature coefficient of surface tension γ, L is the characteristic length of the system, ρ and ν are the density and the dynamic viscosity of the test liquid, respectively. The temperature difference between the end disks, ΔT, is the governing factor to vary the thermocapillary effect. The intensity is also generally described as the Marangoni number $\mathrm{Ma} = |\gamma_T|\Delta TL/(\rho\nu\kappa) = \mathrm{RePr}$, where κ is the thermal diffusivity of the test liquid, and $\mathrm{Pr} = \nu/\kappa$ indicates the Prandtl number. Schwabe et al. [1] found such unique phenomenon in the traveling-wave-type oscillatory convections in the liquid bridges of high Pr fluids, and named this as the particle accumulation structure (PAS). In the geometry concerned, the thermocapillary-driven convection exhibits a transition from the two-dimensional time-independent ("steady") convection to the three-dimensional time-dependent ("oscillatory") one [2–5] by so-called hydrothermal-wave instability [6]. There exist two types of oscillatory convection [4,5] before the secondary instability [7]: the standing-wave-type oscillation and the travelling-wave-type oscillation. In the former, one finds a pair of thermal waves propagating over the free surface with the same amplitude and the same propagation speed but in the opposite azimuthal directions. In the latter, on the other hand, one finds a single thermal wave propagating in an azimuthal direction. The PAS is formed in the travelling-wave-type oscillatory convection [1,8,9]. After the onset of the hydrothermal-wave instability, the oscillatory convection emerges in the liquid bridge accompanied by an azimuthal wave number m, depending on the liquid bridge shapes defined as the aspect ratio $\Gamma = H/R$ [2,3,5] and the volume ratio $V/V_0 = V/(\pi R^2 H)$ [10], where H and R are the distances between the disks (or, the height of the liquid bridge) and the disk radius, respectively, and V is the volume of the liquid bridge. Here V_0 indicates the volume of a straight cylinder, of which the end disks correspond to the end surfaces. The PAS has the same azimuthal wave number as the oscillatory convection by the hydrothermal-wave instability [8,9,11–14]. Such characteristics were also well reproduced by the numerical simulations for the PAS of $m = 3$ [15–20].

Tanaka et al. [8] indicated via terrestrial experiments that there exist two-types of PAS of the same m by varying the intensity of the thermocapillary effect; spiral loops 1 (SL1) and 2 (SL2). Typical examples of those are illustrated in Figure 1 for the PAS of $m = 3$. These are the snapshots observed through the transparent top rod; the white dots correspond to the particles suspended in the liquid bridge. The black circle at the centre is an outlet of the channel drilled through the bottom rod for a liquid supply. The PAS consists of the major structure called the "blade" [8]: When one observes the PAS from above in the laboratory frame, it seems to be rotating without changing its shape as the rigid structure. The azimuthal direction of the rotation of the PAS is the same as that of the thermal wave over the free surface. The number of the blades matches to the azimuthal wave number m of the oscillatory convection after the onset of the primal instability. The tip of the blades corresponds to the trajectory of the particles travelling near the free surface toward the cold end from the hot end in the case of $\gamma_T < 0$. After the travel near the free surface, the particles then penetrate into the central region of the liquid bridge, and rise toward the hot end by following the return flow. The particles change their direction toward the free surface near the hot end to form the tip of the adjacent blade. The net azimuthal direction of the particles forming the PAS is opposite to that of the PAS. The SL1-PAS is the basic structure; if one follows the radial position of the blade (r) by varying the azimuthal position (θ), one obtains $\partial r/\partial\theta > 0$ from the central region toward the tip, and $\partial r/\partial\theta < 0$ from the tip toward the central region. The SL2-PAS exhibits an additional loop structure near the tip; when the particles forming the PAS travel near the free surface, they exhibit a sharp additional rotation near the free surface before penetrating into the central region of the liquid bridge. Existing research has dealt mainly with the SL1-PAS, because this type of PAS emerges at lower Ma than that for the SL2-PAS [8,9,11]. That means one has to add larger ΔT to realize the SL2-PAS, which induces a severe and inevitable problem for carrying out the experiments under the normal gravity condition; that is, the evaporation of the test liquid. When one conducts terrestrial experiments, one has to prepare the liquid bridge with a characteristic length L is small enough in order to prevent the deformation of the liquid bridge due to the static pressure [21]. In order to realize high Re for small L, however, a large ΔT must be imposed between the both ends of the liquid bridge. This might be one of the critical reasons why the amount of research on the SL2-PAS has been limited compared to those on the SL1-PAS. One can find some research via numerical simulation on SL2-PAS [20].

In this article, the author introduces the experimental results accumulated in terrestrial environments, which lead to future microgravity experiments on the International Space Station (ISS) by the project "Japanese European Research Experiments on Marangoni Instability (JEREMI)" [22,23].

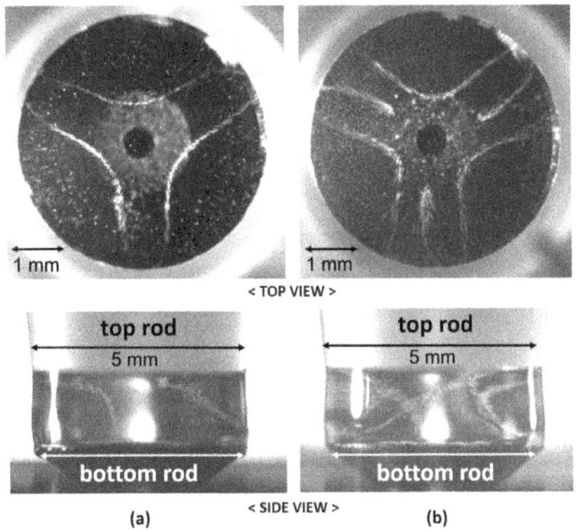

Figure 1. Snapshots of particle accumulation as viewed from above (top row) and from the side (bottom row) in a liquid bridge of Pr $= 28.6$, $\Gamma = 0.68$ ($R = 2.5\,$mm), and $V/V_0 = 1.0$: (**a**) spiral loop (SL)1-particle accumulation structures (PAS) under Ma $= 4.7 \times 10^4$ and (**b**) SL2-PAS under Ma $= 5.8 \times 10^4$. The direction of the PAS rotation or the travelling-wave-type oscillatory convection by the hydrothermal wave instability is counterclockwise for both cases in the top row, and is from right to left in the bottom row. Note that Ma is evaluated by applying H for the characteristic length L. (Figure 2 in Toyama et al. [11]).

2. Materials and Methods

The geometry of the interests is quite simple as aforementioned; one prepares a set of cylindrical disks placed face-to-face with a designated distance between the both end surfaces. There exist a fair amount of research dealing with the half-zone liquid bridges [2,5,24–38]. Note that the transparent rod enables one to observe the particles' behaviours through it. Such a technique was introduced by Hirata et al. [29,30], and has been widely spread to the various research groups. Here an example of the experimental set up is introduced in Figure 2 according to Gotoda et al. [13]. A liquid is sustained between the top rod made of sapphire and the bottom one made of aluminium. End surfaces of the both rods are finished to have the same radius R. The top rod is heated by the electric heater winding the rod. The temperature of the rod is measured by the thermocouple, and is controlled by the PID controller. The bottom rod is connected to the base block with the drilled channels. The block is cooled by the flow of coolant pumped from the constant temperature bath. In order to suppress undesignated natural convection around the liquid bridge, an external shield is often installed coaxially to the liquid bridge. An infrared (IR) camera is commonly used to measure the surface temperature. One might have to prepare a tiny window for the IR camera installed on the external shield. The particles' behaviours are observed by camera(s). If one has two cameras, simultaneous observation is possible through the top rod for the top view and through the external shield for the side view. Oba et al. [14] applied three-dimensional particle tracking velocimetry (3-D PTV) with a conventional algorithm; a cubic beam splitter was installed above the top rod in order to realize simultaneous observation via multiple cameras with a different incident angle to the central axis of the liquid bridge through the top rod. It is noted that the characteristic time for the oscillatory convection becomes shorter as the characteristic length becomes smaller; thus high-speed camera(s) might have to be installed for precise observations.

A variety of test liquids have been employed as high Pr fluids. Silicone oil has been widely used as in Chun and Wuest [24] because of its transparency for visible light and well-known thermal properties including surface tension that exhibits almost linear variation against temperature. Due to the high temperature needed in order to realize thermocapillary-driven convection under high Ma, evaporation of the test liquid becomes a common and severe problem to conducting the experiment as aforementioned. A liquid supply system through the rod has been installed in order to keep V/V_0 constant [13,14,39–41]. One finds apparatus with a similar mechanism used for microgravity experiments [9,42–44].

The existing research has dealt with high Pr fluids in a range of $1 \lesssim \mathrm{Pr} \lesssim 200$. In the following, the effect of Pr on the coherent structures and the thermal flow field will be introduced, especially in the cases of $\mathrm{Pr} \lesssim 30$. Microgravity experiments [21,43–50] dealt with higher Pr than those in terrestrial experiments thus far. Further research is indispensable to discussing the Pr effect on the coherent structures and the oscillatory convection induced by the hydrothermal wave instability in the case of $\mathrm{Pr} > 100$.

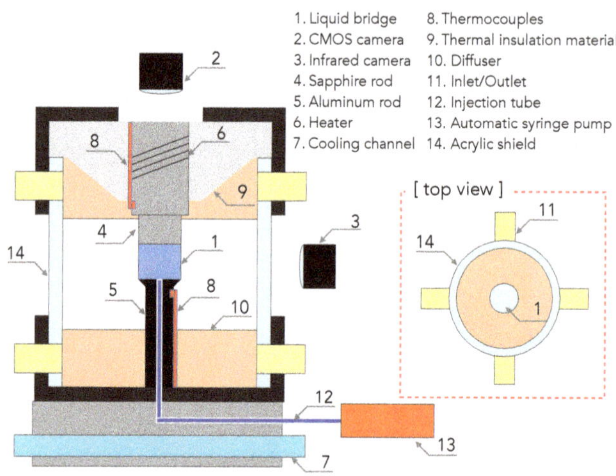

Figure 2. Cross-sectional view of the experimental apparatus (not to scale). The liquid bridge is heated from above in this configuration. A narrow ZnSe window is placed at a portion of the external shield for the infrared camera to detect the surface temperature through the shield. The CMOS camera for the side view and the displacement sensor are omitted. (Figure 3 in Gotoda et al. [13]).

3. Thermocapillary-Driven Convection and Coherent Structures

3.1. General View

The physical mechanism for particle accumulation has been discussed over a decade; three major models have been proposed so far. At the early stage after the PAS discovery, Schwabe et al. [9] proposed that the particles were gathered near the free surface due to the Marangoni effect. Through fine experimental observation by using the liquid of $\mathrm{Pr} = 15$, they found that the tip of the PAS blade is located on the colder region of the thermal wave over the free surface. They considered that there exists thermocapillary-driven flow toward the colder region due to the negative temperature coefficient of the surface tension, which drives the particles approaching the free surface toward the low-temperature band. When the particles return to the internal region of the liquid bridge, they follow a narrow band of the stream lines to form the PAS. It was indicated, however, that such spatial correlation between the PAS tip and the relatively cold band over the free surface is not always realized; Toyama et al. [11] illustrated the spatial correlations between the SL1- and SL2-PASs and the thermal wave over the free surface (see Figure 3). They conducted a simultaneous observation of the surface temperature from the side and the particle motion inside the liquid bridge of Pr = 28.6 through the top rod, and reconstructed those distributions in the rotating frame of reference. It was indicated that the tip of the PAS blade locates between the cold and hot bands in the liquid bridge of higher Pr than that in Schwabe et al. [9]. This experimental work revealed that the particles are not collected by the thermocapillary effect over the free surface.

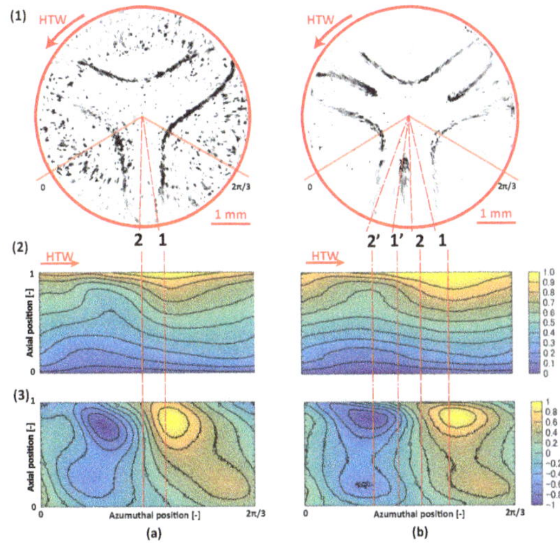

Figure 3. Correlation between PAS and surface temperatures for (**a**) SL1-PAS under Ma $= 4.7 \times 10^4$ and (**b**) SL2-PAS under Ma $= 5.8 \times 10^4$ in a liquid bridge of Pr = 28.6, $\Gamma = 0.68$ ($R = 2.5$ mm), and $V/V_0 = 1.0$ (the same condition as shown in Figure 1): Row (**1**) illustrates the top views of the PASs, and rows (**2**) and (**3**) the absolute temperature and the temperature deviation over the free surface in a range of $0 \leq \theta \leq 2\pi/3$ (as defined in row (**1**)), respectively. Ma is evaluated by applying H for the characteristic length L. (Figure 11 in Toyama et al. [11]).

Two other models are called the (i) "phase locking model" [51] and (ii) "particle–free-surface interaction model" [17,52]. Pushkin et al. [51] proposed a model based on the "phase locking" between the flow field and the particle motion; they suggested that a PAS would be formed by "synchronization" between the particle turnover motion due to the basic flow in the liquid bridge and the azimuthal convective motion due to the hydrothermal-wave instability. Hofmann and Kuhlmann [17] proposed a model by considering density-matched particles: the PAS would be formed through the transfer of particles that "collide" with the free surface to specific streamlines. They illustrated the presence of closed stream tubes in the flow in the rotating frame, and that particles accumulate on these stream tubes. It was considered that an effect of the finite particle size would force transfers of the particle from one streamline to another. After the proposals of these models, there has been active discussion in this decade [20,39,53–57]. It is indispensable and of great importance to accumulate comprehensive knowledge from experimental and numerical approaches with fine spatial–temporal resolution to lead comprehensive understandings of the PAS formation.

When one pays attention to the formation process of the PAS, especially in order to evaluate the formation time [8,9], we have to measure the accumulation; how much the accumulation is realized. Some measures have been introduced such as "contrast" [41] and "K(t) parameter" [58]. It was indicated that the formation time of SL1-PAS of $m = 3$ is of the order of the thermal diffusion time in terrestrial experiments [59]. It must be noted, however, that it is rather impossible in the experiments to disperse the particles without disturbing the thermal-flow field in the liquid bridge [59,60] to evaluate the time it takes for the particles to form the PAS. We need supportive data from fine numerical simulation precisely solving the particles' behaviour even close to the free surface and walls [57].

In the following, knowledge on the PAS in fully developed states is introduced.

3.2. Particle Path Lines to Form Coherent Structures

3.2.1. Path Lines

The PAS seems to rotate azimuthally without changing its shape as a rigid structure at a constant angular velocity in the laboratory frame [8], as aforementioned. When one observes the PAS in the rotating frame of reference, the particles exhibit a coherent structure inside the liquid bridge [13,14,61]. Figure 4 illustrates the typical examples of the path lines of the particles forming (a) a triangle zone of the particle depletion and (b) the SL1-PAS observed from above (i) in the rotating frame of reference and (ii) in the laboratory frame [13]. The images are obtained by accumulating the snapshots for 7 s or 10 periods. When one observes the particles in the liquid bridge in the laboratory frame (column (ii)), the particles almost scatter inside the liquid bridge except for the particle depletion zone. At an appropriate ΔT, the particles gather along closed orbits to form coherent structures in the frame of reference rotating with the traveling-wave-type oscillatory convection due to the hydrothermal-wave instability ((b) in column (i)). Mukin and Kuhlmann [55] indicated that the PAS emerges after the particle transfer from the chaotic streamlines to other streamlines located inside of Kolmogorov–Arnold–Moser (KAM) tori, and that such particle transfer is realized by the interaction with the free surface. These KAM tori are three-dimensional closed stream tubes in the liquid bridge, and the spatial structures of the particle accumulations match well to those of the KAM tori such as T_3^3, T_3^9, and T_{core} [61].

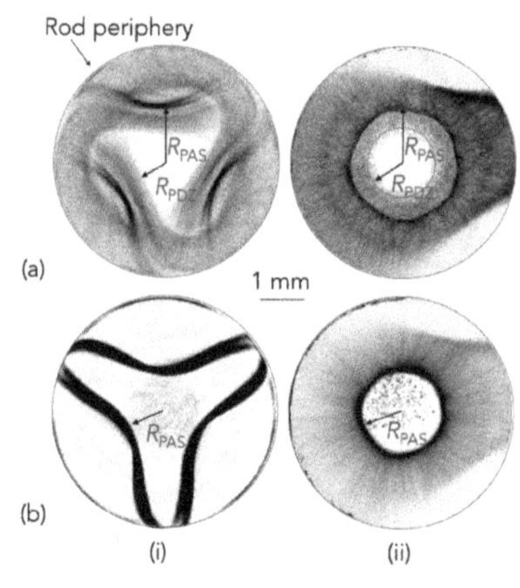

Figure 4. Top views of path lines of particles of 15 µm in diameter (**a**) before and (**b**) after PAS formation in (**i**) the rotating frame of reference and in (**ii**) teh laboratory frame by averaging over 7 s (10 times fundamental periods of hydrothermal wave (HTW)) in a liquid bridge of Pr = 28.6, $\Gamma = 0.64$ ($R = 2.5$ mm), and $V/V_0 = 1.0$: (**a**) Ma = 3.13×10^4 and (**b**) Ma = 4.30×10^4. Note that white regions at the top and bottom right in the images in the column (**ii**) are the regions where the light from the light source never irradiates in the liquid bridge: the light is supplied to the liquid bridge through the free surface from the left in these images, and one cannot avoid the refraction of the light due to the curvature of the free surface. Those regions become vague and are not apparent by integrating images for a long period in the rotating frame of reference as shown in the column (**i**). Ma is evaluated by applying H for the characteristic length L. (Figure 6 in Gotoda et al. [13]).

3.2.2. Pattern Map

The particle accumulation depends on the intensity of the thermocapillary effect; the occurring condition is described as a function of Ma [9,11]. Figure 5 illustrates a typical example of the variation of the particle distribution in the rotating frame of reference for the particles of different sizes [13]. One notices the formation, the development, and the decay of the coherent structures as Ma varies. After the formation of the coherent structures, the width of the PAS varies as a function of Ma; such variation can be seen for the KAM tori predicted by employing the modelled convection field [58]. Note that, even under the same Ma, the shape of the coherent structure depends on the particle size or corresponding Stokes number $\mathrm{St} = \varrho d_\mathrm{p}^2/(18H^2)$ [17], where ϱ is the density ratio between the particle ρ_p and the test fluid ρ_f or $\varrho = \rho_\mathrm{p}/\rho_\mathrm{f}$, and d_p is the particle diameter. In the present study, the value of St is of the order of 10^{-5}. Dependence of the coherent structure on St were examined via numerical simulation by considering a modeled traveling flow in the straight liquid bridge [58]. It has been indicated that the spatial structure of the particle accumulation is defined by the combination of the turnover motion in the r–z plane and the azimuthal motion [8,60,62]. In the following, individual particle behaviour is focused upon.

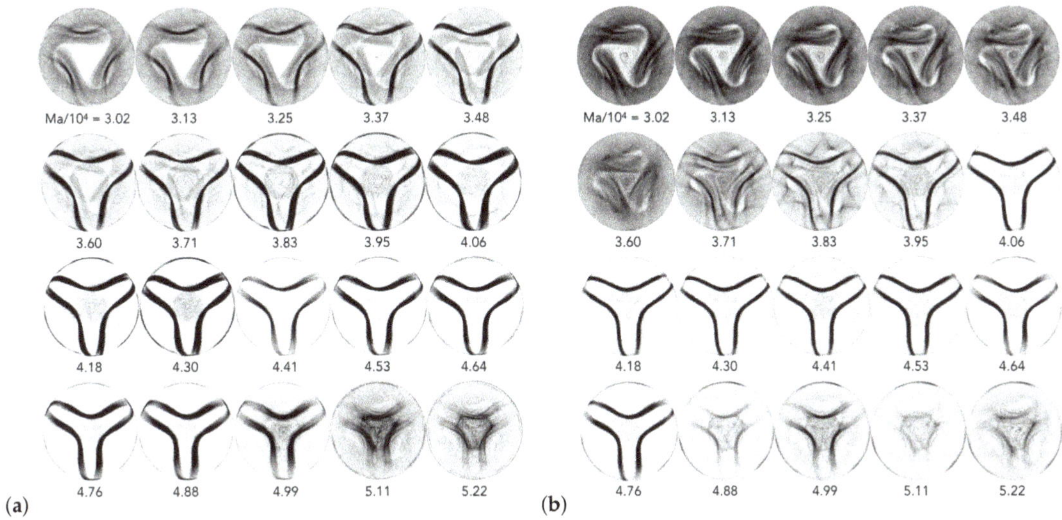

Figure 5. Variations of coherent structures by the particles as a function of Ma for (**a**) $d_\mathrm{p} = 15\,\mu\mathrm{m}$ and (**b**) $d_\mathrm{p} = 30\,\mu\mathrm{m}$. Conditions are the same as indicated in Figure 4. The azimuthal direction of the thermal wave over the free surface due to the HTW is counterclockwise for all images. Ma is evaluated by applying H for the characteristic length L. (Figure 5 in Gotoda et al. [13]).

3.2.3. Particle Behaviours

When one pays their attention to a single particle motion, one finds its characteristics to form the PAS. In addition to the basic turnover motion in the r–z plane due to the thermocapillary-driven convection, the particle exhibits azimuthal motion in the oscillatory convection [8]; the net azimuthal direction of the particle motion is opposite to that of the PAS itself. When the particle forms the PAS, the turnover motion itself varies as Ma; Gotoda et al. [13] indicated the variation of the particle trajectory from (i) strictly periodic to (ii–iii) quasi-periodic and (iv) period doubled by varying Ma (Figure 6a). Such variation is clearly illustrated by monitoring the particle position in r (frame (b)) and its Fourier spectrum (frame (c)). When attention is paid to the absolute minimum of the particle position in r (corresponding to R_PAS as indicated in Figure 4), the variation of the minimum position of the particle in r depends on Ma. In the case of (i), the minimum position is almost constant. By increasing Ma for (ii) and (iii), it exhibits periodic modulations. By further increasing Ma to realize the period-doubled PAS [55], the minimum position of the travelling particle in r exhibits a large oscillation, showing alternative values of 0.7 and 0.8 mm each time in this case, as the particles penetrate to the deepest position in the liquid bridge. One finds a fundamental frequency $f_0^{(\mathrm{p})} = 2.4$ Hz and its subharmonics $f_0^{(\mathrm{p})}/2 = 1.2$ Hz (the frame (c)-(iv)) for the turnover motion of the particle; this subharmonic frequency corresponds to the period-doubled trajectory of the PAS in the rotating frame of reference. It must be emphasized that the oscillatory convection itself is strictly periodic and not modulated under these conditions, which was also indicated by Toyama et al. [11] under Ma for SL1- and SL2-PASs, and even under slightly higher Ma with no stable PASs. Through the tracking of a single particle forming the PAS, the frequency ratio between the particle turnover motion $f_0^{(\mathrm{p})}$ and the HTW $f_0^{(\mathrm{HTW})}$ remains almost constant (about 1.6 in this series of the experiments) under the condition where the PAS is fully formed; the ratio is independent of Ma.

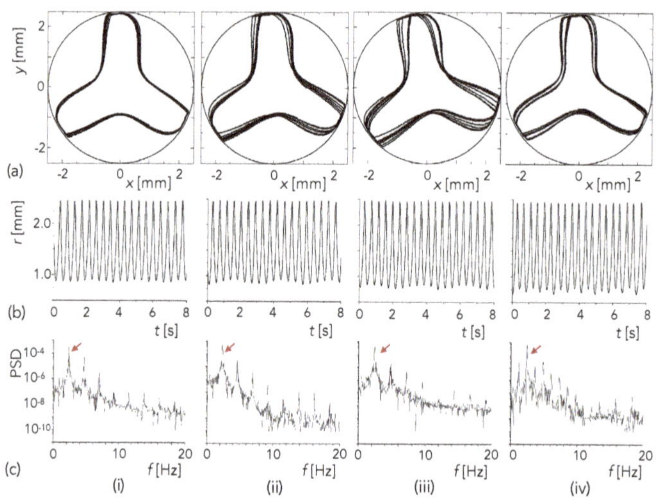

Figure 6. (**a**) Reconstructed SL1-PAS configurations in a rotating frame of reference by tracking a single particle forming PAS, (**b**) time series of particle radial positions, and (**c**) its power spectral density (PSD) for Ma of (**i**) 3.3×10^4, (**ii**) 3.7×10^4, (**iii**) 4.2×10^4, and (**iv**) 4.6×10^4, respectively. Conditions of the liquid bridge are the same as indicated in Figure 4. The particle diameter is 15 µm. Arrows in row (**c**) indicate the fundamental frequency of the particle turnover motion $f_0^{(\mathrm{p})}$. Ma is evaluated by applying H for the characteristic length L. (Figure 9 in Gotoda et al. [13]).

In order to depict the particles' trajectory to form the PAS, one can monitor the Poincaré section. Figure 7 illustrates some example of the Poincaré section at $z \sim H/2$. When one tracks a single particle forming the PAS, the particle passes several fixed areas in the r–θ plane. In the case of the PAS of $m = 3$, one finds three pairs of the area; a pair consists of the area locating near the free surface ("A" in the frame) and near the liquid-bridge centre ("B"). The area "A" near the free surface corresponds to the trajectory near the tip of the PAS blade passing the plane downward, and the area "B" near the centre corresponds to the trajectory where the particle follows the return flow inside the liquid bridge from the region near the cold disk toward the hot disk. One finds finite sizes of the areas. Note that these experimental results were obtained by employing rather big particles ($d_\mathrm{p} = 30$ µm) and a light sheet of finite thickness. One could measure not only the size but also their shape in the Poincaré section, if one employed finer particles with finer resolution of the observation system, in order to compare with those of the KAM tori [58] .

Through such observation, one can find the particle departing from a certain trajectory to another (see Figure 8); the top frame illustrates an example of the Poincaré section detected at $z = 0.81 \pm 0.09$ mm under $\mathrm{Ma}|_{L=H} = 5.2 \times 10^4$ (the same as row (c) in Figure 7 but in a different experimental run). Frames (a)–(c) indicate the sections obtained in the same experimental run to track a single particle under the same condition but in different successive periods. Frame (d) illustrates all the data obtained in a single experimental run for 73.2 s; that is, the sum of (a) to (c) corresponds to the original section shown as frame (d). From these successive sections, one observes that this particle does not stay to settle at a certain structure, but switches among specific structures. In this case, the particle (a) stays on the core first, then (b) migrates to stay for a while on a structure wrapping the core, and then (c) migrates to finally settle on the PAS. Each structure resembles the KAM torus realized in the thermocapillary-driven convection in the half-zone liquid bridge; the structure known as the toroidal core, the structure wrapping the toroidal core, and the PAS of $m = 3$ resembles T_{core}, T_3^9, and T_3^3 [61], respectively. The middle and bottom frames of the figure illustrate the corresponding temporal variations of the r- and θ-positions of a particle on the Poincaré section, respectively. In the variation of the r-position, the corresponding KAM torus is indicated; section (a) exhibits the paths of the particle on the core, that is, the particle is attracted by T_{core} first. Then, the particle switches to the attractor T_3^9 in section (b). This torus, however, does not keep attracting the particle; one finds that the particle migrates between the attractors T_3^9 and T_{core}. Finally, in this case, the particle is attracted to T_3^3 to keep forming the PAS (the section (c)). This scenario of the particle migration is not thus far predictable; each torus becomes unstable and they keep attracting particle under high Ma. It was indicated that the particle would not be able to stay for a long period especially on the core, which might reflect that the torus T_{core} in the high-Pr liquid bridge becomes weakly chaotic [20]. Further research is needed to describe the attractivity of each structure of KAM tori and the mechanism on the interaction between the particles and the KAM tori.

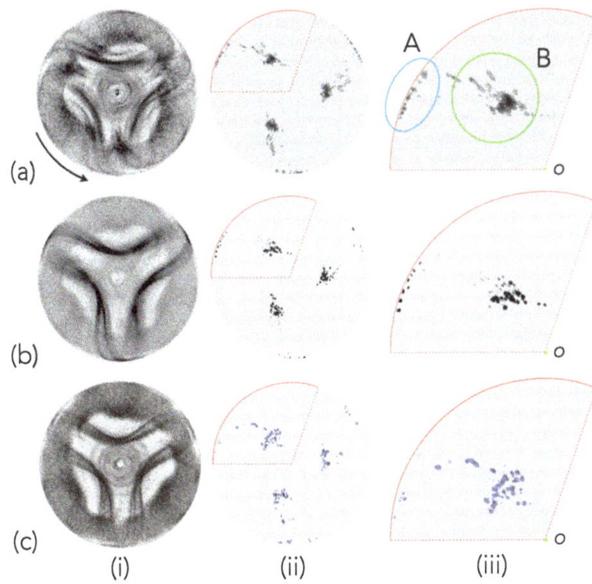

Figure 7. PASs observed through the top rod in a rotating frame of reference under Ma $=$ (**a**) 4.1×10^4, (**b**) 4.7×10^4, and (**c**) 5.2×10^4 in a liquid bridge of Pr $= 28.5$, $\Gamma = 0.68$ ($R = 2.5$ mm), and $V/V_0 = 1.0$ with suspended particles 30 µm in diameter: Columns (**i**,**ii**) indicate the projected images, and Poincaré sections at midheight of the liquid bridge, respectively. Images for column (**i**) are obtained by by integrating 500 frames (for about 8.3 s). Images for (**ii**)-(**a**) and -(**b**) are obtained by integrating 3900 frames (for almost 65 s), and -(**c**) 2340 frames (for almost 39 s). Column (**iii**) shows a zoomed view of the $1/3 = 1/m_0$ region of the image in (**ii**). The fundamental frequencies $f_0^{(HTW)}$ are (**a**) 1.41 Hz, (**b**) 1.42 Hz, and (**c**) 1.44 Hz. The rotating direction of the HTW in this figure is counter-clockwise in the laboratory frame. Ma is evaluated by applying H for the characteristic length L. (Figure 2 in Yamaguchi et al. [12]).

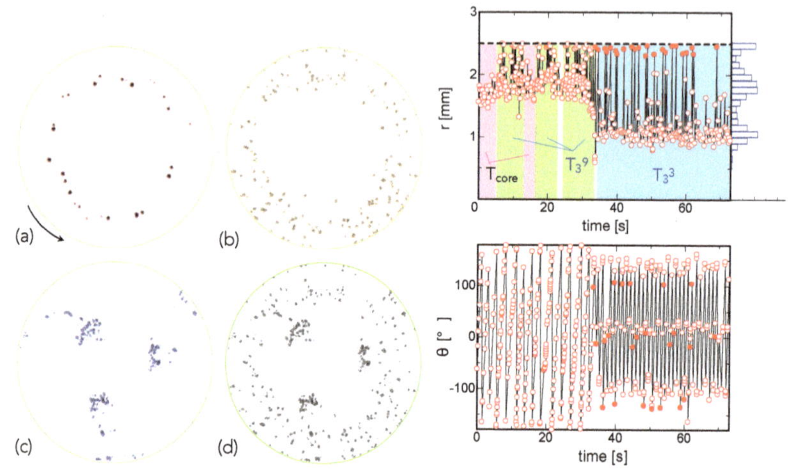

Figure 8. (left) Poincaré section at the midheight of the liquid bridge obtained by integrating (**a**) $0 \leq t$ [s] ≤ 6.1, (**b**) $6.1 \leq t \leq 33.4$, (**c**) $33.4 \leq t \leq 73.2$ and (**d**) total images of the section for 73.2 s detected at $z = 0.81 \pm 0.09$ mm under $\mathrm{Ma} = 5.2 \times 10^4$ (same as row (**c**) in Figure 7 but in a different experimental run). (Middle) and (right) corresponding temporal variations of r- and θ-positions of a particle on the Poincaré section, respectively. The histogram on the right of the top graph illustrates the particle number density measured with a constant interval $\Delta r = 0.1$ mm in the observation period. The direction of the travelling-wave-type oscillation is counter-clockwise in the laboratory frame. The conditions are the same as those in Figure 7. (Figure 4 in Yamaguchi et al. [12]).

It is of great importance and indispensable to detect the three-dimensional behaviour of the particles in order to understand the mechanism of PAS formation and the correlation with the KAM tori. It is, however, quite difficult especially under normal gravity conditions because of the small size of the liquid bridge with the deformed free surface. The time-dependent deformation of the free surface in the oscillatory convection prevents the precise measurement of the particles with the visible light through the free surface. In such a sense, microgravity conditions are an ideal environment because one can employ larger liquid bridges [21,42–50,63–65]. Under normal gravity, some efforts have been made to realize three-dimensional measurements of the particles inside the liquid bridge of $\mathcal{O}(10^{-3}\ \mathrm{m})$ through the transparent top rod [60,62,66]. Oba et al. [14] realized the three-dimensional reconstruction of the particles' position as a function of time by applying 3-D PTV with a classical algorithm (see Figure 9): frames (a) and (b) illustrate the reconstructed trajectories of the particles in the bird's-eye view and the top views, respectively, and frame (c) shows the distributions of all trajectories in the r-ϕ (top) and the z-ϕ (bottom) planes. The figures in frames (a) and (b) consist of the trajectories of the particles on the PAS (14 trajectories in blue), on the toroidal core (16 in red), and on the structure wrapping the core (3 in green). Note that the trajectories were obtained in the 1/3 region of the liquid bridge in the azimuthal direction by the experiments, and the same results are plotted repeatedly with a phase difference of $\pm 2\pi/3$ in the rest region; this process is based on the three-fold rotational symmetric structure of the flow field. Because of the high frequency of the particle motions in the small-size liquid bridge, Oba et al. [14] employed synchronized high-speed cameras. One can track the motion of the particles forming the various coherent structures as a function of time.

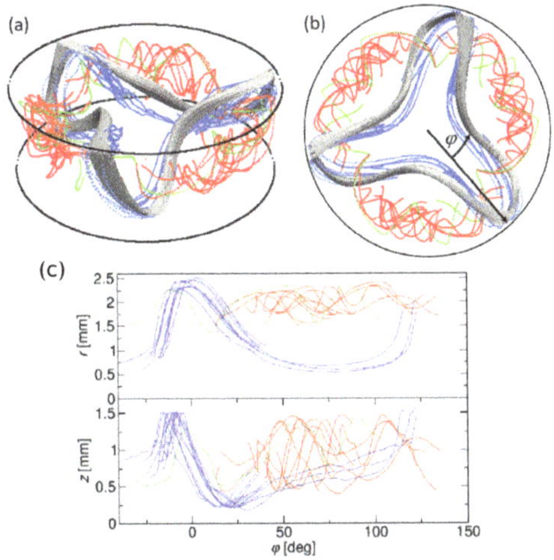

Figure 9. Reconstructed trajectories of particles in (**a**) bird's-eye view and (**b**) top view, respectively, in the liquid bridge of Pr $= 28.6$, $\Gamma = 0.68$ ($R = 2.5$ mm), and $V/V_0 = 1.0$ with suspended particles of 30 µm in diameter. The figures consist of the trajectories of the particles on the PAS (14 trajectories in blue), on the core (16 in red) and on the structure wrapping the core (3 in green). The direction of HTW is clockwise, and the net direction of the particles on the PAS is counter-clockwise (in positive ϕ direction) in the laboratory frame. Bundles of pathlines which form the PAS obtained by the numerical simulation [67] are also drawn in grey. Frame (**c**): distributions of all trajectories in the r-ϕ (top) and the z-ϕ (bottom) planes. (Figure 9 in Oba et al. [14]).

When one obtains the three-dimensional behaviour of the particles, their motions at designated area can be monitored (see Figure 10). It was found that the particles form various structures which resemble various KAM tori such as T_3^3, T_3^9, and T_{core} as indicated by the model flow [58]. As aforementioned, one needs finer tracking with smaller particles in order to make a comprehensive comparison between the KAM tori and the coherent structures in more detail.

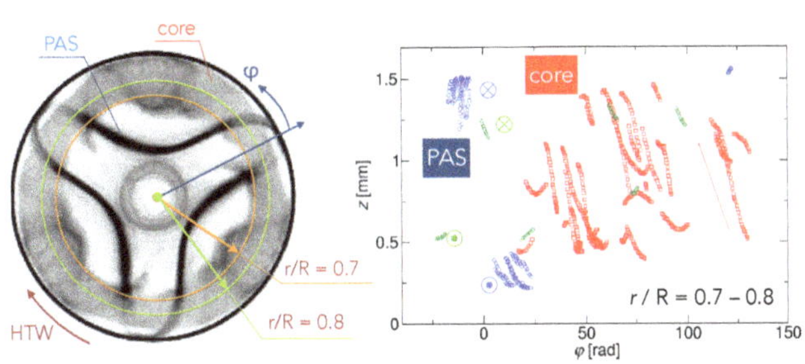

Figure 10. (**left**) Coherent structure under the same condition as shown in Figure 9, and (**right**) spatial distributions of paths in the z–ϕ space in $0.7 \leq r/R < 0.8$. In the right frame, the particles attracted to T_3^3, T_{core} and T_3^9 are drawn in blue, red and green, respectively. The arrow indicates the direction of motion of the particles concerned. The direction in r of the particle motions is also indicated; $\odot$ indicates the motion toward the free surface (or in positive r direction), and $\otimes$ that toward the centre axis of the liquid bridge (or in negative r direction).

3.2.4. On-Going Topics

The azimuthal wave number of the oscillatory convection depends on the shape of the liquid bridge, especially on the aspect ratio Γ. Because of the gravity effect as well as the Rayleigh limit, the main target of the investigation has been the oscillatory convection with $2 \le m \le 4$ [8,9]. Note that the experiments with short liquid bridges for larger m ($m \ge 5$) [68] involve the critical problem of the evaporation; one has to increase ΔT to realize higher Ma despite smaller characteristic length L. Therefore, it is rather difficult to find any knowledge accumulated from the experimental approaches to the PAS of $m = 1$ and $m \ge 5$. As for the PAS of $m = 1$, a small amount of research from terrestrial experiments can be found: Sasaki et al. [69] and Schwabe et al. [9] examined the PAS of $m = 1$ in the liquid bridge of moderate Pr. Despite the fact that Schwabe et al. [9] indicated that the existing range of the PAS of $m = 1$ against Γ and Ma is much narrower than those of the PAS of $2 \le m \le 4$, it was found that two types of PAS of $m = 1$ exist by varying Ma. It must be noted that the PAS of $m \ge 2$ also exhibits two types by increasing Ma (SL1- and SL2-PASs [8] as introduced), but the variation of the PAS of $m = 1$ is different from those: when one observes from above, the particles form the closed path with a spiral structure inside the liquid bridge for the SL-PAS under lower Ma [69]. Such a spiral structure corresponds to the helical motion following the return flow in the central region. As for the SL-PAS under higher Ma, on the other hand, the particles form the closed path without such a spiral structure. That is, this type of PAS realized at higher Ma exhibits a simpler structure. It seems opposite to the correlation for the SL1- and SL2-PASs of $m \ge 2$ against Ma. Schwabe et al. [9] also realized the SL-PAS of $m = 1$ with a spiral structure in the central region of the liquid bridge. The author's group at Tokyo University of Science recently realized the SL-PAS with a spiral structure in the liquid bridge of Pr $= 28.6$, and succeeded in the reconstruction in the rotating frame of reference (see Figure 11a) [70]. This PAS of $m = 1$ is realized under $\mathrm{Ma}|_{L=H} = 1.86 \times 10^4$ or $\Delta T = 20\,\mathrm{K}$ in the liquid bridge of $\Gamma = 2.0$ ($R = 0.75\,\mathrm{mm}$) and $V/V_0 = 0.8$. The induced oscillatory convection seems to correspond to the type of hydrothermal wave (HTW) (b) indicated by the linear stability analysis [71] by judging the spatial distribution of the temperature deviation over the free surface (frame (b)). Recently the SL-PAS of $m = 1$ with a spiral structure was finely reproduced in the liquid bridge of Pr $= 8$ by Capobianchi and Lappa [72] via the numerical simulation. They indicated the correlation between the coherent structure and the KAM torus. It must be noticed, however, that they reproduced the coherent structure only by the particle lighter than the test liquid, or, $\varrho = \rho_p/\rho_l < 1$. The terrestrial experiments [9,69,70] have revealed that the particles of $\varrho > 1$ realize the coherent structure of $m = 1$ as those of $m \ge 2$ [8,9,13,14,59,60,62]. As for the SL-PAS of $m = 1$ without a spiral structure, it was also numerically reproduced by Barmak et al. [73,74]. They illustrated that the attracting orbit exhibits multiple revolutions. Knowledge on the PAS of $m = 1$ has gradually accumulated, but further research is surely needed via experimental approaches for comprehensive understanding of the coherent structures of $m = 1$ in tall liquid bridges.

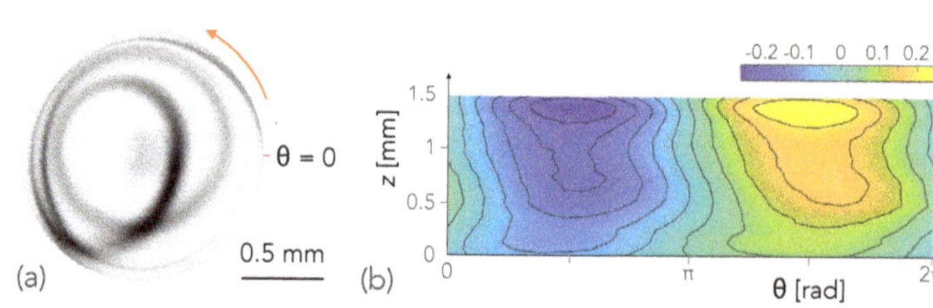

Figure 11. (**a**) Coherent structure in the rotating frame of reference observed through a hot disk and (**b**) temperature-deviation distribution over free surface from a terrestrial experiment with a liquid bridge of $\Gamma = 2.0$ ($R = 0.75\,\mathrm{mm}$) and $V/V_0 = 0.8$ under $\mathrm{Ma}|_{L=H} = 1.86 \times 10^4$. The image of (**a**) the coherent structure is drawn by integrating for 10 fundamental periods or 3.8 s, and the temperature deviation in (**b**) is obtained by averaging the images for 65 fundamental periods or 24.7 s. The arrow in the row (**a**) indicates the direction of the travelling wave.

4. Concluding Remarks

Coherent structures by particles suspended in half-zone thermocapillary liquid bridges through experimental approaches were introduced in this article. After Schwabe et al. [1] found this unique phenomenon called particle accumulation structure (PAS), general knowledge on the shape, the occurring conditions, and their formation processes has accumulated. By focusing on the spatial–temporal behaviour of the particles forming the PAS of the azimuthal wave number $m = 3$, a correlation between the particle behaviour and the ordered flow structures known as the Kolmogorov–Arnold–Moser tori was illustrated as functions of the intensity of the thermocapillary effect and the particle size. Recent works on the PAS of $m = 1$, through the experimental and numerical approaches, were briefly introduced. Knowledge accumulated via ground-based research on this unique phenomenon would lead to future research including microgravity experiments on the International Space Station (ISS) by the project "Japanese European Research Experiments on Marangoni Instability (JEREMI)" [22,23].

Funding: The experimental research on the PAS of $m = 1$ through terrestrial experiments [70] were supported by the Japan Society for the Promotion of Science (JSPS) through Challenging Research (Exploratory) (project number: 20K20977).

Institutional Review Board Statement: Not applicable.

Informed Consent Statement: Not applicable.

Acknowledgments: The author acknowledges Dietrich Schwabe, the professor emeritus of Justus-Liebig-Universität Gießen, Valentina Shevtsova, Denis Melnikov, Aliaksandr Mialdun, and Yury Gaponenko (Université Libre de Bruxelles), Hendrik C. Kuhlmann, Thomas Lemée, and Francesco Romanò (TU Wien), Marcello Lappa (University of Strathclyde), Koichi Nishino (Yokohama National University), Taishi Yano (Kanagawa University), Hiroshi Kawamura (professor emeritus) and Shiho Tanaka who belonged to Tokyo University of Science, for fruitful discussion. The author also acknowledges ex- and current graduate and undergraduate students belonging to the Interfacial Thermo-Fluid Dynamics Lab at Tokyo University of Science; all of experimental results introduced in this article owe to their invaluable contributions.

Conflicts of Interest: The author declares no conflict of interest.

Abbreviations

The following abbreviations are used in this manuscript:

HTW	Hydrothermal wave
JEREMI	Japanese European Research Experiments on Marangoni Instability
PAS	Particle accumulation structure

References and Notes

1. Schwabe, D.; Hintz, P.; Frank, S. New features of thermocapillary convection in floating zones revealed by tracer particle accumulation structure (PAS). *Microgravity Sci. Technol.* **1996**, *9*, 163–168.
2. Preisser, F.; Schwabe, D.; Scharmann, A. Steady and oscillatory thermocapillary convection in liquid columns with free cylindrical surface. *J. Fluid Mech.* **1983**, *126*, 545–567. [CrossRef]
3. Wanschura, M.; Shevtsova, V.M.; Kuhlmann, H.C.; Rath, H.J. Convective instability mechanisms in thermocapillary liquid bridges. *Phys. Fluids* **1995**, *7*, 912–925. [CrossRef]
4. Leypoldt, J.; Kuhlmann, H.C.; Rath, H.J. Three-dimensional numerical simulation of thermocapillary flows in cylindrical liquid bridges. *J. Fluid Mech.* **2000**, *414*, 285–314. [CrossRef]
5. Ueno, I.; Tanaka, S.; Kawamura, H. Oscillatory and chaotic thermocapillary convection in a half-zone liquid bridge. *Phys. Fluids* **2003**, *15*, 408–416. [CrossRef]
6. Xu, J.J.; Davis, S.H. Convective thermocapillary instabilities in liquid bridges. *Phys. Fluids* **1984**, *27*, 1102–1107. [CrossRef]
7. Ogasawara, T.; Motegi, K.; Hori, T.; Ueno, I. Secondary instability induced by thermocapillary effect in half-zone liquid bridge of high Prandtl number fluid. *Mech. Eng. Lett.* **2019**, *5*, 1900014. [CrossRef]
8. Tanaka, S.; Kawamura, H.; Ueno, I.; Schwabe, D. Flow structure and dynamic particle accumulation in thermocapillary convection in a liquid bridge. *Phys. Fluids* **2006**, *18*, 067103. [CrossRef]

9. Schwabe, D.; Mizev, A.I.; Udhayasankar, M.; Tanaka, S. Formation of dynamic particle accumulation structures in oscillatory thermocapillary flow in liquid bridges. *Phys. Fluids* **2007**, *19*, 072102. [CrossRef]
10. Yano, T.; Maruyama, K.; Matsunaga, T.; Nishino, K. Effect of ambient gas flow on the instability of Marangoni convection in liquid bridges of various volume ratios. *Int. J. Heat Mass Transf.* **2016**, *99*, 182–191. [CrossRef]
11. Toyama, A.; Gotoda, M.; Kaneko, T.; Ueno, I. Existence conditions and formation process of second type of spiral loop particle accumulation structure (SL-2 PAS) in half-zone liquid bridge. *Microgravity Sci. Technol.* **2017**, *29*, 263–274. [CrossRef]
12. Yamaguchi, K.; Hori, T.; Ueno, I. Long-term behaviors of a single particle forming a coherent structure in thermocapillary-driven convection in half-zone liquid bridge of high Prandtl-number fluid. *Int. J. Microgravity Sci. Appl.* **2019**, *36*, 360203.
13. Gotoda, M.; Toyama, A.; Ishimura, M.; Sano, T.; Suzuki, M.; Kaneko, T.; Ueno, I. Experimental study of coherent structures of finite-size particles in thermocapillary liquid bridges. *Phys. Rev. Fluids* **2019**, *4*, 094301. [CrossRef]
14. Oba, T.; Toyama, A.; Hori, T.; Ueno, I. Experimental study on behaviors of low-Stokes number particles in weakly chaotic structures induced by thermocapillary effect within a closed system with a free surface. *Phys. Rev. Fluids* **2019**, *4*, 104002. [CrossRef]
15. Takatsuka, M.; Tanaka, S.; Ueno, I.; Kawamura, H. Dynamic particle accumulation structure of Marangoni convection in liquid bridge—2. numerical simulation. In *Proceedings of the Thermal Engineering Conference*; Japan Society for Mechanical Engineers: Tokyo, Japan, 2002; pp. 307–308.
16. Seki, T.; Tanaka, S.; Kawamura, H. Numerical simulation of particle accumulatoin structure in oscillatory thermocapillary convection of a liquid bridge. In *Proceedings of Thermal Engineering Conference*; Japan Society for Mechanical Engineers: Tokyo, Japan, 2005; pp. 169–170.
17. Hofmann, E.; Kuhlmann, H.C. Particle accumulation on periodic orbits by repeated free surface collisions. *Phys. Fluids* **2011**, *23*, 072106. [CrossRef]
18. Melnikov, D.E.; Pushkin, D.O.; Shevtsova, V.M. Accumulation of particles in time-dependent thermocapillary flow in a liquid bridge: Modeling and experiments. *Eur. Phys. J. Spec. Top.* **2011**, *192*, 29–32. [CrossRef]
19. Lappa, M. Assessment of the role of axial vorticity in the formation of particle accumulation structures in supercritical Marangoni and hybrid thermocapillary-rotation-driven flows. *Phys. Fluids* **2013**, *25*, 012101. [CrossRef]
20. Romanò, F.; Kuhlmann, H.C. Finite-size Lagrangian coherent structures in thermocapillary liquid bridges. *Phys. Rev. Fluids* **2018**, *3*, 094302. [CrossRef]
21. Kawamura, H.; Nishino, K.; Matsumoto, S.; Ueno, I. Report on microgravity experiments of Marangoni convection aboard International Space Station. *J. Heat Transf.* **2012**, *134*, 031005. [CrossRef]
22. Shevtsova, V.; Gaponenko, Y.; Kuhlmann, H.C.; Lappa, M.; Lukasser, M.; Matsumoto, S.; Mialdun, A.; Montanero, J.M.; Nishino, K.; Ueno, I. The JEREMI-project on thermocapillary convection in liquid bridges. Part B: Overview on impact of co-axial gas flow. *Fluid Dyn. Mater. Process.* **2014**, *10*, 197–240.
23. Kuhlmann, H.C.; Lappa, M.; Melnikov, D.E.; Mukin, R.V.; Muldoon, F.H.; Pushkin, D.O.; Shevtsova, V.M.; Ueno, I. The JEREMI-Project on thermocapillary convection in liquid bridge. Part A: Particle Accumulation Structures. *Fluid Dyn. Mater. Process.* **2014**, *10*, 1–10.
24. Chun, C.H.; Wuest, W. A micro-gravity simulation of the Marangoni convection. *Acta Astronaut.* **1978**, *5*, 681–686. [CrossRef]
25. Lan, C.W.; Kim, Y.J.; Kou, S. A half-zone study of Marangoni convection in floating-zone crystal growth under microgravity. *J. Cryst. Growth* **1990**, *104*, 801–808. [CrossRef]
26. Cao, Z.H.; Xie, J.C.; Tang, Z.M.; Hu, W.R. The influence of buoyancy on the onset of oscillatory convection in a half floating zone. *Adv. Space Res.* **1991**, *11*, 163–166. [CrossRef]
27. Velten, R.; Schwabe, D.; Scharmann, A. The periodic instability of thermocapillary convection in cylindrical liquid bridges. *Phys. Fluids A* **1991**, *3*, 267–279. [CrossRef]
28. Petrov, V.; Schatz, M.F.; Muehlner, K.A.; VanHook, S.J.; McCormick, W.D.; Swift, J.B.; Swinney, H.L. Nonlinear control of remote unstable states in a liquid bridge convection experiment. *Phys. Rev. Lett.* **1996**, *77*, 3779–3782. [CrossRef] [PubMed]
29. Hirata, A.; Nishizawa, S.; Sakurai, M. Experimental results of oscillatory Marangoni convection in a liquid bridge under normal gravity. *J. Jpn. Soc. Microgravity Appl.* **1997**, *14*, 122–129.
30. Hirata, A.; Sakurai, M.; Ohishi, N. Effect of gravity on Marangoni convection in a liquid bridge. *J. Jpn. Soc. Microgravity Appl.* **1997**, *14*, 130–136.
31. Shevtsova, V.M.; Mojahed, M.; Legros, J.C. The loss of stability in ground based experiments in liquid bridges. *Acta Astronaut.* **1999**, *44*, 625–634. [CrossRef]
32. Kamotani, Y.; Wang, L.; Hatta, S.; Wang, A.; Yoda, S. Free surface heat loss effect on oscillatory thermocapillary flow in liquid bridges of high Prandtl number fluids. *Int. J. Heat Mass Transf.* **2003**, *46*, 3211–3220. [CrossRef]
33. Shiomi, J.; Kudo, M.; Ueno, I.; Kawamura, H.; Amberg, G. Feedback control of oscillatory thermocapillary convection in a half-zone liquid bridge. *J. Fluid Mech.* **2003**, *496*, 193–211. [CrossRef]
34. Irikura, M.; Arakawa, Y.; Ueno, I.; Kawamura, H. Effect of ambient fluid flow upon onset of oscillatory thermocapillary convection in half-zone liquid bridge. *Microgravity Sci. Technol.* **2005**, *16*, 176–180. [CrossRef]
35. Wang, A.; Kamotani, Y.; Yoda, S. Oscillatory thermocapillary flow in liquid bridges of high Prandtl number fluid with free surface heat gain. *Int. J. Heat Mass Transf.* **2007**, *50*, 4195–4205. [CrossRef]

36. Tiwari, S.; Nishino, K. Effect of confined and heated ambient air on onset of instability in liquid bridges of high Pr fluids. *Fluid Dyn. Mater. Process. (FDMP)* **2010**, *6*, 109–136.
37. Ueno, I.; Kawazoe, A.; Enomoto, H. Effect of ambient-gas forced flow on oscillatory thermocapillary convection of half-zone liquid bridge. *Fluid Dyn. Mater. Process. (FDMP)* **2010**, *6*, 99–108.
38. Schwabe, D.; Mizev, A.I. Particles of different density in thermocapillary liquid bridges under the action of travelling and standing hydrothermal waves. *Eur. Phys. J. Spec. Top.* **2011**, *192*, 13–27. [CrossRef]
39. Melnikov, D.E.; Pushkin, D.O.; Shevtsova, V.M. Synchronization of finite-size particles by a traveling wave in a cylindrical flow. *Phys. Fluids* **2013**, *25*, 092108. [CrossRef]
40. Melnikov, D.E.; Watanabe, T.; Matsugase, T.; Ueno, I.; Shevtsova, V. Experimental study on formation of particle accumulation structures by a thermocapillary flow in a deformable liquid column. *Microgravity Sci. Technol.* **2014**, *26*, 365–374. [CrossRef]
41. Gotoda, M.; Melnikov, D.E.; Ueno, I.; Shevtsova, V. Experimental study on dynamics of coherent structures formed by inertial solid particles in three-dimensional periodic flows. *Chaos* **2016**, *26*, 073106. [CrossRef]
42. Yano, T.; Nishino, K.; Ueno, I.; Matsumoto, S.; Kamotani, Y. Sensitivity of hydrothermal wave instability of Marangoni convection to the interfacial heat transfer in long liquid bridges of high Prandtl number fluids. *Phys. Fluids* **2017**, *29*, 044105. [CrossRef]
43. Yano, T.; Nishino, K.; Matsumoto, S.; Ueno, I.; Komiya, A.; Kamotani, Y.; Imaishi, N. Overview of "Dynamic Surf" project in Kibo—Dynamic behavior of large-scale thermocapillary liquid bridges in microgravity. *Int. J. Microgravity Sci. Appl.* **2018**, *35*, 350102.
44. Yano, T.; Nishino, K.; Matsumoto, S.; Ueno, I.; Komiya, A.; Kamotani, Y.; Imaishi, N. Report on microgravity experiments of dynamic surface deformation effects on Marangoni instability in high-Prandtl-number liquid bridges. *Microgravity Sci. Technol.* **2018**, *30*, 599–610. [CrossRef]
45. Yano, T.; Nishino, K.; Kawamura, H.; Ueno, I.; Matsumoto, S.; Ohnishi, M.; Sakurai, M. Space experiment on the instability of Marangoni convection in large liquid bridge—MEIS-4: Effect of Prandtl number. *J. Phys. Conf. Ser.* **2011**, *327*, 012029. [CrossRef]
46. Yano, T.; Nishino, K.; Kawamura, H.; Ueno, I.; Matsumoto, S.; Ohnishi, M.; Sakurai, M. 3-D PTV measurement of Marangoni convection in liquid bridge in space experiment. *Exp. Fluids* **2012**, *53*, 9–20. [CrossRef]
47. Sato, F.; Ueno, I.; Kawamura, H.; Nishino, K.; Matsumoto, S.; Ohnishi, M.; Sakurai, M. Hydrothermal wave instability in a high-aspect-ratio liquid bridge of Pr > 200 On-orbit experiments in the Japanese Experiment Module 'Kibo' aboard the International Space Station. *Microgravity Sci. Technol.* **2013**, *25*, 43–58. [CrossRef]
48. Yano, T.; Nishino, K.; Kawamura, H.; Ueno, I.; Matsumoto, S. Instability and associated roll structure of Marangoni convection in high Prandtl number liquid bridge with large aspect ratio. *Phys. Fluids* **2015**, *27*, 024108. [CrossRef]
49. Nishino, K.; Yano, T.; Kawamura, H.; Matsumoto, S.; Ueno, I.; Ermakov, M.K. Instability of thermocapillary convection in long liquid bridges of high Prandtl number fluids in microgravity. *J. Cryst. Growth* **2015**, *420*, 57–63. [CrossRef]
50. Matsugase, T.; Ueno, I.; Nishino, K.; Ohnishi, M.; Sakurai, M.; Matsumoto, S.; Kawamura, H. Transition to chaotic thermocapillary convection in a half zone liquid bridge. *Int. J. Heat Mass Transf.* **2015**, *89*, 903–912. [CrossRef]
51. Pushkin, D.O.; Melnikov, D.E.; Shevtsova, V.M. Ordering of small particles in one-dimensional coherent structures by time-periodic flows. *Phys. Rev. Lett.* **2011**, *106*, 234501. [CrossRef] [PubMed]
52. Kuhlmann, H.C.; Hofmann, E. The mechanics of particle accumulation structures in thermocapillary flows. *Eur. Phys. J. Spec. Top.* **2011**, *192*, 3–12. [CrossRef]
53. Kuhlmann, H.C.; Muldoon, F.H. Particle-accumulation structures in periodic free-surface flows: Inertia versus surface collisions. *Phys. Rev. E* **2012**, *85*, 046310. [CrossRef] [PubMed]
54. Kuhlmann, H.C.; Muldoon, F.H. On the different manifestations of particle accumulation structures (PAS) in thermocapillary flows. *Eur. Phys. J. Spec. Top.* **2013**, *219*, 59–69. [CrossRef]
55. Mukin, R.V.; Kuhlmann, H.C. Topology of hydrothermal waves in liquid bridges and dissipative structures of transported particles. *Phys. Rev. E* **2013**, *88*, 053016. [CrossRef] [PubMed]
56. Muldoon, F.H.; Kuhlmann, H.C. Origin of particle accumulation structures in liquid bridges: Particle-boundary-interactions versus inertia. *Phys. Fluids* **2016**, *28*, 073305. [CrossRef]
57. Romanò, F.; Kuhlmann, H.C.; Ishimura, M.; Ueno, I. Limit cycles for the motion of finite-size particles in axisymmetric thermocapillary flows in liquid bridges. *Phys. Fluids* **2017**, *29*, 093303. [CrossRef]
58. Muldoon, F.H.; Kuhlmann, H.C. Coherent particulate structures by boundary interaction of small particles in confined periodic flows. *Phys. D Nonlinear Phenom.* **2013**, *253*, 40–65. [CrossRef]
59. Gotoda, M.; Sano, T.; Kaneko, T.; Ueno, I. Evaluation of existence region and formation time of particle accumulation structure (PAS) in half-zone liquid bridge. *Eur. Phys. J. Spec. Top.* **2015**, *224*, 299–307. [CrossRef]
60. Abe, Y.; Ueno, I.; Kawamura, H. Dynamic particle accumulation structure due to thermocapillary effect in noncylindrical half-zone liquid bridge. *Ann. N. Y. Acad. Sci.* **2009**, *1161*, 240–245. [CrossRef] [PubMed]
61. Kuhlmann, H.C.; Mukin, R.V.; Sano, T.; Ueno, I. Structure and dynamics of particle-accumulation in thermocapillary liquid bridges. *Fluid Dyn. Res.* **2014**, *46*, 041421. [CrossRef]
62. Niigaki, Y.; Ueno, I. Formation of particle accumulation structure (PAS) in half-zone liquid bridge under an effect of thermo-fluid flow of ambient gas. *Trans. JSASS (Japan Soc. Aeronaut. Space Sci. Aerosp. Technol.)* **2012**, *10*, Ph33–Ph37. [CrossRef]
63. Kang, Q.; Wu, D.; Duan, L.; Hu, L.; Wang, J.; Zhang, P.; Hu, W. The effects of geometry and heating rate on thermocapillary convection in the liquid bridge. *J. Fluid Mech.* **2019**, *881*, 951–982. [CrossRef]

64. Kang, Q.; Wu, D.; Duan, L.; Zhang, J.; Zhou, B.; Wang, J.; Han, Z.; Hu, L.; Hu, W. Space experimental study on wave modes under instability of thermocapillary convection in liquid bridges on Tiangong-2. *Phys. Fluids* **2020**, *32*, 034107.
65. Wang, J.; Wu, D.; Duan, L.; Kang, Q. Transition to chaos of buoyant-thermocapillary convection in large-scale liquid bridges. *Microgravity Sci. Technol.* **2020**, *32*, 217–227. [CrossRef]
66. Nishimura, M.; Ueno, I.; Nishino, K.; Kawamura, H. 3D PTV measurement of oscillatory thermocapillary convection in half-zone liquid bridge. *Exp. Fluids* **2005**, *38*, 285–290. [CrossRef]
67. Romanò, F.; Kuhlmann, H.C. Private communications.
68. Yazawa, S.; Kawamura, H. Experiment of Marangoni convection in a liquid bridge with a low to medium aspect ratio. In Proceedings of the National Heat Transfer Symposium of Japan (In Japanese), Hokkaido, Japan, 26–28 May 2010; p. SP409.
69. Sasaki, Y.; Tanaka, S.; Kawamura, H. Particle accumulation structure in thermocapillary convection of small liquid bridge. In Proceedings of the 6th Japan/China Workshop on Microgravity Sciences, Saga, Japan, 22–26 October 2005.
70. Sakata, T.; Ueno, I. The University of Tokyo, Tokyo, Japan, private communications.
71. Fujimoto, S.; Ogasawara, T.; Ota, A.; Motegi, K.; Ueno, I. Effect of heat loss on hydrothermal wave instability in half-zone liquid bridges of high Prandtl number fluid. *Int. J. Microgravity Sci. Appl.* **2019**, *36*, 360204.
72. Capobianchi, P.; Lappa, M. On the influence of gravity on particle accumulation structures in high aspect-ratio liquid bridges. *J. Fluid Mech.* **2021**, *908*, A29. [CrossRef]
73. Barmak, I.; Romanò, F.; Kuhlmann, H.C. Particle accumulation in high-Prandtl-number liquid bridges. *Proc. Appl. Math. Mech.* **2019**, *19*, e201900058. [CrossRef]
74. Barmak, I.; Romanò, F.; Kannan, P.K.; Kuhlmann, H.C. Coherent particle structures in high-Prandtl-number liquid bridges. *Microgravity Sci. Technol.* **2021**, *33*, 19. [CrossRef]

fluids

Article

Coating Flow Near Channel Exit. A Theoretical Perspective

Roger E. Khayat * and Mohammad Tanvir Hossain

Department of Mechanical and Materials Engineering, University of Western Ontario, London, ON N6A 5B9, Canada; mhossa69@uwo.ca
* Correspondence: rkhayat@uwo.ca; Tel.: +1-519-661-2111

Received: 13 September 2020; Accepted: 9 October 2020; Published: 15 October 2020

Abstract: The planar flow of a steady moving-wall free-surface jet is examined theoretically for moderate inertia and surface tension. The method of matched asymptotic expansion and singular perturbation is used to explore the rich dynamics near the stress singularity. A thin-film approach is also proposed to capture the flow further downstream where the flow becomes of the boundary-layer type. We exploit the similarity character of the flow to circumvent the presence of the singularity. The study is of close relevance to slot and blade coating. The jet is found to always contract near the channel exit, but presents a mild expansion further downstream for a thick coating film. We predict that separation occurs upstream of the exit for slot coating, essentially for any coating thickness near the moving substrate, and for a thin film near the die. For capillary number of order one, the jet profile is not affected by surface tension but the normal stress along the free surface exhibits a maximum that strengthens with surface tension. In contrast to existing numerical findings, we predict the existence of upstream influence as indicated by the nonlinear pressure dependence on upstream distance and the pressure undershoot (overshoot) in blade (slot) coating at the exit.

Keywords: coating flow; free surface; boundary layer; stress singularity; matched asymptotic expansions

1. Introduction

We examine the free-surface flow of a planar moving-wall jet at moderate Reynolds and capillary numbers near and far from the channel exit as encountered in coating flow. With the advent of high-speed coating and the use of low-viscosity liquids, inertia is becoming increasingly important but has been traditionally neglected in the modelling of coating flow. General aspects on the classification and analyses of coating flows can be found in the reviews by Ruschak [1] and Weinstein and Ruschak [2]. The current work focuses on slot and blade coating flows which, in addition to the substrate movement, typically involve an adverse or a favorable pressure gradient.

In slot coating, the liquid is forced into the slot die, and distributes through the narrow slot before it emerges onto the moving substrate. A low-pressure area or vacuum is imposed upstream of the die to facilitate a faster and more stable coating process. Consequently, part of the flow is driven downstream by the moving substrate and part of it circulates upstream in the low-pressure area, which causes a streamwise adverse pressure gradient to act inside the channel formed between the downstream die and the moving substrate. In blade coating, the flow lies between a fixed blade of a prescribed shape and a substrate moving parallel to itself. The coating liquid is dragged inside the channel by the moving substrate, which causes a hydrodynamic pressure rise at the upstream of the blade. The pressure rise causes the blade to reject most of the liquid, and only a fraction passes into the narrow channel. Since the drag flow can only carry half of the coating liquid, a streamwise favorable pressure gradient is generated to carry the rest [3]. In this case, the pressure gradient forces the coating liquid in the same direction as the movement of the substrate inside the channel. In both slot and blade coating, the moving substrate drags the flow out of the channel in the form of a free-surface wall jet, and a thin layer of liquid film is obtained. In the present work, only the pressure gradient immediately upstream of the blade or die exit is accounted for, which is found in reality to be very close to constant. Consequently, the flow inside the channel will be assumed to be a superposition of Couette (velocity driven) and Poiseuille (pressure driven) flows.

Aside from its industrial importance, coating flow is fundamentally important as it spans various flow regimes. In an effort to clearly identify these regimes, de Ryck and Quere [4] carried out different experiments on fibre coating and used dimensional arguments, which illustrate the situation of coating flow in general. The visco-capillary range corresponds to slow coating flow and negligible Weber number, obeying the well-known Landau equation [5]. In this case, the film thickness results from a balance between capillarity and viscous forces. For coating flow of liquids of low viscosity (such as water) at higher velocity (about 1 m/s), the measurements of de Ryck and Quere were shown to deviate considerably from the Landau law, even if the capillary number remained negligible ($Ca < 0.05$). This is the visco-inertial regime, corresponding to a Weber number of order 1. At larger velocities ($We > 1$), the coating flow does not seem to depend any longer on the surface tension. This observation corroborates well findings for coating flow in general, as the thickness measurements reported by Lee et al. [6] and Becker and Wang [7] indicate for slot coating. De Ryck and Quere [4] also point out that the boundary-layer regime should be relevant to most industrial fibre-coating processes. Similarly, the more recent thickness measurements of Chang et al. [8] for low-viscosity slot-coating films show that viscous and surface tension effects become important only below a critical Reynolds number. At low speed, the minimum wet thickness increases with increasing capillary number but becomes independent of the capillary number for $Ca > 0.3$. Above the critical Reynolds number, fluid inertia becomes dominant. In this region, the minimum wet thickness decreases as the Reynolds number increases. Chang et al. [8] also established that inertial forces cannot be neglected in slot coating for Re larger than 10, which roughly corresponds to a substrate speed of 1 m/s. The earlier measurements of Carvalho and Kheshgi [9] show that the visco-capillary model remains valid at capillary numbers below unity. Their measurements also indicate that the model becomes less and less useful as the capillary number and the Reynolds number rise beyond unity. According to Romero et al. [10], high capillary number ($Ca \approx 5$) and Reynolds number ($Re \approx 3$) occur in higher-speed coating operations. It is this inertial regime that is of primary interest in the present study.

Although inertia has the desirable effect of lowering the flow limit and stabilizing the coating process [9], existing studies have predominantly focused on the role of surface tension. Lee et al. [6] and Chin et al. [11] showed that reducing the viscosity lowers the minimum thickness and increases the maximum coating speed. They found that a higher viscosity of the coating solution tends to be destabilizing, and a lower coating speed under high viscosity induces coating defects such as air entrainment and ribbing. Lee et al. [6] measured the minimum wet thickness for extrusion slot coating. In their experiment, the range of Reynolds number was low, from 0.2 to 24. They noticed that there exists a critical capillary number beyond which the minimum wet thickness becomes constant regardless of the value of Ca. Following the work of Lee et al. [6], and using a highly viscous fluid in the slot coater, Yu et al. [12] found that a minimum wet thickness can be reached if a low viscous fluid is used as a bottom carrier layer. Chang et al. [8] visualized the slot coating process for a low viscosity fluid, and observed that beyond a critical Reynolds number ($Re = 20$), both viscous and surface tension effects become negligible, with inertia dominating the flow.

Although extensive theoretical work has been devoted to the analysis of coating flow, the focus has been primarily on inertialess flow of a non-Newtonian fluid [13]. To a much lesser extent, numerical studies accounting for inertia were also carried out. An overview of high-speed blade coating was given by Aidun [14], with direct relevance to paper coating. One of the earlier numerical work was done by Saito and Scriven [15], who carried out a finite-element analysis coupled to an iterative scheme to examine the capillary number effect on the curved meniscus close to the static contact line in slot coating. They showed that the downstream meniscus no longer remains attached to the slot die at higher capillary number and lower flow rate. They found that the rate of convergence is highly dependent on the capillary number, and became exceedingly slow for Ca > 10. Carvalho and Kheshgi [9] and Jang and Song [16] examined the low-flow limit for slot coating in the inertial-capillary regime. Iliopoulos and Scriven [17] examined numerically the influence of particle suspensions on the blade and substrate. They found that the blade coating thickness increased with the Reynolds number, for any capillary number, but the coating thickness tapered off at the higher capillary numbers. In their study, the approximate ranges were, 0 < Re < 55 and 10 < Ca < 30. Lin et al. [18] carried out a comparison between numerical predictions and experiment, assessing operating windows for slot coating for their Reynolds number ranging from 0 to 100.

Perhaps the scarcity of theoretical studies on inertial coating flow is due to the stronger singularity at the contact point (line) at higher Reynolds number. Bajaj et al. [19] reported that for surface-tension dominated flow, a circulation appears upstream of the exit. This recirculation zone was found to gradually shrink with increasing Reynolds number. Consequently, a geometric singularity emerges in slot coating flow, resulting from the discontinuous slope at the contact line, which is increasingly exposed in relatively stronger flows. The singularity is also physical as a result of the change in the boundary condition from adherence at the die wall to the shear-free condition on the free surface (see also below). Inertia can be used to counteract the receding action of the downstream meniscus and the flow reversal, and delay the onset of the flow limit [9]. However, this gain will be offset by the strengthening of the singularity, which is no longer attenuated by a dominant surface tension. In this case, the circulation just upstream of the contact line weakens and eventually disappears altogether with increasing inertia, causing a more abrupt drop in the shear stress and a stronger singularity. We recall that the flow limit corresponds to a minimum coating thickness for a substrate speed or a maximum substrate speed for a given coating thickness. Inertia helps reduce the flow limit when, for instance, lower viscosity liquids are used, which tend to spread more easily [6]. The measurements and theoretical predictions of Carvalho and Keshghi [9] indicate that thinner films can be obtained at faster web speeds. Finally, the imposition of appropriate boundary conditions for free-surface inlet and outlet flows remains an open issue, particularly for small or vanishing surface tension [20].

Coating flow is modelled as a free-surface wall jet emerging from a channel bounded by a semi-infinite stationary (die or blade) wall and an infinite moving substrate. We assume inertia to remain relatively important, allowing the asymptotic development of the flow field in terms of some inverse power of the Reynolds number. The jet is assumed to be subject to a constant favorable or adverse pressure gradient far upstream where the flow acquires a Couette–Poiseuille (CP) character, typically as encountered in blade and slot coating. The assumption of fully developed upstream flow and the channel flow geometry are generally reasonable and commonly adopted in numerical simulation [15,19,21,22]. In roll coating, for instance, it is established that when the radius of the roller is large compared to the capillary length $\sqrt{\frac{\sigma}{\rho g}}$, then the effect of the curvature of the cylinder is very weak and results in only small perturbations to the uniform film thickness on the roller [23]. Here σ is the surface tension coefficient, ρ is the fluid density and g is the gravitational acceleration.

The stress singularity constitutes the major difficulty in a computational approach. The incorporation of the singularity point and its immediate vicinity is unavoidable in this case since the entire flow domain must be considered (discretized). The neighborhood region around the singularity, which is crucial to the rest of the flow domain, is difficult to handle numerically. In a numerical approach, the singularity is typically smoothed out or smeared over. The present asymptotic approach represents a viable alternative, at least for the flow in the vicinity of the singularity, which is avoided altogether as a result of the flow similarity in the free-surface layer and the boundary layer near the wall. In a numerical approach, the mesh is typically refined near the singularity, thus capturing more closely the singular behavior; the numerical difficulty resides in handling the resulting stronger flow gradients. Mitsoulis [24] conducted a finite-element analysis of blade coating. From his figures, we can see that both the shear stress as well as the pressure and the normal stress become singular at the channel exit. Finally, although the flow is primarily dominated by the stress singularity close to the exit, this eventually influences the accuracy of the numerical predictions of the flow further downstream and upstream.

Saito and Scriven [15] used an iterative numerical scheme to determine the free surface profile. They reported on earlier studies [25] where convergence difficulties were encountered. This was particularly the case for slot coating, with severe bending of the free surface. Convergence difficulties were also encountered when inertia grows large at a fixed flow rate and dominates surface tension. Saito and Scriven [15] were able to overcome the convergence difficulties by combining simple coordinate parametrizations of parts of the free surface, and computing accurately the derivatives of finite-element residuals with respect to free-surface locations along the spines. Alternatively, mapping techniques have also been used more recently, where the governing equations and boundary conditions are transformed to an equivalent set defined in a known reference frame [9,21,22]. The volume-of-fluid approach has also been employed [16].

The present asymptotic approach circumvents the singularity altogether and does not require an iterative scheme to determine the shape of the meniscus. Perhaps more importantly, the present formulation provides a deeper insight on the flow structure and the rich dynamics near the singularity (see Section 7 for further discussion on this point and the relation with the triple-deck approach). We observe that the use of the singular perturbation technique also provides a procedure by which higher-order terms could be determined to the desired accuracy [26]. The formulation is essentially analytical, providing the steady flow solution in a manner that is completely amenable to the implementation of a linear or nonlinear stability analysis. Generally, asymptotic analyses have been successfully adopted for flows in the visco-capillary range [27,28]. Timoshin [29] developed a high-Reynolds-number asymptotic theory and examined the stability of boundary-layer flow over a coated surface. More recently, Tsang et al. [30] implemented a high-Reynolds-number asymptotic approach to study the interaction of a boundary layer on a solid plate and the free surface above. Studies of closer relevance to the present flow in the visco-inertial range were also conducted, but to a much lesser extent. Tillett [31] analyzed the laminar symmetric free-jet flow near the channel exit using the method of matched asymptotic expansions. Miyake et al. [32] carried out a similar analysis on a vertical jet of inviscid fluid, taking into account gravity effect. Philippe and Dumargue [33] applied an analysis similar to Tillett's for viscous axisymmetric vertical jets, emphasizing the interplay between the effects of gravity and inertia on the free surface shape and the velocity profile. Their approach, which is similar to the one in the present study, was validated against experiment. A local similarity transformation was carried out by Wilson [34] for the axisymmetric viscous-gravity jet emerging from a tube, but, unlike the present formulation, Wilson neglected the upstream influence. Khayat and co-workers examined the flow of a jet emerging from a channel of Newtonian [35,36] and non-Newtonian [37,38] liquids. We refer the reader to the book by Sobey [39] on interactive boundary layer and triple-deck theory for a perspective on asymptotic analyses, their applications and historic development.

Asymptotic analysis has also been applied to study coating flow, but the majority of the studies focused on dominant surface tension. Ruschak [26] performed a theoretical analysis based on the thin-film theory of Landau and Levich [5] to determine the effects of different parameters on the flow limit of extrusion slot coating for a Newtonian fluid. Ruschak considered a very small capillary number by setting the coating speed close to zero, and determined the film thickness in the slow flow (negligible inertia) limit by carrying out a singular perturbation method. Ruschak observed that film thickness becomes thinner for smaller capillary number, and noticed that gravity does not affect the flow much in the limit considered in the study. Higgins and Scriven [40] extended Ruschak's analysis by incorporating the viscous effect in the coating bead (inside the channel) with variable meniscus location. They concluded that as the coating speed increases, the dynamic contact angle increases, resulting in an altered coating behavior. Christodoulou and Scriven [41] provided an asymptotic analysis based on a thin-film approach in slide coating. Carvalho and Kheshgi [9] assessed the low-flow limit for slot coating, which is defined as the maximum substrate speed possible without any coating defects at a fixed minimum thickness or vice versa. By modifying the visco-capillary model, Carvalho and Kheshgi [9] developed a 2D numerical tool to study the effect of higher capillary numbers. They observed that at higher capillary number (Ca > 0.3), inertia started to dominate the flow, delaying the onset of the low-flow limit. Of close relevance to the present formulation is the matched asymptotic approach used by Rushack and Scriven [42], focusing on low flow rate and high surface-tension limits for an inertialess jet.

Other coating configurations were also analyzed. Savage [43] and, later, Gaskell et al. [44] analyzed the meniscus roll-coating flow of an inertialess fluid. Kelmanson [45] examined the effect of inertia in the small- and large- surface-tension limits for a coating film flowing around a rotating cylinder. Blythe and Simpkins [46] developed an asymptotic analysis using the Landau–Levich equation to examine the inertialess falling coating flow on a fibre. Later, Jang et al. [16] proposed a model for slot coating by modifying the visco-capillary model to accommodate the high capillary and to some extent inertia effects. Unlike the visco-capillary model, they accounted for the pressure variation under the slot die, which enabled them to predict the coating thickness at high capillary and Reynolds numbers. They observed that when Re > 10 the film thickness decreased with increasing inertia, and visco-capillary model did not show good agreement with the simulation for Ca > 0.2.

Finally, the present work is reminiscent of the early treatment of Goldstein [47] of the flow near the trailing edge of a semi-infinite plate, where the flow downstream of the singular edge was examined. Later, Stewrtson [48] and Messiter [49] developed a more comprehensive treatment of the flow very near the singularity, laying down the foundation for the triple-deck theory [39]. They established a rational theoretical approach for the investigation of separation and other non-linear features emerging in external flow situations. The present wall-jet flow differs fundamentally from the external separation flow due to the absence of an external inviscid flow. The present development is somewhat of the same level as the asymptotic treatment of Goldstein. The more complete analysis of the full Navier–Stokes equations and the implementation of a triple-deck approach may be envisageable in a future extension.

2. Problem Formulation and the Physical Domain

We consider the planar flow of an incompressible Newtonian fluid of density ρ, viscosity μ and surface tension σ, flowing between the semi-infinite stationary blade or slot die and the translating substrate, separated by a distance D. The flow is induced by the simultaneous forward translation velocity V of the substrate and an applied adverse or favorable pressure gradient g far upstream, where fully-developed Couette–Poiseuille (CP) flow is assumed to prevail. The assumption of fully-developed flow far upstream is commonly adopted in numerical studies on slot coating [10,15,21]. We also assume the CP flow to hold for blade coating as the pressure gradient is sensibly constant over a sufficiently long upstream distance [17]. Far downstream, the flow becomes uniform with speed V and film thickness T.

Non-dimensional variables are introduced by measuring the coordinates (x, z), the velocity components (u, w), the stream function ψ and the pressure p in units of D, V, VD and ρV^2, respectively. There result three dimensionless parameters appearing in the problem, namely the Reynolds number Re, the capillary number Ca, and the thickness-to-gap ratio Q, which are expressed as

$$\mathrm{Re} = \frac{\rho V D}{\mu}, \qquad \mathrm{Ca} = \frac{\mu V}{\sigma}, \qquad Q = \frac{T}{D}. \tag{1}$$

A parameter related to Q, which is conveniently introduced here:

$$G \equiv 3 - 6Q, \tag{2}$$

with 2G being the dimensionless applied pressure gradient scaled by $\frac{\mu V}{D^2}$. Clearly, Q < 1 for coating flow. Consequently, $-3 < G = O(1) < 3$. The final thickness T is typically greater than half the gap for blade coating where $G < 0$. That the pressure gradient can be constant far upstream is easily observed in blade coating for a long flat or even angled blade. We refer the reader to the numerical studies of Iliopoulos and Scriven [17], particularly their figures for different blade angles, and Mitsoulis and Athanasopoulos [50]. For slot coating, the pressure gradient is commonly directly related to the flow rate, which can be independent of the substrate speed [10]. In this case, G can be positive or negative, and is typically assumed to be constant in numerical studies of slot coating [15,22,25]. Finally, the value of G varies typically from −0.05 to −0.20 in blade coating [17,24,50], and from 0.05 to 0.20 in slot coating [11,18].

Figure 1 illustrates the flow configuration schematically where dimensionless notations are used. We conveniently choose the stationary wall (slot die or blade) to lie along the x axis at z = 0, and the moving substrate at z = 1. The z axis lies across the channel, with the origin coinciding with the contact line ($x = z = 0$) between the stationary wall and the free surface $z = \zeta(x)$. The figure depicts the two possibilities for the flow far upstream that typically correspond to a thin or a thick coating film, involving an adverse ($G > 0$) or a favorable ($G < 0$) pressure gradient, respectively. In both cases, the pressure gradient is generally expected to be constant. Figure 1 also illustrates the four regions or layers that constitute the physical domain, which will be discussed in Section 2.2.

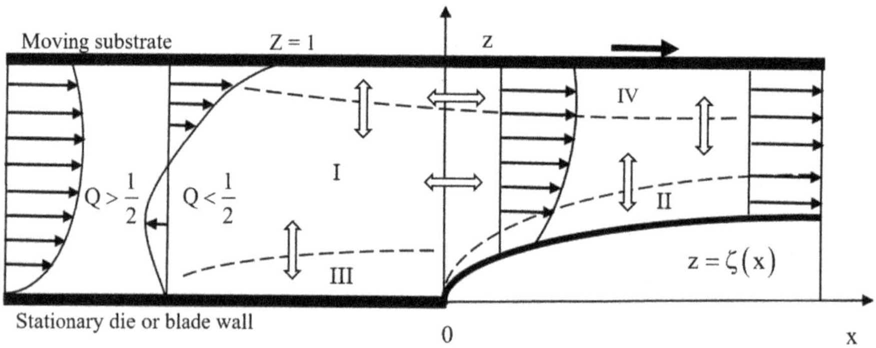

Figure 1. Schematic illustration of the coating flow domain. The two possible Couette–Poiseuille profiles correspond to a favorable pressure gradient $\left(Q > \frac{1}{2}, G < 0\right)$ and an adverse pressure gradient $\left(Q < \frac{1}{2}, G > 0\right)$. The domain is subdivided in four layers: (I) the core layer, (II) the free-surface layer ($x > 0$), (III) the stationary wall slip layer ($x < 0$), and (IV) the substrate slip layer. Double arrows indicate matching directions.

2.1. Governing Equations and Boundary Conditions

The problem is formulated in terms of the stream function ψ, with $u = \psi_z$ and $w = -\psi_x$. The stream function and velocity corresponding to the CP flow far upstream turn out to be the leading-order solution in the core layer I, and are conveniently introduced here in terms of G as

$$\psi_0(z) = \frac{1-G}{2}z^2 + \frac{G}{3}z^3, \quad u_0(z) = (1-G)z + Gz^2. \tag{3}$$

In the current study, the Reynolds number, Re, is assumed to be moderately large and $G = O(1)$ at most. If $Q = 1/2$ ($G = 0$), the flow is only driven by the forward substrate translation, resulting in drag flow. For steady laminar planar flow, the non-dimensional Navier–Stokes equations take the following form:

$$\begin{aligned} \psi_z\psi_{xz} - \psi_x\psi_{zz} &= -p_x + \tfrac{1}{Re}(\psi_{xxz} + \psi_{zzz}), \\ \psi_x\psi_{xz} - \psi_z\psi_{xx} &= -p_z - \tfrac{1}{Re}(\psi_{xxx} + \psi_{xzz}). \end{aligned} \tag{4}$$

Except for the stress components, a subscript with respect to x or z denotes partial differentiation. For $x > 0$, the kinematic and dynamic boundary conditions at the free surface $z = \zeta(x)$ are

$$\psi(x > 0, z = \zeta) = 0, \tag{5}$$

$$p(x>0,z=\zeta)\zeta' - \frac{1}{Re}(2\zeta'\psi_{xz}+\psi_{xx}-\psi_{zz})\Big|_{z=\zeta} = -\frac{1}{ReCa}\frac{\zeta'\zeta''}{1+\zeta'^2}, \tag{6}$$

$$p(x>0,z=\zeta) + \frac{1}{Re}[2\psi_{xz}+\zeta'(\psi_{zz}-\psi_{xx})]_{z=\zeta} = -\frac{1}{ReCa}\frac{\zeta''}{1+\zeta'^2}. \tag{7}$$

Here, a prime denotes total differentiation. In addition, the following conditions must be satisfied at the walls and far upstream:

$$\psi_z(x,z=1)=1, \qquad \psi(x,z=1)=1-\frac{G}{6}, \tag{8}$$

$$\psi_z(x<0,z=0)=0, \tag{9}$$

$$\psi(x<0,z=0)=0, \tag{10}$$

$$\psi(x\to-\infty,z) \sim \frac{1-G}{2}z^2+\frac{G}{3}z^3. \tag{11}$$

It is worth noting that in the coating literature, the pressure scale is taken as $\mu V/D$ instead of the current ρV^2 scale. In the former case, the pressure becomes Rep(x, z), which we shall use as such. The overall solution strategy of problem (4)–(11) along with the flow structure is summarized next.

2.2. The Physical Domain and the Flow Structure

The physical domain is shown schematically in Figure 1. As the fluid emerges out of the channel, the shear stress vanishes at the channel exit, causing *free-surface layer* II to develop along the free surface. The *core layer* I is influenced by the thin free-surface layer and the jet contraction. It is important to recall that the core layer remains predominantly of inviscid rotational character. However, this inviscid layer cannot extend to the stationary die or blade wall where significant viscous shearing occurs upstream of the stress singularity ($x < 0$) at the exit, forcing the formation of the *slip layer* III in the vicinity of the die or blade. Similarly, the core flow must also adjust to the shearing viscous flow in the *slip layer* IV upstream and downstream. Consequently, the predominantly inviscid core flow profile cannot satisfy adherence at the rigid walls, thus causing a slip layer to form with thickness that is not expected to be significant at high Reynolds number, but can be large otherwise. It is therefore fundamentally important to examine the flow in the slip layer and the extent of upstream influence. In fact, the thickness of the wall layers will turn out to be of $O(Re^{-1})$. The layer IV along the moving substrate is expected to grow rapidly with distance downstream of the channel exit.

The flow in each layer of the domain (Figure 1) is dominated by different physical mechanisms with corresponding characteristic length scales. The flow in layers II, III and IV, near the free surface, the die or blade and the moving substrate, respectively, is shear dominated, and is of the boundary-layer type. In the core layer I, both shear and elongation are in balance as a result of the predominance of the Couette–Poiseuille character of the flow upstream and downstream of the channel exit. At the channel exit, $x = 0$, the shear stress undergoes a step change from a non-zero value $\tau_{xz}(x<0,z=0) \approx 1-G$ at the die or blade to zero at the free surface $z = \zeta(x)$. The effect of this drop diffuses upstream inside the channel ($x < 0$) over a certain distance where fully developed Couette–Poiseuille flow is recovered, and downstream ($x > 0$), toward the moving substrate. In the limit of infinite Reynolds number, the flow retains the Couette–Poiseuille profile (3), which will turn out to be the lowest-order solution in the core layer I. In this limit, the jet height remains horizontal, with $\zeta(x) = 0$. In this case, the jet does not contract, and mass remains conserved throughout the upstream channel and downstream jet. For finite Reynolds number, the Couette–Poiseuille profile is modified when the fluid leaves the channel in the form of the wall jet. As the fluid detaches itself from the stationary wall, the removal of the wall stress causes the free-surface layer II to develop, where the Couette–Poiseuille profile adjusts itself so as to satisfy the stress-free (zero-traction) condition at the free surface. At infinite Reynolds number, the zero-traction condition would not be imposed; there is no viscous mechanism for the stress singularity to dissipate. In this limit, all the conditions of the problem would be satisfied since the Couette–Poiseuille profile continues undisturbed downstream from the exit. However, as the inviscid flow is governed by the Euler equations with no unique solution, the fully developed Couette–Poiseuille flow can be assumed to be everywhere the proper inviscid limit. Consequently, the core flow is, not affected by the in the free-surface layer II to lowest order; this layer, however, induces perturbations to the Couette–Poiseuille flow when higher-order terms are included, with some upstream influence. This assumption is similar to the one made for channel and tube flows with mild [51] and severe [52] constrictions, where the flow field in the core region, to leading order, satisfies the inviscid equations of motion. The reader is again referred to the book by Sobey [39] for other asymptotic analyses on interactive boundary layers.

The analysis of the flow in the core layer I is conducted separately upstream ($x < 0$) and downstream ($x > 0$) of the exit. The flow is obtained as a perturbation to the Couette–Poiseuille flow by solving Equation (4) subject to the far-upstream condition (11) and the matching condition with the free-surface layer II as the vertical double-arrow between layers I and II indicates in Figure 1. The upstream and downstream flows in layer I are matched at $x = 0$ as the horizontal double-arrow indicates. Boundary-layer solutions are sought in layers II, III and IV near the free surface and the walls, but not all layers admit a similarity solution. For the flow downstream of the exit, a similarity solution is possible in the free-surface layer II.

This determines completely the flow in layers I and II. The details will be given in Sections 3 and 4. We observe that no matching is required at $x = 0$ between the free-surface layer II and the slip layer III. Consequently, the singularity is entirely circumvented in the solution process, which constitutes a major advantage of the present approach. In Section 3, we also use a thin-film approach to analyze the flow far downstream. This is an appropriate approach since the flow is completely of the boundary-layer type at some distance downstream from the exit. In Section 5, we examine the flow near the stationary wall or slip layer III. The flow in layer III is matched onto the flow in the core layer I for $x < 0$ as the vertical double-arrow between layers I and III indicates. Finally, the flow in the slip layer IV near the moving substrate is examined in Section 6. The non-similarity solution is obtained separately for $x < 0$ and $x > 0$, and matched at $x = 0$.

3. The Flow in the Free-Surface Layer II and the Coating Film Profile Near and Far from the Exit

In this section, we examine the flow structure in the free-surface layer II. As the layer grows with downstream distance, it eventually invades the entire film region, at which point we adopt a thin-film approach to capture the flow far from the die or blade exit. The flows near and far downstream from the exit are matched to obtain the profile of the coating film everywhere downstream.

3.1. The Flow in the Free-Surface Layer Close to the Die or Blade Exit

In addition to the kinematic and dynamic conditions (5)–(7) at the free surface, the matching of the flow at the edge of the free-surface layer and the core layer also provides conditions needed to solve for the flow near the free surface as well as the core stream function and pressure terms. The matching process is detailed in Appendix A. For a successful application of the matching rule (A1), it is required that the stretching transformation must be expressed in the canonical form $y = \varepsilon\eta$ where $y = z - \zeta(x)$ is the near boundary transverse coordinate in the free-surface layer II. Here, $\varepsilon(\mathrm{Re}) << 1$ is the small parameter in the problem or the aspect ratio as referred to by Weinstein and Ruschak [2], which will be defined precisely shortly. The core expansion in this case, must be written in terms of y, not z; otherwise (A1) will be satisfied only approximately.

To examine the free-surface layer structure, we let $\varepsilon = \mathrm{Re}^{-\alpha}$, where α is to be determined. Anticipating that the height ζ of the free surface is of the same order of magnitude as the boundary-layer thickness, one can write $\zeta(x) = \varepsilon h(x)$, and henceforth work with $h(x)$, with the matching indicating that $h(x) = O(1)$. The following change of coordinates is introduced, namely,

$$x = \xi, \qquad z = y + \zeta(x) = \varepsilon(\eta + h). \tag{12}$$

The use of ξ instead of x helps emphasize the distinction between the core region where the (x, z) variables are used, and the free-surface layer where the (ξ, η) variables are used, and makes clearer the mathematical development in each region. The aim is to find a solution of the problem in the (ξ, η) plane in the form of a boundary-layer expansion in ε. In order to match this to the core (predominantly) CP flow, it is necessary to have $\psi \sim \left(\frac{1-G}{2}\right)\varepsilon^2\eta^2$ as $\eta \to \infty$ in layer II, to lowest order in ε. Therefore, ψ must be of order ε^2. In order to determine the value of α, the convective and viscous terms in the transformed momentum Equation (4) must balance. This is achieved upon taking the value of $\alpha = 1/3$ similar to the case of a Newtonian jet [31–33] as well as a non-Newtonian jet [37]. The streamwise and transverse velocity components are now expressed in terms of the stream function as $u = \psi_z = \frac{1}{\varepsilon}\psi_\eta$ and $w = -\psi_x = -\psi_\xi + h'\psi_\eta$, respectively. Considering the fact that the streamwise velocity u in layer II must match the velocity in the core layer I, that is $u \sim (1 - G)\varepsilon\eta$ as $\eta \to \infty$, it is concluded that u is of order ε. From continuity, w is of order ε^2. Consequently, the momentum conservation Equation (4) become

$$\begin{aligned}\psi_\eta\psi_{\xi\eta} - \psi_\xi\psi_{\eta\eta} &= -\varepsilon^2\left(p_\xi - h'p_\eta\right) \\ &\quad + \varepsilon^2\psi_{\eta\eta\eta} + \varepsilon^4\left(\psi_{\xi\xi\eta} - 2h'\psi_{\xi\eta\eta} + h'^2\psi_{\eta\eta\eta} - h''\psi_{\eta\eta}\right),\end{aligned} \tag{13}$$

$$\begin{aligned}-\psi_\eta\psi_{\xi\xi} + \psi_\xi\psi_{\xi\eta} + h''\psi_\eta^2 \quad &+ h'\left(\psi_\eta\psi_{\xi\eta} - \psi_\xi\psi_{\eta\eta}\right) \\ &= -p_\eta - \varepsilon^2\left(\psi_{\xi\eta\eta} - h'\psi_{\eta\eta\eta}\right) - \varepsilon^4\left(\frac{\partial}{\partial\xi} - h'\frac{\partial}{\partial\eta}\right)^3\psi.\end{aligned} \tag{14}$$

Similarly, the boundary conditions on the free surface, i.e., at $\eta = 0$, can be rewritten as

$$\psi = 0, \tag{15}$$

$$h'p + \psi_{\eta\eta} + \varepsilon^2\left(h''\psi_\eta + h'^2\psi_{\eta\eta}\right) = -Ca^{-1}\varepsilon^4\frac{h'h''}{1 + \varepsilon^2 h'^2}, \tag{16}$$

$$p + \varepsilon^2\left(2\psi_{\xi\eta} - h'\psi_{\eta\eta}\right) + \varepsilon^4 h'\left(2h'\psi_{\eta\xi} + h''\psi_\eta - h'^2\psi_{\eta\eta}\right) = -Ca^{-1}\varepsilon^4\frac{h''}{1 + \varepsilon^2 h'^2}. \tag{17}$$

The expansion for ψ in the free-surface layer II begins with a term in ε^2, and precedes in powers of ε as

$$\psi(\xi, \eta) = \varepsilon^2\Psi_2(\xi, \eta) + \varepsilon^3\Psi_3(\xi, \eta) + \cdots. \tag{18}$$

Similarly, h is expanded as

$$h(\xi) = \varepsilon^{-1}\zeta(\xi) = h_0(\xi) + \varepsilon h_1(\xi) + \cdots. \tag{19}$$

From (18) and (19), it is concluded that the pressure p in the free-surface layer is of order ε^4. Hence,

$$p(\xi, \eta) = \varepsilon^4 P_4(\xi, \eta) + \varepsilon^5 P_5(\xi, \eta) + \cdots. \tag{20}$$

This leaves the leading-order shear-stress term in Equation (16) to balance the surface tension term, indicating that surface tension is of order $\frac{\varepsilon^2}{Ca} = \frac{1}{CaRe^{2/3}}$. We shall see in Section 3.2 that for a thin film at high Reynolds number, surface tension is of order $\frac{1}{CaRe^3}$ (refer also to Weinstein and Ruschak [2]). Consequently, for coating flow, which behaves like a thin film far downstream of the exit, surface tension becomes negligible when $\frac{1}{CaRe^3} < \frac{1}{CaRe^{2/3}} << 1$ or $Ca >> \frac{1}{Re^{2/3}} > \frac{1}{Re^3}$. So, even for a relatively moderate Reynolds number, say $Re = 64$, Ca needs only be greater than 0.06 for surface tension to be negligible. In this work, we assume a moderate surface tension with Ca remaining of order one.

Recalling that $u = \frac{1}{\varepsilon}\psi_\eta$ and $w = -\psi_\xi + h'\psi_\eta$, then the velocity components take the following for:

$$u(\xi,\eta) = \varepsilon U_1(\xi,\eta) + \varepsilon^2 U_2(\xi,\eta) + \cdots, \tag{21}$$

$$w(\xi,\eta) = \varepsilon^2 W_2(\xi,\eta) + \varepsilon^3 W_3(\xi,\eta) + \cdots \tag{22}$$

In this case, $U_1 = \Psi_{2\eta}$, $U_2 = \Psi_{3\eta}$ and $W_2 = -\Psi_{2\xi} + h_0{}'\Psi_{2\eta}$, and so on.

We next consider the leading-order solution of problem (14) and (15). Thus, to leading order in ε, Equation (13) reduce to:

$$\Psi_{2\eta}\Psi_{2\xi\eta} - \Psi_{2\xi}\Psi_{2\eta\eta} = \Psi_{2\eta\eta\eta}. \tag{23}$$

The above problem is similar to the case of symmetric free jet [31–33] with different boundary conditions. The conditions at the free surface are deduced from Equation (15) as

$$\Psi_2(\xi,0) = \Psi_{2\eta\eta}(\xi,0) = 0. \tag{24}$$

To complete the problem for Ψ_2, another boundary condition is required, which is obtained by matching the flow at the edge of the free-surface layer with the flow in the core layer downstream of the channel exit. This is shown in Appendix A. To this order, there is no interaction between the free-surface layer II and the flow in the core layer I, this latter still retaining the Couette-Poiseuille profile. The resulting condition is conveniently written here:

$$\Psi_2(\xi,\eta \to \infty) \sim \left(\frac{1-G}{2}\right)\eta^2. \tag{25}$$

We observe that the growth of the free-surface layer II can be estimated by considering the balance of the viscous and inertial terms in Equation (23) and using Equation (25), thus indicating that $\eta \propto \frac{\xi}{\Psi_2} \sim \left(\frac{2}{1-G}\right)\frac{\xi}{\eta^2}$. This leads to $y \propto \left(\frac{2}{1-G}\right)^{1/3}\left(\frac{x}{Re}\right)^{1/3}$ or

$$\delta \propto \left(\frac{2}{1-G}\frac{x}{Re}\right)^{1/3}, \tag{26}$$

which suggests that the thickness of the free-surface layer for a coating film decreases like $Re^{-1/3}$ with increasing Reynolds number. However, this also suggests that, for the same substrate speed (same Reynolds number), the free-surface layer is expected to be thicker for slot coating ($G > 0$) compared to blade coating ($G < 0$). As an interesting consequence, the surface layer grows faster for the thinner film $\left(Q < \frac{1}{2}\right)$ in slot coating than for the thicker film $\left(Q > \frac{1}{2}\right)$ in blade coating. This fundamental observation has an important practical significance in modelling coating film flow, which will be discussed in Section 3.2 when the near exit flow is matched to the thin-film flow. Finally, as we shall see, the boundary-layer growth is different along the stationary and moving walls.

We now turn to the solution of problem (23)–(25), which admits a similarity profile: $\Psi_2(\xi,\eta) = \xi^{2/3} f_2(\theta)$, where $\theta = \eta\xi^{-1/3}$ is the similarity variable. The problem for $f_2(\theta)$ is given by

$$3f_2''' + 2f_2 f_2'' - f_2'^2 = 0, \tag{27}$$

$$f_2(0) = f_2''(0) = 0, \tag{28}$$

$$f_2(\theta \to \infty) \sim \left(\frac{1-G}{2}\right)(\theta + c_1)^2, \tag{29}$$

where $c_1 = (1-G)^{-1/3} d_1$ is a constant function of G to be determined numerically. Problem (27)–(29) is solved as an initial-value problem using a fourth-order Runge–Kutta scheme (IMSL-DIVERK), coupled with a shooting technique. Equation (27) is integrated subject to conditions (28) and a guessed value $f_2'(0)$ for the slope at the origin. The slope is adjusted until reasonable matching is achieved with the asymptotic form (29) at large θ, or, more precisely, until $f_2'' \sim 1 - G$ is reached. The value of c_1 is then determined upon matching the numerical solution and its asymptotic form. Matching the numerical solution with its asymptotic behavior, we find $d_1 = 0.892$. Moreover, the slope $f_2'(0) = (1-G)^{2/3}e_1$ is related to the velocity at the film surface where we numerically find $e_1 = 1.611$. We observe that the numerical solution was carried out by scaling out the factor 1–G from the problem. The variables were rescaled by letting $f_2(\theta) = (1-G)^{1/3}g_2(t)$ and $t = (1-G)^{1/3}\theta$. In this case, problem (27)–(29) reduces to: $3g_2''' + 2g_2g_2'' - g_2'^2 = 0,\ g_2(0) = g_2''(0) = 0,\ g_2(t \to \infty) \sim \frac{(t+d_1)^2}{2}$. Pursuing the solution and the matching process to next order, the governing equation for $\Psi_3(\xi, \eta)$ becomes

$$\Psi_{2\eta}\Psi_{3\xi,\eta} + \Psi_{3\eta}\Psi_{2\xi,\eta} - \Psi_{2\xi}\Psi_{3\eta\eta} - \Psi_{3\xi}\Psi_{2\eta\eta} = \Psi_{3\eta\eta\eta}, \tag{30}$$

subject to the boundary conditions

$$\Psi_3(\xi, 0) = \Psi_{3\eta\eta}(\xi, 0) = 0, \tag{31}$$

$$\Psi_3(\xi, \eta) \sim \frac{G}{3}\eta^3 \text{ as } \eta \to \infty. \tag{32}$$

A similarity solution is also possible, namely $\Psi_3(\xi, \eta) = \xi f_3(\theta)$, resulting in the following problem for $f_3(\theta)$:

$$3f_3''' + 2f_2f_3'' - 3f_2'f_3' + 3f_2''f_3 = 0, \tag{33}$$

$$f_3(0) = f_3''(0) = 0, \tag{34}$$

$$f_3(\theta \to \infty) \sim \frac{G}{3}\left[(\theta + c_1)^3 - \frac{6}{1-G}\right] + c_2(\theta + c_1). \tag{35}$$

here $c_2 = G(1-G)^{-2/3}d_2$, where the constant $d_2 = 1.654$ is determined numerically. Here again, the numerical solution is obtained for the rescaled equation and boundary conditions:

$$3g_3''' + 2g_2g_3'' - 3g_2g_3' + 3g_2''g_3 = 0, \tag{36}$$

$$g_3(0) = g_3''(0) = 0,\ g_3(t \to \infty) \sim \frac{1}{3}\left[(t+d_1)^3 - 6\right] + d_2(t + d_1). \tag{37}$$

We find it is helpful to summarize in Table 1 the constants and numerical values that are used in the numerical results.

Table 1. Constants arising in the solution of $f_2(\theta)$ and $f_3(\theta)$.

$c_1 = \frac{d_1}{(1-G)^{1/3}}$	$c_2 = \frac{G}{(1-G)^{2/3}}d_2$	$f_2'(0) = (1-G)^{2/3}e_1$	$f_3'(0) = \frac{G}{(1-G)^{2/3}}e_2$
$d_1 = 0.892$	$d_2 = 1.654$	$e_1 = 1.611$	$e_2 = 2.85$

We recall that G is directly related to the flow rate (or coating thickness) through $G = 3(1-2Q)$. The height of the free surface is then obtained to the current order by substituting (A11) and (A18) into Equation (19), yielding

$$\zeta(x) = \varepsilon c_1 x^{1/3} + \varepsilon^2 \frac{c_2}{1-G}x^{2/3} = c_1\left(\frac{x}{Re}\right)^{1/3} + \frac{c_2}{1-G}\left(\frac{x}{Re}\right)^{2/3}. \tag{38}$$

This expression reveals clearly the intricate interplay between inertia and flow rate (coating thickness). In particular, the free-surface height $\zeta(x)$ or the thickness $\chi(x) \equiv 1 - \zeta(x)$ can exhibit an extremum depending on the value and signs of the constants involved. An important observation to make here is the absence of surface-tension effect on the film profile given by Equation (38). This is not surprising since $Ca = O(1)$, which is relatively large. In this range of capillary number, the film thickness remains essentially independent of surface tension as this has been clearly demonstrated experimentally. The reader is referred, among others, to Lee et al. [6] and Becker and Wang [7] where film thickness measurements were reported for slot coating. In their numerical simulation.

Saito and Scriven [15] examined the influence of flow parameters, such as the flow rate, the capillary number and the Reynolds number, on the shape of the meniscus in slot coating. They observed that the effect of inertia was the least evident to interpret given the non-monotonic response in the meniscus profiles as they varied the Reynolds number. A similar non-monotonic response was also reported more recently by Carvalho and Kheshgi [9] who plotted the numerically found film profiles for different capillary numbers for constant Re/Ca ratio. Incidentally, this ratio is the Ohnesorge number $Oh^{-2} \equiv \frac{Re}{Ca}$ [53], and not the "Property" number as sometime referred to in the coating literature. The non-monotonic response becomes particularly clear when the separation angle and the radius of curvature near the contact line are plotted against Re, displaying a maximum and a minimum, respectively, which is illustrated in Saito and Scriven [15].

At the free surface ($z = \zeta$), the velocity becomes

$$u(x, z = \zeta) = \varepsilon x^{1/3} f_2'(0) + \varepsilon^2 x^{2/3} f_3'(0) = \left(\frac{x}{Re}\right)^{1/3} f_2'(0) + \left(\frac{x}{Re}\right)^{2/3} f_3'(0). \tag{39}$$

We note that $f_3'(0) = G(1-G)^{-2/3} e_2$ where $e_2 = 2.85$. Both the initial slope $f_3'(0)$ and c_2 increase with positive G values, reflecting a higher order strengthening effect of the adverse pressure gradient on the flow near the free surface. In contrast, the trend is reversed for negative G, pointing to a higher order weakening effect of the favorable pressure gradient on the flow close to the free surface.

It is interesting to observe that both the free-surface height (or film thickness) and the velocity depend on $\frac{x}{Re}$ rather than x. This is also the case for the pressure along the free surface of the meniscus, which is determined by inserting Equations (18)–(20) into condition (17). To the current order, the pressure may be written as

$$Re^2 p(\xi, 0) = -\frac{2}{3}\left[\left(\frac{x}{Re}\right)^{-2/3} f_2'(0) + 2\left(\frac{x}{Re}\right)^{-1/3} f_3'(0)\right] + \frac{2}{9} We^{-1}\left(c_1\left(\frac{x}{Re}\right)^{-5/3} + \frac{c_2}{1-G}\left(\frac{x}{Re}\right)^{-4/3}\right). \tag{40}$$

Here We = Ca/Re is the Weber number. Thus, near the exit, the pressure behaves like $x^{-2/3}$ in the absence of surface tension, which is different from the $x^{-1/2}$ behavior reported by Aidun [14]. Moreover, the strength of the pressure singularity depends intricately on inertia, surface tension and the flow rate (or upstream pressure gradient). In the absence of surface tension, the singularity weakens with inertia and the flow rate for both slot and blade coating. As mentioned earlier. Chang et al. [8] observed that viscous and surface-tension effects become negligible for Re > 20. One expects the pressure to be vanishingly small along the free surface. In this case, Equation (40) gives an estimate of pressure on the order of $Re^{-4/3}$.

Although surface tension can play a significant role, not only by becoming dominant near the exit, but also by altering the force behavior as we shall see when we discuss the normal stress distribution shortly. For now, we consider the effects of inertia and flow rate on the film height, the velocity and the pressure along the film surface.

Figure 2 depicts the dependence of the film height (Figure 2a), the velocity (Figure 2b) and the pressure (Figure 2c) along the free surface, on the flow rate (or pressure gradient) and inertia. It is evident from the figure that as the dimensionless flow rate Q increases, the film height in Figure 2a drops as expected, signalling a thicker film and a longer relaxation distance. This behavior agrees well with existing numerical predictions as confirmed by comparing with Saito and Scriven [15] and Carvalho and Kheshgi [9]. Very close to the exit, the film height (Figure 2a), the velocity (Figure 2b) as well as the pressure (Figure 2c) increase sharply and monotonically with distance, for any flow rate, indicating a strong contraction. This behavior is the result of the dominance of the $x^{1/3}$ term in Equations (38) and (39) near the origin. Although the height and the streamwise velocity components are continuous at the exit (as they remain equal to zero immediately before x = 0), the flow behavior is intrinsically singular through the surface slope, transverse velocity and stress components, and is well illustrated in the pressure behavior near x = 0 (Figure 2c). We observe from Figure 2a that the separation angle is consistently 90 degrees for the order of capillary and Reynolds numbers considered here. The separation angle is the angle between the normal to the stationary slot or blade wall and the normal to the surface at the contact line (x = z = 0). Further downstream, only the film at low flow rate, typically depicting the behavior of a film in slot coating, continues to grow, resulting in a smaller coating thickness. While the film height grows monotonically with distance for a drag film (Q = 0.5), the free-surface height and velocity for blade coating (Q > 0.5) exhibit a maximum before decaying, indicating a local expansion. However, the presence of the maximum should be interpreted with some caution. The non-monotonic behavior suggests a relatively strong influence of the higher-order $x^{2/3}$ term, which should not dominate the leading-order $x^{1/3}$ term if the expansions (38) and (39) are to remain uniformly valid. However, this does not seem to be the case for the range of flow rates reported in Figure 2. The expansion displayed for the thicker film in blade coating (Q > 0.5) appears to be physically real as can be established from the numerical values of the various terms. However, the validity of (38) and (39) also depends on the thickness of the free-surface layer II (boundary layer along the free surface), which is assumed to be small compared to the free-surface height. The curves in Figure 2a are based on the near-exit solution (38). The error is $O\left(\frac{x}{Re}\right)^{2/3}$, except for the Q = 0.5 (G = 0), which is $O\left(\frac{x}{Re}\right)^{1/3}$ since $c_2(G=0) = 0$ as per Table 1. Despite the relatively large error, the case Q = 0.5 is included to illustrate the trend as Q increases. The more accurate prediction is given in Section 3.2 when the far field is examined. The pressure profiles in Figure 2c are essentially monotonic; the pressure increases and decays asymptotically to zero. A slight overshoot ahead of the decay is observed for the thicker coating films. Despite the clear dependence of the elongation rate, the second term on the left of (17) that yields the wild non-monotonic dynamics in Figure 2b, the monotonicity of the pressure is mainly caused by surface tension. Although this behavior is reported for We = 1, the same trend is predicted essentially for any moderate level of surface tension. This behavior sharply contrasts that of the normal stress, which depends strongly on surface tension as we shall now see.

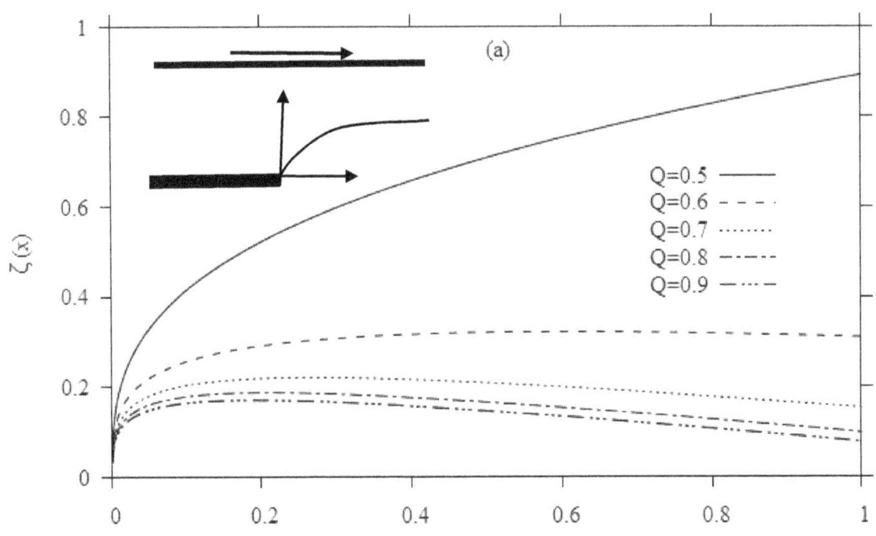

Figure 2. *Cont.*

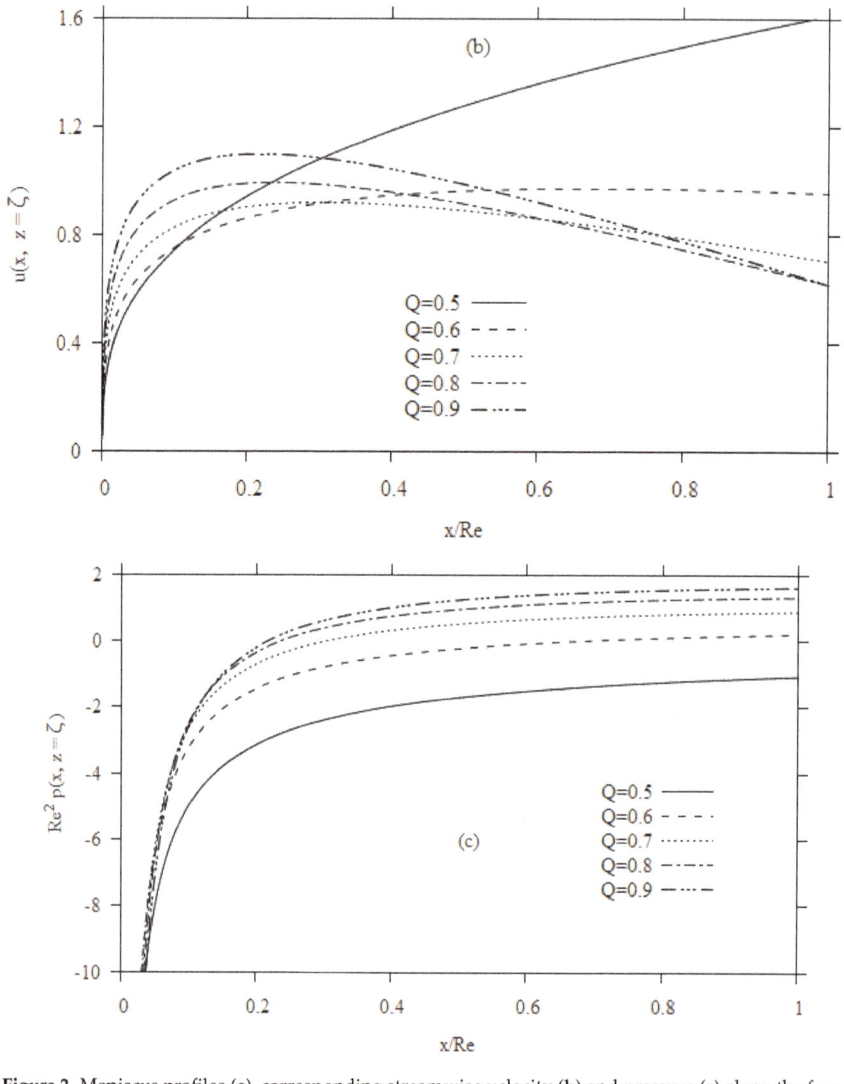

Figure 2. Meniscus profiles (**a**), corresponding streamwise velocity (**b**) and pressure (**c**) along the free surface for various coating thickness or flow rates for We = 1. The drag-out case (Q = 0.5) is also included for reference.

The influence of flow rate and surface tension on the (negative) primary normal stress difference $\tau_{tt} - \tau_{nn}$ (Hoop stress) along the free surface is displayed in Figure 3 for Re = 10. Both the cases of slot coating (Q < 0.5) and blade coating (Q > 0.5) are illustrated in the presence (Figure 3a) and the absence (Figure 3b) of surface tension. Intricate non-monotonic behavior can be expected for blade coating, resulting from the dynamics already observed in Figure 2a,b, thus allowing the possibility of extrema, and the waviness displayed in the free surface and velocity profiles. Indeed, Figure 3a shows that the normal stress exhibits a relatively strong localized maximum followed by an asymptotic decay towards zero further downstream. This behavior is reminiscent of that reported by Bajaj et al. [19] for viscoelastic creeping flow. They examined numerically the influence of Weissenberg number and viscosity ratio on the normal stress along the free surface for dilute polymer solutions. Their figures also show a local maximum for $\tau_{tt} - \tau_{nn}$ close to the die exit followed by a decay similar to Figure 3a. A major difference, however, is worth noting here: the plots of Bajaj et al. [19] do not clearly display the singularity at the origin. They do assess, on the other hand, the existence and strength of the singularity by examining the behavior of the rate-of-strain components. As to the crucial role of surface tension, especially near the exit, it is demonstrated in Figure 3b, where the Hoop stress is plotted in the absence of surface tension. Comparison between Figure 3a,b clearly indicates that the dynamics exhibited in the stress near the origin in Figure 3a are caused by surface tension. Figure 3b shows that the Hoop stress increases essentially monotonically, exhibiting a very weak overshoot for the thicker coating film before decaying to zero.

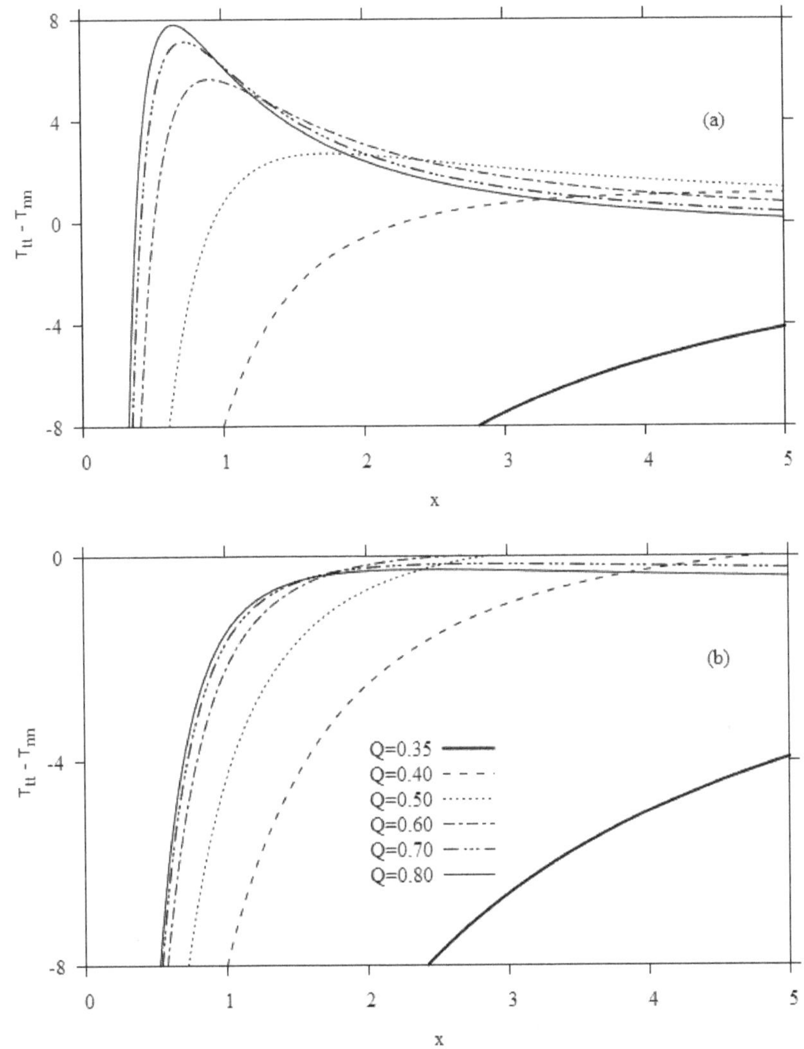

Figure 3. Influence of flow rate on the normal-stress difference along the film surface as function of distance at Re = 10, and (**a**) We = 5 and (**b**) in the absence of surface tension.

Figure 4 displays the variation with the inclination angle for Re =10 and We = 0.1 of the free-surface curvature (Figure 4a) and pressure along the free surface (Figure 4b). The abscissa is the angle of inclination of the normal to the free surface from the z direction, as depicted in the inset of Figure 4b. We see that the curvature rises with increasing flow rate. This is expected as the meniscus in slot coating contracts more than in blade coating. The results in Figure 4 are overall qualitatively similar to the numerical results of Saito and Scriven [15] and Lee et al. [21], when compared to their figures for the slot coating flow. The agreement is particularly obvious when comparison is made with the high Ca curves of Saito and Scriven [15]. Similar to their curve for Ca = 2, we also find that the surface curvature exhibits a change in concavity at low inclination angle for any flow rate. At low capillary number, Saito and Scriven [15] as well as Lee et al. [21] predict a linear growth, which remains monotonic for very low capillary number. Saito and Scriven observed that a large portion of the curvature is constant when the capillary number is very small; but the curvature varies more and more rapidly, displaying a maximum, as Ca increases, a behavior that is also reflected in the rapid variation and the maxima we see in Figure 4a, especially for blade coating (Q > 0.4). Simultaneously, the pressure in Figure 4b displays a similar linear growth for small inclination angle as in the earlier numerical studies, but tends to increase sharply for larger angles as the singularity is approached near $\phi = 90$ degrees. This behavior is also captured by Lee et al. [21].

The similarity between our high-Re results and the high-Ca results at Re = 0 of Saito and Scriven [15] highlights the crucial role of surface tension. In this regard, it is helpful to examine the current predictions for the curvature relative to a flow with higher surface tension. For Ca < 1, the shape of the meniscus near the separation line has often been approximated as a static meniscus, which was deduced by Ruschack (1976) using a quasi-static approach where the shape of the meniscus is the arc of a circle of constant curvature as predicted by Landau and Levich (1942). The theory imposes an upper bound on the meniscus curvature as it cannot exceed $\frac{2}{1-Q}$ (when using our dimensionless notations). Based on the results in Figure 4a, this criterion appears to be plausible at best for the lowest flow rate considered here (Q = 0.4). This observation corroborates well that of Saito and Scriven [15] who found that the quasi-static assumption is valid only for large surface tension, which is not the case here. The convective (dynamic) effects are simply too dominant for Ruschak's approximate to approximation hold.

3.2. *The Coating Profile near and Far from the Exit*

In order to capture the coating profile at any location downstream of the exit, we exploit the flow structure as the free-surface layer II grows and invades the entire film region, at a critical location $x = x_c$. Downstream of this location, the flow becomes of the boundary-layer or thin-film type. We exploit this simplification and formulate the flow using a thin-film approach, and match it at $x = x_c$ with the flow obtained earlier near the exit. Far from the exit for $x > x_c$, the conservation equations, adherence and no-penetration conditions at the moving substrate z = 1, and the kinematic and dynamic conditions at the free surface $z = \zeta(x)$ reduce to

$$u_x + w_z = 0,\ \mathrm{Re}(uu_x + wu_z) = -\mathrm{Re}p_x + u_{zz},\ p_z = 0, \tag{41}$$

$$u(x, z = 1) = 1,\ \ w(x, z = 1) = 0, \tag{42}$$

$$w(x, \zeta) = u(x, \zeta)\zeta'(x), \tag{43}$$

$$u_z(x, \zeta) = 0\ p(x, \zeta) = \frac{\zeta'(x)}{\mathrm{ReCa}}. \tag{44}$$

We proceed by using a Karman–Pohlhausen approach, and integrate the momentum equation across the film, using condition (43) and noting that $p_x(x) = \frac{\zeta''(x)}{ReCa}$, to obtain

$$\frac{d}{dx}\int_{\zeta}^{1} u^2 dz = -\frac{\zeta''(1-\zeta)}{ReCa} + Re^{-1}u_z(x, z = 1). \tag{45}$$

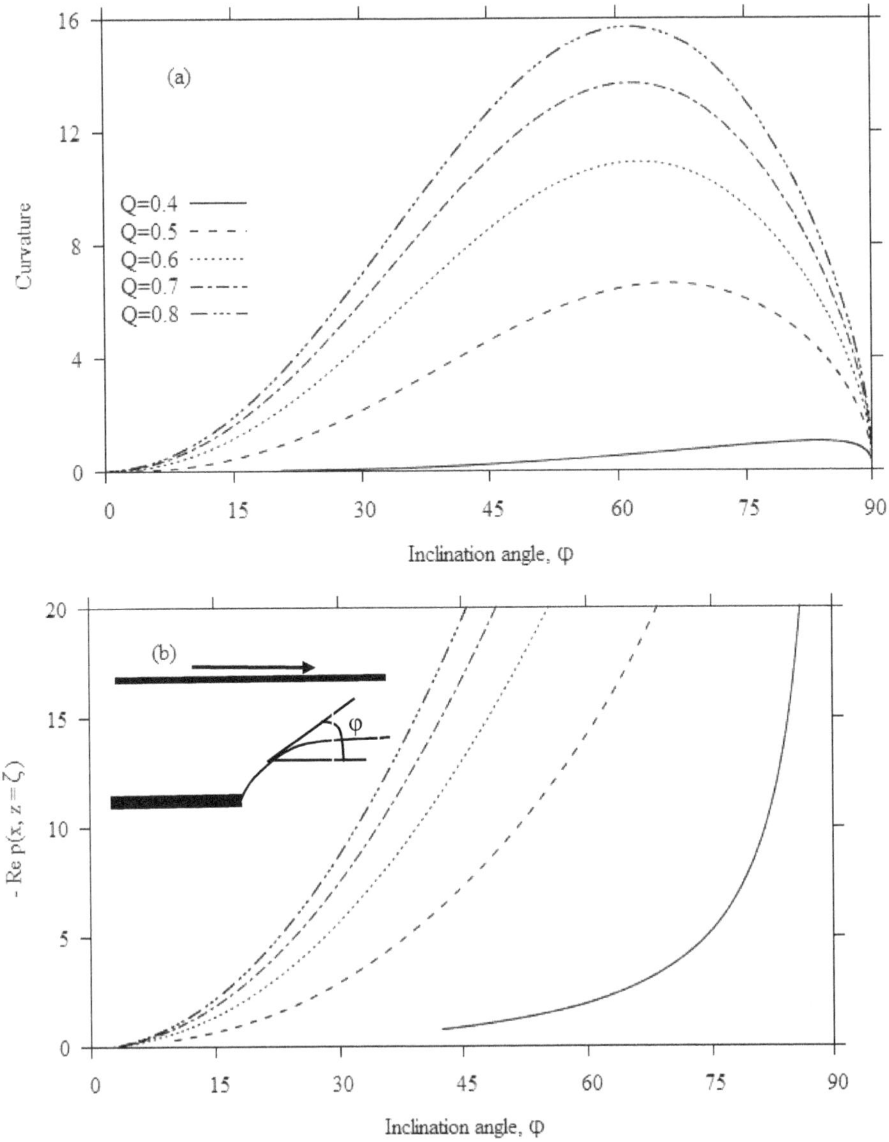

Figure 4. Variation of the meniscus curvature (**a**) and the free-surface pressure (**b**) with the free-surface inclination angle for various flow rates for slot coating (Q ≤ 0.5) and blade coating (Q ≥ 0.5), for Re = 10 and We = 0.1.

Different levels of accuracy have been adopted in the literature for the velocity profile across the film in the presence of inertia, ranging from the simplest parabolic, cubic and quartic profiles [2] to spectral expansions [54,55]. Another alternative would be the use of the asymptotic approach of Higgins and Scriven [40] for visco-capillary flow, which is based on a small departure from the plug flow that prevails far downstream. For our purpose here, we adopt the parabolic profile, which is the simplest and most commonly used, to obtain the following velocity distribution in terms of the thickness $\chi(x) \equiv 1 - \zeta(x)$ as

$$u(x,z) = -\frac{3}{2}\frac{\chi_\infty - \chi(x)}{\chi^3(x)}\left[(1-z)^2 - 2\chi(x)(1-z)\right] + 1, \tag{46}$$

where $\chi_\infty = \chi(x \to \infty) = Q$ is the coating film thickness far downstream. The velocity along the free surface is then $u(x, z = \zeta) = \frac{1}{2}\left(3\frac{\chi_\infty}{\chi} - 1\right)$. Substituting (45) into Equation (46) leads to the following equation for the thickness of the coating meniscus:

$$\frac{5}{\mathrm{ReCa}}\chi^3\frac{d^2\chi}{dx^2} - \left(\chi^2 - 6\chi_\infty^2\right)\frac{d\chi}{dx} + \frac{15}{\mathrm{Re}}(\chi - \chi_\infty) = 0. \tag{47}$$

This equation requires two boundary conditions, which can be imposed as initial conditions at the location $x = x_c$, where the near-exit profile given by (38) matches the far-exit profile dictated by (47). For this, we choose to match the surface height and its slope. This leaves x_c as unknown, which we determine by matching the concavity. In the absence of surface tension, only the height and slope need to be matched. In this case, (47) admits an analytical solution:

$$\chi^2 - \chi_c^2 + 2\chi_\infty(\chi - \chi_c) - 10\chi_\infty^2 \ln\left(\frac{\chi - \chi_\infty}{\chi_c - \chi_\infty}\right) = \frac{30}{\mathrm{Re}}(x - x_c), \tag{48}$$

where $\chi_c \equiv \chi(x = x_c)$ is the thickness at x_c. In this case, upon matching the thickness and the slope, x_c and χ_c are determined by solving the following two equations:

$$\chi_c = 1 - c_1\left(\frac{x_c}{\mathrm{Re}}\right)^{1/3} - \frac{c_2}{(1-G)}\left(\frac{x_c}{\mathrm{Re}}\right)^{2/3}, \tag{49}$$

$$45\frac{\chi_c - \chi_\infty}{\chi_c^2 - 6\chi_\infty^2} + c_1\left(\frac{x_c}{\mathrm{Re}}\right)^{-2/3} + 2\frac{c_2}{(1-G)}\left(\frac{x_c}{\mathrm{Re}}\right)^{-1/3} = 0. \tag{50}$$

It is not difficult to deduce that the matching location turns out to be simply proportional to Re, whereas the corresponding thickness is independent of Re. Both quantities are of course functions of the flow rate (minimum coating thickness).

Figure 5 depicts the composite thickness (Figure 5a) and the free-surface height (Figure 5b) profiles with downstream position for a typical minimum blade coating thickness range ($0.5 \le Q = \chi_\infty \le 0.9$) in the absence of surface tension. Figure 5a depicts the typical monotonic decrease in thickness with distance. The profiles appear to saturate as Q approaches one. The matching location (see Table 2) x_c for each profile coincides with the intersection of the free-surface layer with $z = 1$, which suggests that matching occurs further downstream for a thicker film, which is reflected in Figure 5b where the free-surface (boundary) layer is plotted along with the surface height. Again, the matching location coincides with the free-surface layer II reaching the moving substrate ($z = 1$). We observe that G drops from 0 to −2.4 when Q increases from 0.5 to 0.9, indicating that a drop in pressure in the blade region yields a thicker coating film and thinner free-surface layer. The thinning of the free-surface layer with (negatively) increasing pressure gradient is expected as the favorable pressure difference effectively contributes additional flow inertia in blade coating, as illustrated by Iliopoulos and Scriven [17].

Table 2. Influence of flow rate on matching location and corresponding film thickness between the flow near the exit and the flow far from the exit.

x_c/Re	χ_c	χ_∞=Q	G
0.0173	0.7690	0.5	0
0.0206	0.825	0.6	−0.6
0.0442	0.824	0.7	−1.2
0.1539	0.814	0.8	−1.8
0.5824	0.865	0.9	−2.4

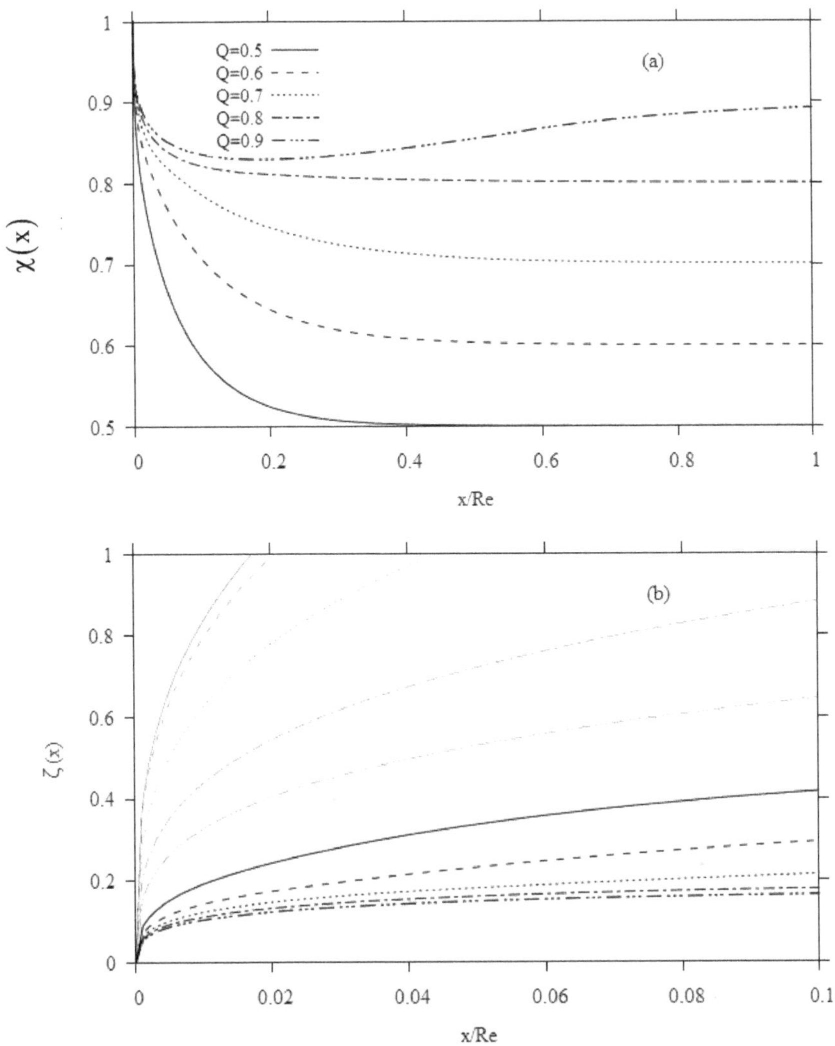

Figure 5. Composite thickness profiles for near and far distance downstream of the exit (**a**), and corresponding free-surface heights close to the exit (**b**) for various minimum coating thickness (Q ≥ 0.5). Also shown in (**b**) are the corresponding free-surface layer heights in lighter plots.

An estimate of the error resulting from the use of the parabolic profile can be obtained by examining the shear stress at the moving substrate. The shear stress along the wall far downstream is generally deduced from

$$\tau_{xz}(x \geq x_c, z = 1) = 3\frac{\chi(x) - \chi_\infty}{\chi^2(x)}. \tag{51}$$

We consider the case of drag flow, in the absence of pressure gradient. In this case, $\tau_{xz}(x, z = 1) = \frac{3}{2}\frac{2\chi(x)-1}{\chi^2(x)}$. Matching the shear stress yields the following equation for the film thickness at $x = x_c$: $2\chi_c^2 - 6\chi_c + 3 = 0$, leading to the only admissible root: $\chi_c = \frac{3-\sqrt{3}}{2} = 0.634$, which is close to 0.769 (from Table 2) but not exactly the same as a result of the parabolic approximation (46), suggesting an error of 20%. Consequently, the thin-film Equation (47) is based on a crude approximation of the parabolic velocity profile (see discussion by Weinstein and Ruschak [2]). Higher-order polynomials or spectral representation may be used for a more accurate description [54,55]. Our aim here, however, is to demonstrate how the current asymptotic theory, which is valid upstream of the exit and downstream close the exit, can be matched at some location with the thin-film flow to provide a formulation to predict the flow anywhere. A more thorough approach involves adopting the asymptotic flow as initial condition for a computational (finite-element) implementation, thus avoiding having to deal numerically with the singularity at the exit.

4. The Flow in the Core Layer I

The flow in the core layer (region I in Figure 1) remains predominantly of inviscid rotational character. For this reason, the flow retains predominantly the CP profile since there is little viscous mechanism for it to develop as it approaches the exit and moves beyond the exit. The core layer is central in the current formulation as it is matched at the edge of each of the three boundary layers: the free-surface layer (region II), the stationary wall layer near a slot die or blade (region III), and the moving-wall layer near the substrate or web (region IV). Since the treatment is similar to earlier studies on the symmetric jet, only a summary of the formulation is outlined here. In fact, the present core formulation reduces to that of Tillett [31] by setting $G = -1$. The formulation of Khayat [37] is also recovered in the limit of a Newtonian fluid. Since the core flow does not satisfy adherence, the core formulations for the current wall jet and the symmetric jet are not affected by the nature of the boundary at $z = 1$, irrespective of whether it is a solid line (for the wall jet) or a symmetry line (for the symmetric jet).

The core layer is divided into two different regions: upstream ($x < 0$) and downstream ($x > 0$) of the exit at $x = 0$. The solution of problems (4) is sought in each region separately. As in the case of the symmetric free jet [31,37], we determine ψ and p by considering a correction to the base Couette-Poiseuille flow resulting from its interaction with the free-surface layer (II). In this case, the dominant corrections for both the stream function and the pressure turn out to be of $O\left(\mathrm{Re}^{-1}\right)$ so that

$$\psi(x,z) = \psi_0(x,z) + \mathrm{Re}^{-1}\overline{\psi}(x,z), \quad p(x,z) = \mathrm{Re}^{-1}\overline{p}(x,z). \tag{52}$$

Here, we recall that $\psi_0 = \left(\frac{1-G}{2}\right)z^2 + \frac{G}{3}z^3$ is the Couette-Poiseuille stream function given in (1). In this case, $\overline{\psi}(x,z)$ represents the deviation from the base flow due to its interaction between the core and free-surface layer. The character of the base flow is similar to the laminar flow of a free jet [31] and channel or tube flow with constriction [51,52,56] at high Reynolds number. In such cases, as well, the fully developed (Poiseuille) profile is the flow to leading order. When examining the flow with severe constriction, Smith [56] expanded (1) as in (52) above. Although Smith's leading order terms in (1) still satisfy the inviscid equations of motion, they do not exactly correspond to fully developed flow as in (52). Smith adopted the free-streamline theory, and obtained the proper inviscid limiting form of the Navier–Stokes equations [56].

We observe that, given the ellipticity of the governing equations in the streamwise direction, the departure from the Couette–Poiseuille profile extends upstream to include the region $x < 0$, causing the upstream influence. Substituting (52) into (4) leads to

$$u_0\overline{\psi}_{xz} - u_{0z}\overline{\psi}_x = -\overline{p}_x + 2G,\ u_0\overline{\psi}_{xx} = \overline{p}_z. \tag{53}$$

Upon eliminating the pressure and noting that $\overline{w} = -\overline{\psi}_x$, one deduces the following problem for $\overline{w}(x,z)$ over the ranges $-\infty < x < \infty$ and $0 \leq z \leq 1$:

$$\overline{w}_{xx} + \overline{w}_{zz} - \frac{u_0''}{u_0}\overline{w} = 0, \tag{54}$$

$$\overline{w}(x \to -\infty, z) \sim 0, \tag{55}$$

$$\overline{w}(x < 0, z = 0) = \overline{w}(-\infty < x < \infty, z = 1) = 0, \tag{56}$$

$$\overline{w}(x > 0, z \to 0) \sim \frac{2G}{1-G}. \tag{57}$$

In addition, $\overline{w}$ must remain bounded as $|x| \to \infty$. Condition (57) is deduced from (A22), which is obtained from the matching between the core and the free-surface layers. The solution of problem (54)–(57) will now be examined separately for $x < 0$ and $x > 0$.

For $x < 0$, it is not difficult to see that problem (54)–(57) admits a general solution of the following form [31,37]:

$$\overline{w}(x < 0, z) = -\sum_{n=1}^{\infty} A_n e^{\beta_n x} V_n(z). \tag{58}$$

Here, $V_n(z)$ and β_n are orthogonal shape functions and corresponding eigenvalue governed by:

$$V_n'' + \left(\beta_n^2 - \frac{u_0''}{u_0}\right)V_n = 0. \tag{59}$$

$$V_n(0) = 0, \tag{60}$$

$$V_n(1) = 0. \tag{61}$$

The solution of (59)–(61) is obtained numerically subject to the additional normalization condition $V_n'(0) = 1$. Figure 6 depicts the profiles of the shape functions for the first five modes for Q = 0.35. We observe that β_n increases with n (see Appendix B). For large n, problem (59)–(61) tends to a Sturm-Liouville problem with eigenvalues $\beta_n \sim n\pi$ and trigonometric shape functions $V_n(z) \sim \frac{\sin(n\pi z)}{n\pi}$. The profiles in Figure 6 clearly acquire the trigonometric character for larger n values. The profiles for other Q values present the same trend.

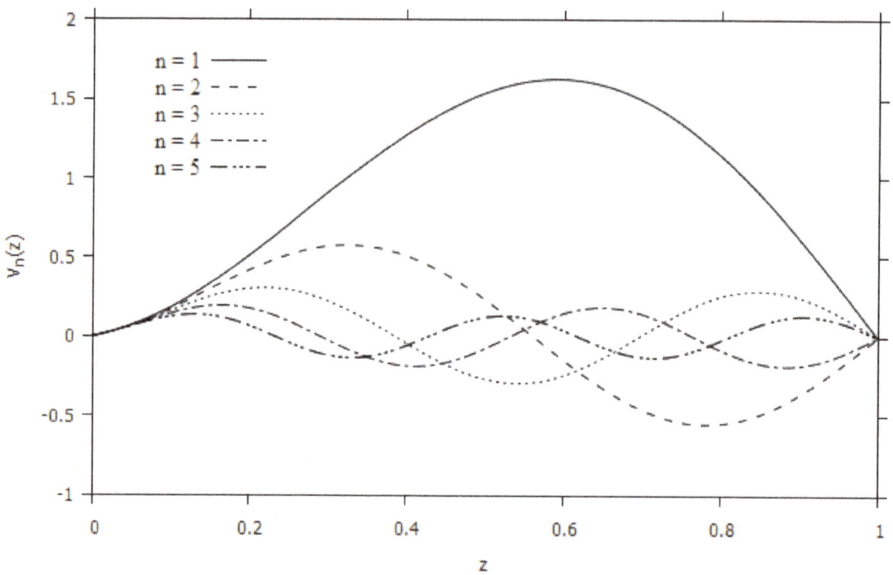

Figure 6. Eigenfunction distributions for the first five modes at Q = 0.35.

The stream function, velocity components and pressure to $O(Re^{-1})$ are then given by

$$\psi(x<0,z) = (1-G)\frac{z^2}{2} + G\frac{z^3}{3} + Re^{-1}\sum_{n=1}^{\infty}\frac{A_n}{\beta_n}e^{\beta_n x}V_n(z), \tag{62}$$

$$u(x<0,z) = (1-G)z + Gz^2 + Re^{-1}\sum_{n=1}^{\infty}\frac{A_n}{\beta_n}e^{\beta_n x}V'_n(z), \tag{63}$$

$$w(x<0,z) = -Re^{-1}\sum_{n=1}^{\infty}A_n e^{\beta_n x}V_n(z), \tag{64}$$

$$p(x\le 0,z) = Re^{-1}\left[2Gx - \sum_{n=1}^{\infty}\frac{A_n}{\beta_n}e^{\beta_n x}\left(\left[(1-G)z+Gz^2\right]V'_n - \left[(1-G)+2Gz\right]V_n\right)\right]. \tag{65}$$

We observe that the pressure at the stationary die or blade (z = 0) is simply $Rep(x \le 0, z = 0) = 2Gx$, corresponding to fully-developed flow. Additional correction is determined when we examine the boundary layer near the stationary wall in Section 5. In this regard, we emphasize that the solution in the core layer is predominantly inviscid rotational. Consequently, the flow field given by (62)–(65) is not expected to satisfy adherence at the moving substrate (z = 1) and the stationary slot or blade wall (z = 0). This becomes particularly evident when examining the velocity at the walls: $u(x<0,z=0) = Re^{-1}\sum_{n=1}^{\infty}\frac{A_n}{\beta_n}e^{\beta_n x} \neq 0$ and $u(x<0,z=1) = 1 + Re^{-1}\sum_{n=1}^{\infty}\frac{A_n}{\beta_n}e^{\beta_n x}V'_n(z=1) \neq 1$. This signals the presence of the slip layers III and IV in Figure 1, which will be dealt with in Sections 5 and 6, respectively.

The coefficients A_n are determined by matching the inside and outside flows at the exit, x = 0, and will therefore be determined after the outside flow is examined next. Thus, for x > 0, the solution of problem (54)–(57) takes the following form:

$$\overline{w}(x>0,z) = 2\frac{G}{1-G}V_0(z) + \sum_{n=1}^{\infty} A_n e^{-\beta_n x} V_n(z). \tag{66}$$

The additional term $V_0(z)$ contributes to the matching of the core flow with the flow in the free-surface layer. It satisfies the following equation and boundary conditions:

$$V_0'' - \frac{u_0''}{u_0}V_0 = 0,\ V_0(0) = 1,\ V_0(1) = 0, \tag{67}$$

which admits an analytical solution:

$$V_0(z) = 2kz(1+kz)\ln\left|\frac{z}{1+kz}\right| + 1 + 2kz + z(1+kz)\left[2k\ln|1+k| - \frac{1+2k}{1+k}\right], \tag{68}$$

where $k = \frac{G}{1-G}$. It is useful to examine the first contribution in (66) as it is the term that survives with increasing downstream distance. The profiles of $\frac{G}{1-G}V_0(z)$ are shown in Figure 7 for $0.4 \le Q \le 0.8$. A couple of interesting distinctions can be made between slot coating ($Q \le 0.5$) and blade coating ($Q \ge 0.5$). For slot coating, $\frac{G}{1-G}V_0(z)$ decreases monotonically with z and remains positive. In contrast, it exhibits a minimum and remains negative for blade coating. This behavior suggests that the vertical flow is predominantly towards the moving substrate in slot coating, and away from it in blade coating.

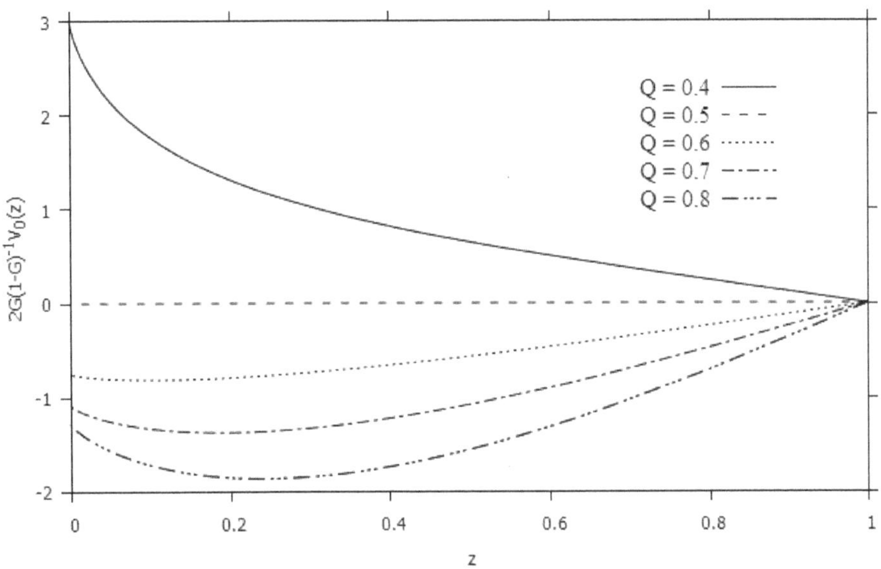

Figure 7. Influence of flow rate on the profile of the dominant term in (66) or $V_0(z)$.

As mentioned above, the coefficients A_n are found by matching the flow at the exit. For this purpose, equating the expressions (58) and (66) at $x = 0$ yields

$$V_0(z) = -\frac{1-G}{G}\sum_{n=1}^{\infty} A_n V_n(z), \tag{69}$$

which is a spectral representation of $V_0(z)$ in terms of the orthogonal shape functions $V_n(z)$. Upon projecting and noting that $\langle V_n V_0 \rangle = \frac{1}{\beta_n^2}$, where the brackets < > denote integration over the interval $0 \le z \le 1$, we have

$$A_n = -\left(\frac{G}{1-G}\right)\frac{\langle V_n V_0 \rangle}{\langle V_n^2 \rangle} = -\left(\frac{G}{1-G}\right)\frac{1}{\beta_n^2 \langle V_n^2 \rangle}. \tag{70}$$

A useful relation is obtained by first noting from (68) that $V_0'(1) = -(1-G)$. Consequently, upon differentiating (69) and evaluating it at z = 1, we have

$$\sum_{n=1}^{\infty} A_n V_n'(1) = G. \tag{71}$$

Finally, the flow field downstream of the die or blade exit become

$$\psi(x>0,z) = (1-G)z^2 + G\frac{z^3}{3} + Re^{-1}\left[-2\frac{G}{1-G}xV_0(z) + \sum_{n=1}^{\infty}\frac{A_n}{\beta_n}e^{-\beta_n x}V_n(z)\right], \tag{72}$$

$$u(x>0,z) = (1-G)z + Gz^2 + Re^{-1}\left[-2\frac{G}{1-G}xV_0'(z) + \sum_{n=1}^{\infty}\frac{A_n}{\beta_n}e^{-\beta_n x}V_n'(z)\right], \tag{73}$$

$$w(x>0,z) = Re^{-1}\left[2\frac{G}{1-G}V_0(z) + \sum_{n=1}^{\infty}A_n e^{-\beta_n x}V_n(z)\right], \tag{74}$$

$$p(x \ge 0,z) = -Re^{-1}\sum_{n=1}^{\infty}\frac{A_n}{\beta_n}e^{-\beta_n x}\left(\left[(1-G)z + Gz^2\right]V_n' - \left[(1-G) + 2Gz\right]V_n\right). \tag{75}$$

Here again, we see that the core solution does not ensure adherence at the moving substrate. This is clearly reflected in the value of the streamwise velocity at the moving substrate: $u(x>0,z=1) = 1 + Re^{-1}\left[2Gx + \sum_{n=1}^{\infty}\frac{A_n}{\beta_n}e^{-\beta_n x}V_n'(1)\right] \neq 1$, indicating the existence of a slip layer near the substrate, which will be discussed in Section 6. Interestingly, we can further see that the streamwise elongation rate component is also not zero since $u_x(x>0,z=1) = Re^{-1}\left[2G - \sum_{n=1}^{\infty}A_n e^{-\beta_n x}V_n'(1)\right] \neq 0$. In particular, using (71), we see that the exit $u_x(x=0,z=1) = GRe^{-1}$. These expressions will help estimate the thickness of the slip layer IV (see Section 6).

Unlike the velocity, expressions (65) and (75) for the core pressure turn out to be valid everywhere: in the core layer as well as at the stationary and the moving walls and near the free surface. The uniform validity of the pressure can be shown by determining the composite solution across the film [31,37]. Physically, the reason for the validity of the core pressure in the boundary layers is the result of the hydrostatic nature of the pressure in each layer and its invariability across each layer. Interestingly, while the core pressure given by (65) and (75) yields a correction to the Poiseuille level at any location, this correction is not felt at the stationary slot or blade wall to the current order. In fact, setting z = 0 in (65) and recalling that $V_n(0) = 0$ from (60), we obtain the linear behavior: $p(x \le 0, z=0) = 2Re^{-1}Gx$. The departure from the Poiseuille level will be established once we examine the pressure in the slip layer III near the stationary wall in the next Section 5. In contrast, the departure becomes increasingly evident for z > 0 as one approaches the moving substrate for a given x position. Indeed, evaluating the pressure at the moving substrate (75) gives

$$Rep(x \le 0,z=1) = 2Gx - \sum_{n=1}^{\infty}\frac{A_n}{\beta_n}e^{\beta_n x}V_n'(1), \tag{76}$$

$$\text{Rep}(x \geq 0, z = 1) = -\sum_{n=1}^{\infty} \frac{A_n}{\beta_n} e^{-\beta_n x} V_n'(1). \tag{77}$$

These expressions clearly signal the departure from the Poiseuille level. Figure 8 displays the pressure distribution along the moving substrate, typically illustrating the case for slot (Figure 8a) and blade (Figure 8b) coating. For each flow rate, the figure shows the linear behavior of the pressure starting far upstream, dictated by the 2Gx term in (76). Upon approaching the exit, the pressure experiences an exponential deviation, which is sustained beyond the exit as the pressure continues to decay and relaxes eventually to zero. The rate of relaxation is slower for a thinner film in slot coating and a thicker film in blade coating. A similar behavior is also predicted for the pressure distribution along the centerline of a symmetric planar jet as shown by Khayat [37] for the Newtonian limit (n = 1). Incidentally, the pressure continues to decrease for a shear-thinning fluid.

The departure from the Poiseuille level may be estimated by examining the pressure at the exit from (76)–(77), namely $\text{Rep}(x = 0, z = 1) = -\sum_{n=1}^{\infty} \frac{A_n}{\beta_n} V_n'(1)$. This expression, however, is not particularly illuminating as the dependence on the flow rate is only implicit through the value of A_n and β_n, as well as $V_n'(1)$, as indicated in the table in Appendix B. In contrast, the influence of Q on the departure in the pressure gradient at the exit can be determined explicitly. Indeed, since the pressure gradient must match at the exit, then upon equating $p_x(x = 0, z = 1)$ from (76) and (77), we find that

$$\text{Rep}_x(x = 0, z = 1) = \sum_{n=1}^{\infty} A_n V_n'(1) = G = 3(1 - 2Q), \tag{78}$$

which explicitly reflects the dependence of the departure from the constant upstream slope $2G = 6(1 - 2Q)$. Thus, the pressure gradient along the moving substrate evolves from $2G = 6(1 - 2Q)$ far upstream and reduces to $G = 3(1 - 2Q)$ at the exit, which is also reflected Figure 8. Only the drag flow does not experience any change in the (zero) magnitude of the pressure gradient. Incidentally, (78) shows that (71) is equivalent to matching the pressure gradient at x = 0.

Interestingly, while the core pressure near the moving substrate exhibits a departure from the Poiseuille level, it retains the same behavior near the stationary die or blade. Another interesting observation worth making is the absence of transverse pressure gradient near both the stationary and moving walls. This is easily confirmed by showing from (65) and (75) that $p_z(x, z)$ vanishes at z = 0 an z = 1. The absence of a transverse pressure gradient suggests, in turn, the existence of a boundary or slip layer near each wall. The flow structure in each slip layer is examined in the next two sections.

5. The Flow Near the Stationary Slot and Blade Walls (Slip Layer III)

The flow structure in the boundary or slip layer III near the die or blade wall, the lower-wall layer, is examined in this section. In this layer, similar to the free-surface layer I, the transverse coordinate near the boundary will be taken as $y = z = \varepsilon\bar{\eta}$, where ε is the same small parameter used before. To examine the structure of layer III upstream of the channel exit, the near-wall coordinates are introduced as $x = \xi$, and $z = \varepsilon\bar{\eta}$. Similar to the free-surface layer, matching with the core flow upstream of the channel exit shows that $\bar{\psi} \sim (1 - G)\frac{y^2}{2} = (1 - G)\frac{\varepsilon^2\bar{\eta}^2}{2}$ as $\bar{\eta} \to \infty$. Therefore, $\bar{\psi}$ must be of $O(\varepsilon^2)$ close to the lower wall. In this case, the transformed Equation (4), along with conditions (10) and (11) become (dropping the overbar):

$$\begin{gathered}
\psi_\eta \psi_{\xi\eta} - \psi_\xi \psi_{\eta\eta} = -\varepsilon^2 p_\xi + \varepsilon^2 \psi_{\eta\eta\eta} + \varepsilon^4 \psi_{\xi\xi\eta}, \\
\psi_\eta \psi_{\xi\xi} - \psi_\xi \psi_{\xi\eta} = p_\eta + \varepsilon^2 \psi_{\xi\eta\eta} + \varepsilon^4 \psi_{\xi\xi\xi}, \\
\psi(\xi, \eta = 0) = \psi_\eta(\xi, \eta = 0) = 0, \\
\psi(\xi \to -\infty, \eta) \sim (1 - G)\frac{\varepsilon^2\eta^2}{2} + G\frac{\varepsilon^3\eta^3}{3}.
\end{gathered} \tag{79}$$

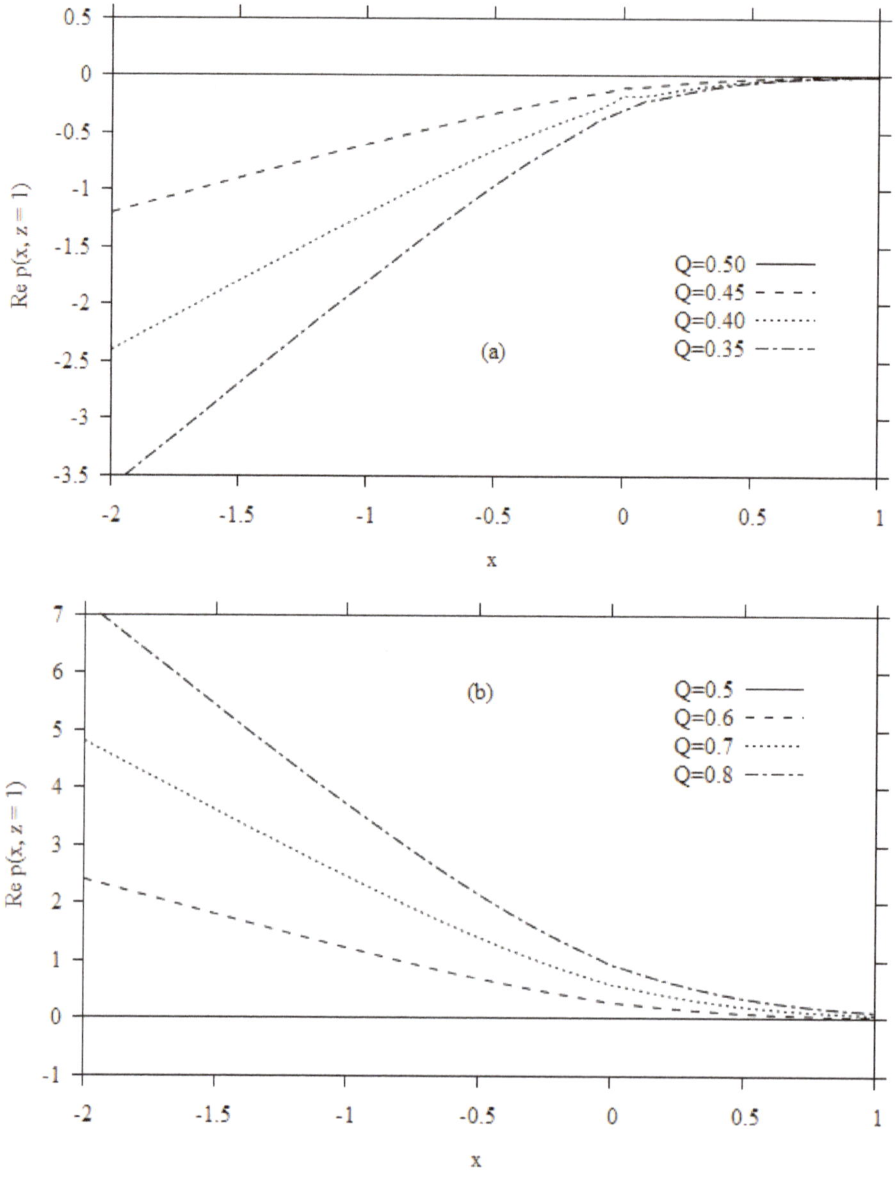

Figure 8. Influence of the flow rate on the pressure distribution along the moving substrate for (**a**) $Q \leq 0.5$ and (**b**) $Q \geq 0.5$. Drag flow corresponds to $Q = 0.5$.

The streamwise and transverse velocity components become $u = \frac{1}{\varepsilon}\psi_\eta$ and $w = -\psi_\xi$, respectively. Noting from the expressions (79) that p is of order ε^2, the flow field expansion reads

$$\psi(\xi,\eta) = \varepsilon^2\Psi_2(\xi,\eta) + \varepsilon^3\Psi_3(\xi,\eta) + \cdots, \tag{80}$$

$$p(\xi,\eta) = \varepsilon^2 P_2(\xi,\eta) + \varepsilon^3 P_3(\xi,\eta) + \cdots. \tag{81}$$

On inserting (80) and (81) in (79), we obtain a hierarchy of problems to different orders, each problem requiring only one boundary condition in the x direction. The problem is therefore well posed by imposing the far-upstream flow as condition. Consequently, the matching of the upstream flow with the flow in the free-surface layer II at the channel exit (x = 0) is not necessary. Moreover, the rescaling in the streamwise direction is not required unless one wants to capture the flow structure very close to the exit, and the distance x is assumed to remain at least of O(1). In contrast to the flow in a constricted (dilated) channel [51], where the streamwise direction x is rescaled in terms of the inverse power of the indentation slope (and the Reynolds number), the flow is not captured very close to the origin in the present formulation.

We determine the additional boundary conditions to solve the problem (79) by matching between the lower-wall and the core layers. Proceeding as in Section 2 and Appendix A, we find that $\Psi_2 \sim (1-G)\frac{\eta^2}{2}$ and $\Psi_3 \sim G\frac{\eta^3}{3}$ for large η. It is not difficult to show that the solution is trivial in both cases. Thus,

$$\Psi_2(\xi,\eta) = (1-G)\frac{\eta^2}{2},\ P_2(\xi,\eta) = 0, \tag{82}$$

$$\Psi_3(\xi,\eta) = G\frac{\eta^3}{3}, P_3(\xi,\eta) = 2G\xi. \tag{83}$$

Thus, up to $O\left(\varepsilon^3 = \mathrm{Re}^{-1}\right)$, the flow retains its Couette–Poiseuille character near the stationary die or blade wall. The correction is established when considering the next order. The equations for Ψ_4 and P_4 are

$$\Psi_{2\eta}\Psi_{4\xi\eta} - \Psi_{4\xi}\Psi_{2\eta\eta} = -P_{4\xi} + \Psi_{4\eta\eta\eta},\ P_{4\eta} = 0. \tag{84}$$

Matching with the core flow by equating $E_3H_4\psi = H_4E_3\psi$ yields the desired matching condition $\Psi_4(\xi,\eta\to\infty) \sim \eta\psi_{3z}(x,z=0)$ where, upon recalling that $V_n'(z=0) = 1$, we obtain $\psi_{3z}(\xi,z=0) = \sum\limits_{n=1}^{\infty}\frac{A_n}{\beta_n}e^{\beta_n\xi}$ from (62). By using (82) and eliminating the pressure from (84), the problem for $\Psi_4(\xi,\eta)$ becomes,

$$\begin{gathered}(1-G)\eta\Psi_{4\xi\eta\eta} = \Psi_{4\eta\eta\eta\eta},\\ \Psi_4(\xi,\eta=0) = \Psi_{4\eta}(\xi,\eta=0) = 0, \qquad \Psi_4(\xi\to-\infty,\eta) \sim 0,\\ \Psi_4(\xi,\eta\to\infty) \sim \eta\sum_{n=1}^{\infty}\frac{A_n}{\beta_n}e^{\beta_n\xi}.\end{gathered} \tag{85}$$

The solution of problem (85) is not trivial, signaling the departure from the C-P flow. We seek a similarity solution of the form $\Psi_4(\xi,\eta) = \sum\limits_{n=1}^{\infty}\frac{A_n}{\beta_n}e^{\beta_n\xi}F_n(\eta)$, where the functions $F_n(\eta)$ are governed by

$$\begin{gathered}F_n'''' = (1-G)\eta\beta_nF_n'',\\ F_n(0) = F_n'(0) = 0, \qquad F_n'(\infty) \sim 1, \qquad F_n''(\infty) \sim 0.\end{gathered} \tag{86}$$

Although an analytical solution is possible in terms of Airy functions, problem (86) can be more conveniently solved numerically. The problem is further simplified by eliminating the explicit dependence on G and β_n by introducing the transformation $\eta \to (1-G)^{1/3}\beta_n^{1/3}\eta$ and $F_n \to (1-G)^{1/3}\beta_n^{1/3}F_n$.

Figure 9 illustrates the profiles of F_n and its derivatives plotted against η, reflecting a monotonic behavior of $F_n(\eta)$ and its derivatives. Of particular physical significance are the values $F_n''(\eta=0) = c_3(1-G)^{1/3}\beta_n^{1/3}$ and $F_n'''(\eta=0) = c_4(1-G)^{2/3}\beta_n^{2/3}$, which, as we shall see, are directly related to the shear stress at the wall and pressure, respectively. Here $c_3 = 1.0651$ and $c_4 = -0.7765$.

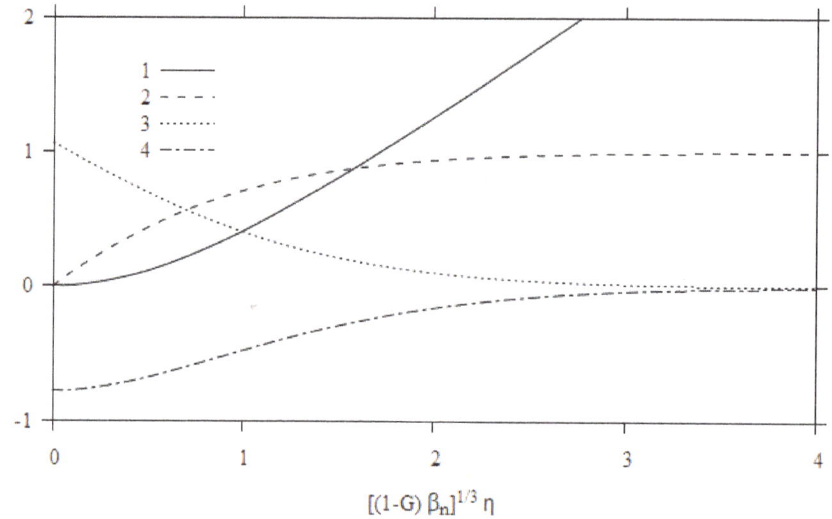

Figure 9. The function F_n and its derivatives plotted against η. Shown are: $(1-G)^{1/3}\beta_n^{1/3}F_n$ as curve (1), F_n' as curve (2), $(1-G)^{-1/3}\beta_n^{-1/3}F_n''$ as curve (3) and $(1-G)^{-2/3}\beta_n^{-2/3}F_n'''$ as curve (4).

To the current order, we determine the flow field in the slip layer III by inserting (82) and (83) as well as Ψ_4 into (80), yielding

$$\psi(x,z) = \frac{1-G}{2}z^2 + \frac{G}{3}z^3 + Re^{-4/3}\sum_{n=1}^{\infty}\frac{A_n}{\beta_n}e^{\beta_n x}F_n\left(Re^{1/3}z\right). \tag{87}$$

Of particular interest are the pressure and the shear stress at the wall:

$$Rep(x) = 2Gx + Re^{-1/3}c_4(1-G)^{2/3}\sum_{n=1}^{\infty}\frac{A_n}{\beta_n^{1/3}}e^{\beta_n x}, \tag{88}$$

$$\tau_{xz}(x,z=0) = 1-G+c_3(1-G)^{1/3}Re^{-2/3}\sum_{n=1}^{\infty}\frac{A_n}{\beta_n^{2/3}}e^{\beta_n x}. \tag{89}$$

We recall that the pressure is hydrostatic across the slip layer III, which is reflected by the absence of the z dependence in (88). Figure 10 illustrates the influence of the flow rate on the pressure (Figure 10a) and the wall shear stress (Figure 10b) at Re = 10. The range Q < 0.5 is taken to correspond typically to slot coating, and Q = 0.5 corresponds to drag flow. In this case, the adverse pressure gradient in Figure 10a opposes the action of the forward translation of the substrate, exhibiting the constant positive gradient $2G = 6(1-2Q)$ far upstream. However, the pressure rises rather rapidly as the flow approaches the die exit after becoming positive at a point that depends on the flow rate. However, this dependence is not monotonic. As Q departs (decreases) from the drag-flow level, Q = 0.5, where there is no change in the pressure, the pressure rises at the exit, but only to reach a maximum around Q = 0.4 and drops again as illustrated for Q = 0.35. This non-monotonicity is a consequence of flow separation, which will be discussed shortly. More generally, the rise in pressure depends also on inertia; its behavior is estimated from (88) to be $Rep(x=0,z) \sim -Re^{-1/3}$. We observe that the explicit dependence on Q is not evident from (88) since the coefficients A_n and eigenvalues β_n also depend on Q (or G as shown in Appendix B). The shear stress plots in Figure 10b show that the stress experiences a drop near the exit. We have also included the shear stress curves that correspond to the Couette–Poiseuille level for each flow rate as asymptotes, which roughly help locate the inception of the slip layer. The rise in adverse pressure above the Poiseuille level (Figure 10a) is caused by the flow acceleration as the film approaches the exit; the flow converges sharply as the film contracts at the exit (see Figure 2). The rise in the pressure is accompanied by a drop in the shear stress, resulting in a loss in forward flow momentum; the increase in pressure in the direction of the flow is, of course, akin to an increase in the potential energy of the fluid, leading to a reduced kinetic energy and a deceleration of the fluid.

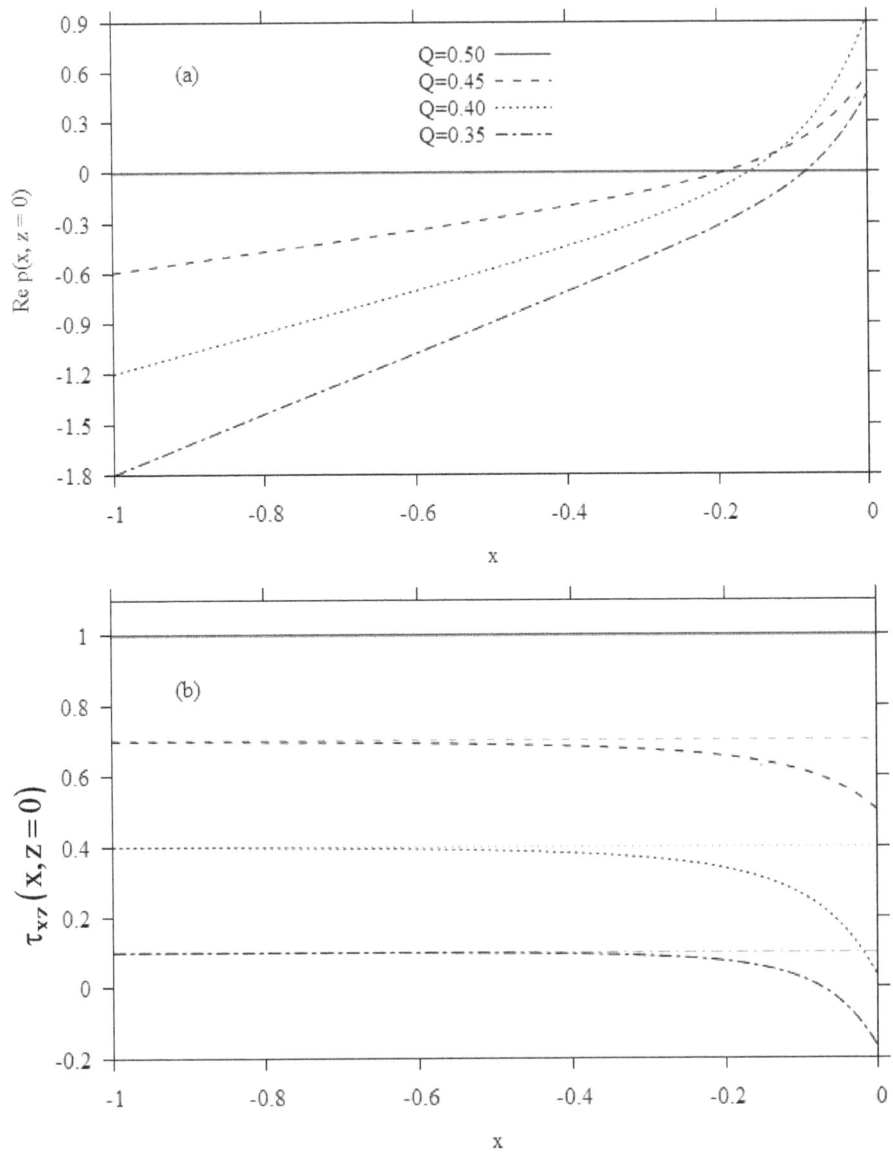

Figure 10. Pressure (**a**) and shear stress (**b**) distributions for slot coating along the stationary die wall for Re = 10 and various coating thicknesses (flow rates). Also added are the Couette–Poiseuille levels in (**b**) to show the leading edge of the slip layer III far upstream.

We observe that the flow in the wall layer is slower than in the core layer, and therefore expect a greater influence of the increasing pressure gradient. This seems to be particularly the case for a thinner coating film, here typically illustrated by Q = 0.35 for slot coating in Figure 10. The adverse pressure gradient is sufficiently large for the shear stress to vanish and a separation to eventually occur, with flow reversal occurring as it separates from the die wall.

We consider next the case of blade coating, which is typically illustrated in Figure 11, where we display the pressure (Figure 11a) and the wall shear stress (Figure 11b) for Q > 0.5 and Re = 10. The drag flow (Q = 0.5) is again included for reference. The pressure at the blade exit drops below atmospheric for the same reason as the rise in pressure in slot coating, namely as a result of the meniscus curvature and the film acceleration while moving along curved streamlines (see Figure 4a). For a higher flow rate, the fall in pressure is sharper, with a corresponding sharper rise in the wall shear stress (Figure 11b). Thus, a thicker film has to adjust more rapidly in height and velocity as Figure 2 indicates. Although the profiles in Figure 11 somewhat mirror those in slot coating, there are two important distinctions to observe. While the deviation from the Poiseuille level is non-monotonic with respect to Q for slot coating, the response is monotonic for blade coating (Figure 11a). This monotonicity is also reflected in the location of the inception of the slip layer, which coincides with the location of departure of the shear stress from the Couette-Poiseuille level, as illustrated in Figure 11b. More importantly, in contrast to slot coating, there is no possibility for separation in blade coating since there is no adverse pressure gradient (Figure 11a) and the wall shear stress remains positive (Figure 11b). Finally, the drop in pressure below atmospheric has been reported in the literature, which we will elucidate further next.

In their computational analysis of high-speed blade coating, Iliopoulos and Scriven [17] reported on pressure drop below atmospheric, similar to the drop predicted by the present formulation (Figure 11a). Direct full quantitative comparison is difficult since Iliopoulos and Scriven included the effect of forces such as shear thinning, the elastic deformation of the substrate and blade, as well as the wear on the blade from particle collisions, which are not accounted for in our study. Nevertheless, we will see that the comparison reveals important fundamental agreement and discrepancies when inertia is involved. In any case, the effect of the various additional forces do not seem to make any significant qualitative and quantitative difference, which can be confirmed by referring to figures of Iliopoulos and Scriven [17]. These figures indicate that the pressure upstream of the blade rises gradually from ambient to a sharp peak at the entrance to the blade region. The adverse pressure gradient causes the deflection of the excess liquid. The pressure then decreases linearly across the channel between the blade and the moving substrate, and drops below atmospheric at the blade exit. As mentioned earlier, this further drop is caused by the meniscus curvature and the film acceleration while moving along curved streamlines. The pressure then rises gradually back to atmospheric level as the film tends to uniform (plug) flow conditions.

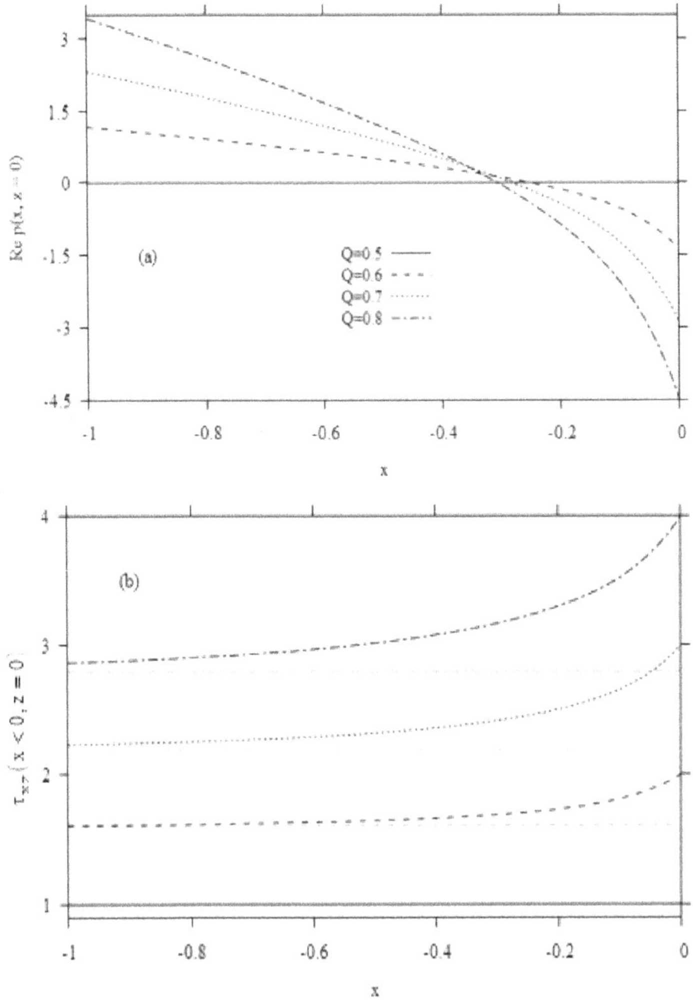

Figure 11. Pressure (**a**) and shear stress (**b**) distributions along the stationary blade or die wall for Re = 10 and various coating thicknesses ($Q \geq 1/2$). Also added are the Couette–Poiseuille levels in (**b**) to show the leading edge of the wall layer far upstream.

This behavior is also captured by the present theoretical analysis, which is depicted from the pressure profiles in Figure 12 for the three cases considered by Iliopoulos and Scriven [17], namely for Re = 1, 5 and 12 and corresponding flow rates of 0.58, 0.6 and 0.62, respectively. We have also included Figure 9 of Iliopoulos and Scriven as the top-right inset for reference, which depicts the pressure profiles for the trailing stiff blade geometry. We note that we used a much narrower streamwise range in Figure 12 than in the inset. In addition, our ambient pressure is 0 as opposed to 1 in the inset. Although the same flow rates are used by Iliopoulos and Scriven (inset), the present pressure gradients are higher. This is expected since the shear stress is higher for a Newtonian fluid. The flow rates used in Figure 12 were deduced from Figure 10 of Iliopoulos and Scriven [17] where the coating thickness is plotted against the Reynolds number. Incidentally, as they report: "Figure 10 suggests that for a stiff trailing blade with deformable substrate, raising the Reynolds number, or equivalently increasing the substrate speed, the liquid density, or decreasing the viscosity, leads to a thicker coating film. For Re around 55, the coated thickness becomes equal to the gap between the blade and the substrate". This response is in line with the basic premise of the present theory which stipulates that the thickness increases with inertia, resulting from the drop in viscous effects responsible for the onset of the free-surface layer and the ensuing film contraction. In fact, the film thickness is equal to the channel gap as Re tends to infinity. In this case, the flow retains its CP profile across the exit region and further downstream. We emphasize that the CP is a solution of the Euler's equations. Another important agreement between the theoretical and numerical predictions is the dependence of the pressure drop below atmospheric on inertia. This is reflected in the inset at the bottom left of Figure 12, where we plot $\mathrm{Rep}(x=0)$ against Re based on the three cases reported in the main figure. Clearly, the trend in the numerical and theoretical data points is essentially the same. The pressure at the exit increases rapidly in the small Re range and tapers (decaying to zero) for large Re. More precisely, expression (88) indicates that the pressure at the exit reduces to $\mathrm{Rep}(x=0)=\mathrm{Re}^{-1/3}c_4(1-G)^{2/3}\sum_{n=1}^{\infty}\frac{A_n}{\beta_n^{1/3}}$, which suggests that the pressure drops below atmospheric is of order $\mathrm{Re}^{-1/3}$. However, we note that if the flow rate (or G) varies with the Reynolds number as the data of Iliopoulos and Scriven [17] seem to suggest for a flexible substrate, then the dependence of the pressure drop on Re may differ slightly. We recall that A_n and β_n are G dependent (see Appendix B). In fact, it seems that $\mathrm{Rep}(0)$ behaves like $\mathrm{Re}^{-1/7}$, a behavior which is surprisingly the same for both the theoretical and numerical predictions

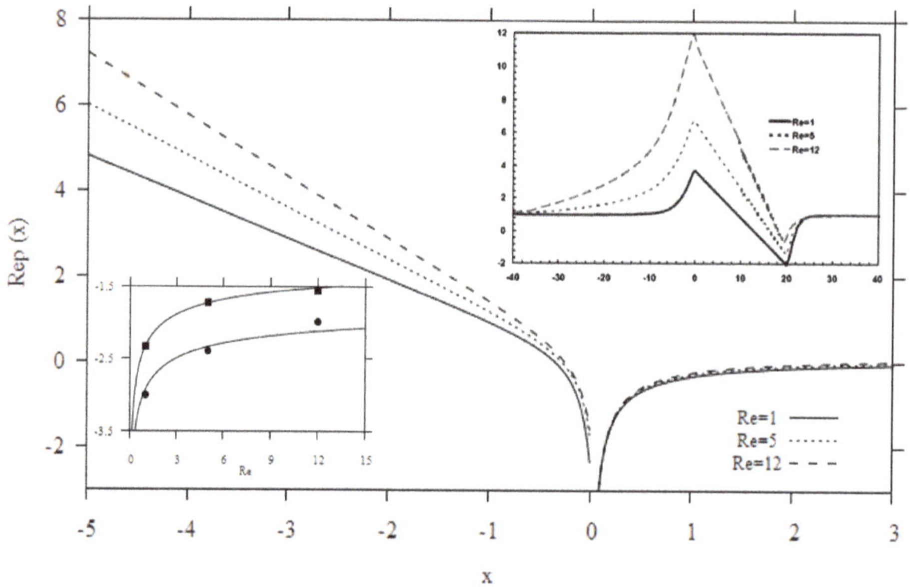

Figure 12. Pressure profiles along the stationary blade (x < 0) and the free surface (x > 0) for Re = 1, 5 and 12 (Q = 0.58, 0.6 and 0.62). Inset on top right shows Figure 13 from Iliopoulos and Scriven [17], and inset on bottom left shows the values of Rep(0) against Re from current theory (squares) and the numerical results of Iliopoulos and Scriven [17] (circles).

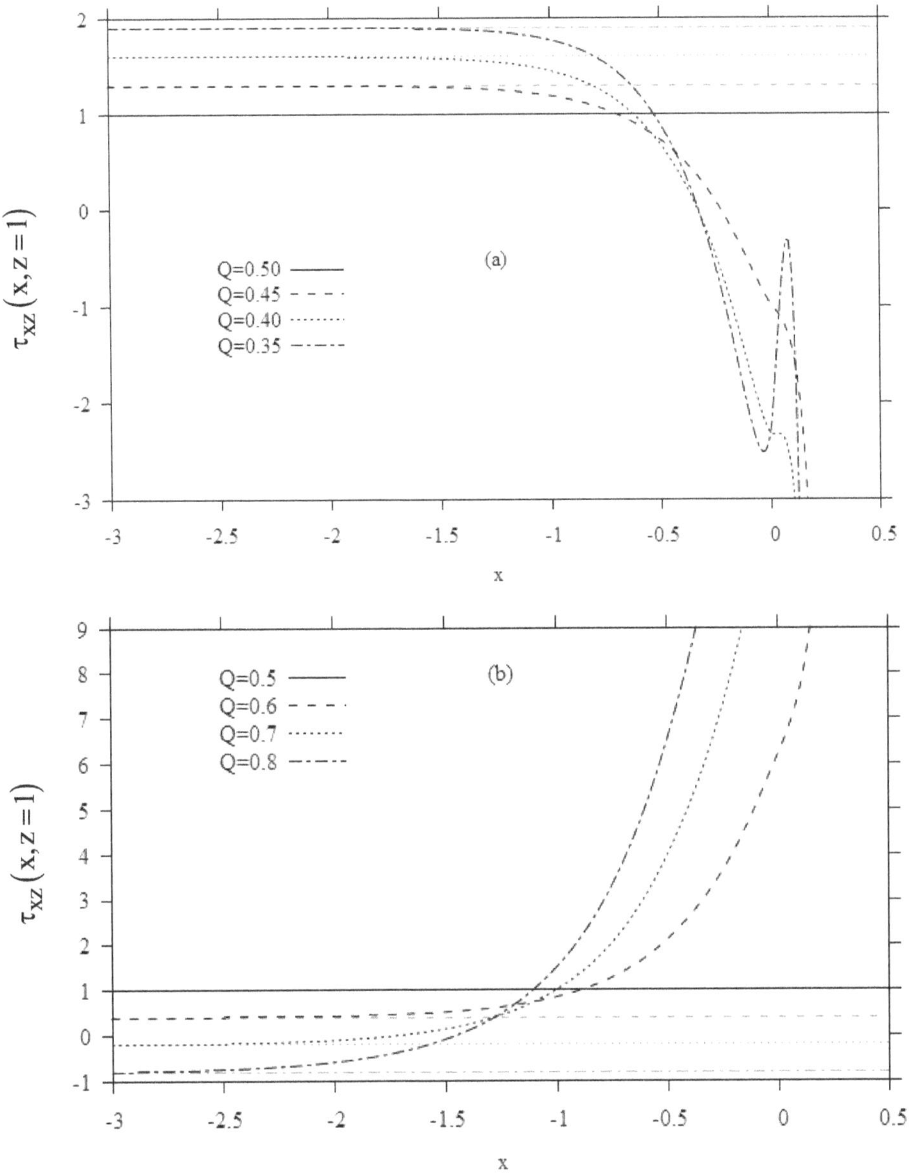

Figure 13. Influence of the flow rate on the shear stress distribution along the moving substrate at Re = 10 for (**a**) $Q \leq 1/2$ and (**b**) $Q \geq 1/2$. Drag flow corresponds to $Q = 1/2$. Corresponding fully-developed profiles are included in lighter lines.

Figure 12 reveals, however, two fundamental features that do not seem to be captured by the numerical simulation (top-right inset). The first is the accelerated drop below atmospheric near the exit (main Figure 12) as opposed to the continuous linear drop in the inset. This means that the numerical simulation does not indicate the existence of any upstream influence, which is rather inaccurate given the elliptic nature of the governing equations. The second discrepancy concerns the stress singularity at the exit, which should inevitably be reflected in the pressure singularity as shown in figure but not in the inset. The pressure singularity is a consequence of elongational effect; near x = 0, the streamwise momentum equation reduces to the balance between the pressure gradient and the gradient of the excess normal stress, thus yielding $Rep_x = u_{xx}$. Consequently, the jump in u_x from zero to a large positive value (see Figure 2b) leads to the jump in the pressure. The pressure singularity can manifest itself numerically in the form of spikes as reported in Figure 11b of Mitsoulis and Athanasopoulos [50]. We suspect that the singularity was smoothed over in the finite-element calculation of Iliopoulos and Scriven [17]. As we shall see next, the flow is very different in the slip layer IV along the moving substrate. We have already reported on the pressure profiles in Figure 8, which turn out to be smooth. The shear stress profiles require further development as we shall see in the next section.

6. The Flow Near the Moving Substrate (Slip Layer IV)

To examine the structure of the slip layer IV near the moving substrate, we let $y = 1 - z > 0$. The scaling in the transverse direction is changed by writing $y = \gamma\overline{\eta} > 0$, where γ is the small parameter in the problem, defined as $\gamma = Re^{-\beta}$ or $Re = \gamma^{-1/\beta}$, where β is to be determined. Similar to the analysis of the free-surface layer II and the slip layer III, the following change of coordinates is introduced, namely $x = \xi, z = 1 - \gamma\overline{\eta}$. Letting $\psi(\xi,\overline{\eta}) = 1 - \frac{G}{6} + \overline{\psi}(\xi,\overline{\eta})$ and proceeding as before, we see that in order to match the flow at the edge of the layer IV to the core layer I, the stream function and pressure must tend to $\overline{\psi}(\xi,\overline{\eta} \to \infty) \sim -\gamma\eta$ and $\overline{p}(\xi,\overline{\eta} \to \infty) \sim \varepsilon^3\left[2Gx - \sum_{n=1}^{\infty} \frac{A_n}{\beta_n} e^{\beta_n x} V_n'(1)\right]$, respectively, to lowest order in γ. In this case, for the inertial and viscous terms to balance in the transformed momentum equations, we must have $\beta = 1/2$. Therefore, $Re = \gamma^{-2} = \varepsilon^{-3}$ and $\gamma = \varepsilon^{3/2}$. In this case, the problem for $\overline{\psi}(\xi,\overline{\eta})$ and $\overline{p}(\xi,\overline{\eta})$ becomes (dropping the bar):

$$\begin{gathered}\psi_\eta\psi_{\xi\eta} - \psi_\xi\psi_{\eta\eta} = -\varepsilon^3 p_\xi - \varepsilon^{3/2}\psi_{\eta\eta\eta} - \varepsilon^{9/2}\psi_{\xi\xi\eta},\\ \psi_\eta\psi_{\xi\xi} - \psi_\xi\psi_{\xi\eta} = p_\eta - \varepsilon^{3/2}\psi_{\xi\eta\eta} - \varepsilon^{9/2}\psi_{\xi\xi\xi},\\ \psi(\xi,\eta = 0) = 0, \qquad \psi_{\overline{\eta}}(\xi,\eta = 0) = -\varepsilon^{3/2},\\ \psi(\xi \to -\infty,\eta) \sim -\varepsilon^{3/2}\eta + \varepsilon^3\eta^2\left(\tfrac{1+G}{2}\right) - G\varepsilon^{9/2}\tfrac{\eta^3}{3}.\end{gathered} \tag{90}$$

Additional conditions for the stream function and the pressure are established from matching with the core flow. However, these conditions are not the same upstream (x < 0) and downstream (x > 0) of the exit. The two problems will be treated separately in each region. However, for any x, the expansions of the stream function and the pressure take the forms:

$$\psi(\xi,\eta) = 1 - \frac{G}{6} + \varepsilon^{3/2}\Psi_2(\xi,\eta) + \varepsilon^3\Psi_3(\xi,\eta) + \varepsilon^{9/2}\Psi_{9/2}(\xi,\eta) + \ldots, \tag{91}$$

$$p(\xi,\eta) = \varepsilon^3 P_3(\xi,\eta) + \varepsilon^4 P_4(\xi,\eta) + \ldots. \tag{92}$$

Proceeding as in the previous sections, we find that the first two terms are contributions to the Couette-Poiseuille flow:

$$\Psi_{3/2}(\xi,\eta) = -\eta, \tag{93}$$

$$\Psi_3(\xi,\eta) = \left(\frac{1+G}{2}\right)\eta^2. \tag{94}$$

The correction occurs at the next order, $O(\varepsilon^6)$, and the problem is governed by

$$\Psi_{9/2\xi\eta} = P_{3\xi} + \Psi_{9/2\eta\eta\eta}, \tag{95}$$

$$P_{3\eta} = 0, \tag{96}$$

$$\Psi_{9/2}(\xi,\eta = 0) = \Psi_{9/2\overline{\eta}}(\xi,\eta = 0) = 0, \tag{97}$$

$$\Psi_{9/2}(\xi \to -\infty, \eta) \sim -\frac{G}{3}\eta^3,\ \Psi_{9/2}(\xi, \eta \to \infty) \sim -\frac{G}{3}\eta^3. \tag{98}$$

We consider the solution of problem (95)–(98) separately upstream and downstream of the exit. For $x < 0$, the pressure remains dictated by its core value from (65):

$$p_3(x \le 0, \eta \to \infty) \sim 2Gx - \sum_{n=1}^{\infty} \frac{A_n}{\beta_n} e^{\beta_n x} V_n'(1), \tag{99}$$

which, when inserted in (95), leads to the following equation for $\Psi_{9/2}(\xi < 0, \eta)$:

$$\Psi_{9/2\xi\eta} = -\sum_{n=1}^{\infty} A_n e^{\beta_n x} V_n'(1) + \Psi_{9/2\eta\eta\eta} + 2G. \tag{100}$$

The solution may be written as $\Psi_{9/2}(\xi < 0, \eta) = -\frac{G}{3}\eta^3 + \sum\limits_{n=1}^{\infty} A_n e^{\beta_n x} V_n'(1) G_n(\eta)$, where the coefficients $G_n(\eta)$ are governed by

$$G_n''' - \beta_n G_n' - 1 = 0,\ G_n(0) = G_n'(0),\ G_n'''(\infty) \sim 0. \tag{101}$$

This problem admits an analytical solution $G_n(\eta) = -\eta + \frac{1}{\sqrt{\beta_n}}\left(1 - e^{-\sqrt{\beta_n}\eta}\right)$, and the stream function near the moving substrate becomes

$$\begin{aligned} \Psi(\xi < 0, \eta) &= 1 - \tfrac{G}{6} - Re^{-1/2}\eta + Re^{-1}\left(\tfrac{1+G}{2}\right)\eta^2 - Re^{-3/2}\tfrac{G}{3}\eta^3 \\ &- Re^{-3/2}\sum_{n=1}^{\infty} \tfrac{A_n}{\beta_n} e^{\beta_n x} V_n'(1)\left[\eta - \tfrac{1}{\sqrt{\beta_n}}\left(1 - e^{-\sqrt{\beta_n}\eta}\right)\right]. \end{aligned} \tag{102}$$

When expressed in terms of z, this expression leads to the following expression for the velocity and the shear stress near the moving substrate:

$$u(x < 0, z) = u_0(z) + Re^{-1}\sum_{n=1}^{\infty} \frac{A_n}{\beta_n} e^{\beta_n x} V_n'(1)\left(1 - e^{-\sqrt{\beta_n Re}(1-z)}\right), \tag{103}$$

$$\tau_{xz}(x < 0, z) = 1 - G + 2Gz - Re^{-1/2}\sum_{n=1}^{\infty} \frac{A_n}{\sqrt{\beta_n}} e^{\beta_n x} V_n'(1) e^{-\sqrt{\beta_n Re}(1-z)}. \tag{104}$$

The exponential term that constitutes the main correction and ensures adherence at the moving substrate. Finally, the skin drag coefficient along the wall for $x < 0$ becomes $\tau_{xz}(x < 0, z = 1) = 1 + G - Re^{-1/2}\sum\limits_{n=1}^{\infty} \frac{A_n}{\sqrt{\beta_n}} e^{\beta_n x} V_n'(1)$. This expression can be used to estimate the starting point of the slip layer IV, a point where the flow begins to deviate from the Couette-Poiseuille limit. Thus, the commencement of the slip layer would correspond, in practice, to a position x_0 satisfying $\left|\frac{1+G-Re^{-1/2}\sum\limits_{n=1}^{\infty} \frac{A_n}{\sqrt{\beta_n}} e^{\beta_n x_0} V_n'(1)}{1+G}\right|$ equal to a small tolerance. Of course, the starting point depends on Q and Re (see below). The slip layer IV should not be confused with the boundary layer examined by Carvalho and Kheshgi [9], which emanates at the dynamic contact line of the upstream meniscus in slot coating. Obviously, the upstream meniscus is not accounted for in the present formulation.

For $x > 0$, the problem is much more complicated, but remains mathematically manageable. The difficulty stems from the core pressure (65), which takes a different form than (99):

$$P_3(x > 0, \eta \to \infty) \sim -\sum_{n=1}^{\infty} \frac{A_n}{\beta_n} e^{-\beta_n x} V_n'(1), \tag{105}$$

which, when inserted in (95), leads to the following equation for $\Psi_{9/2\eta}(\xi > 0, \eta)$:

$$\Psi_{9/2\xi\eta} = \sum_{n=1}^{\infty} A_n e^{-\beta_n x} V'_n(1) + \Psi_{9/2\eta\eta\eta}. \quad (106)$$

In addition to conditions (97), Equation (106) must be solved subject to flow matching at the channel exit (x = 0):

$$\Psi_{9/2\eta}\left(\xi = 0^+, \eta\right) = \Psi_{9/2\eta}\left(\xi = 0^-, \eta\right) = -G\eta^2 - \sum_{n=1}^{\infty} \frac{A_n}{\beta_n} V'_n(1)\left(1 - e^{-\sqrt{\beta_n}\eta}\right). \quad (107)$$

The solution may be conveniently written in the form

$$\Psi_{9/2\bar{\eta}}(\xi, \bar{\eta}) = -G\eta^2 - \sum_{n=1}^{\infty} A_n V'_n(1)\left[U_n(\xi, \eta) + \frac{1}{\beta_n}\left(1 - e^{-\sqrt{\beta_n}\eta}\right)\right], \quad (108)$$

where the coefficients $U_n(\xi, \eta)$ are governed by the following problem

$$\begin{aligned} U_{n\xi} &= U_{n\eta\eta} - e^{-\sqrt{\beta_n}\eta} + 2 - e^{-\beta_n x}, \\ U_n(\xi, \eta = 0) &= U_n(\xi = 0^+, \eta) = 0, \end{aligned} \quad (109)$$

which is of the transient heat conduction type with a source term, admitting the solution:

$$\begin{aligned} U_n(\xi, \eta) &= \int_0^{\xi}\int_0^{\infty} \frac{2 - e^{-\beta_n s} - e^{-\sqrt{\beta_n} y}}{\sqrt{4\pi(\xi - s)}}\left[\exp\left(-\frac{(\eta - y)^2}{4(\xi - s)}\right) - \exp\left(-\frac{(\eta + y)^2}{4(\xi - s)}\right)\right] dy ds \\ &= \int_0^{\xi}\left(2 - e^{\beta_n(\xi - t)}\right)\mathrm{erf}\left(\frac{\eta}{\sqrt{4t}}\right)dt - \frac{1}{\beta_n}\left(1 - e^{\beta_n \xi}\right)\sinh\left(\sqrt{\beta_n}\eta\right) \\ &\quad - \frac{1}{2}\int_0^{\xi} e^{\beta_n t}\left[e^{\sqrt{\beta_n}\eta}\mathrm{erf}\left(\frac{2\sqrt{\beta_n}t + \eta}{2\sqrt{t}}\right) - e^{-\sqrt{\beta_n}\eta}\mathrm{erf}\left(\frac{2\sqrt{\beta_n}t - \eta}{2\sqrt{t}}\right)\right]dt. \end{aligned} \quad (110)$$

This, in turn, yields the following expression for the velocity to $O\left(Re^{-1}\right)$ near the moving substrate:

$$u(x > 0, z) = (1 - G)z + Gz^2 + Re^{-1}\sum_{n=1}^{\infty} A_n V'_n(1)\left[\frac{1}{\beta_n}\left(1 - e^{-\sqrt{\beta_n}\eta}\right) + U_n(\xi, \eta)\right], \quad (111)$$

which is easily evaluated numerically. The shear stress or drag at the substrate takes the following form to $O\left(Re^{-1/2}\right)$:

$$\tau_{xz}(x > 0, z = 1) = 1 + G - Re^{-1/2}\sum_{n=1}^{\infty} A_n V'_n(1)\left[4\sqrt{\frac{x}{\pi}} + \frac{1 - e^{\beta_n x}}{\sqrt{\pi}\beta_n} + \frac{e^{\beta_n x}\mathrm{erfc}\left(\sqrt{\beta_n x}\right)}{\sqrt{\beta_n}}\right]. \quad (112)$$

We recall again that the pressure is hydrostatic across the slip layer IV, and therefore remains as in Figure 8. Figure 13 illustrates the influence of the flow rate on the wall shear stress at Re = 10. The ranges Q < 0.5 and Q > 0.5 in Figure 13a,b are again taken to correspond typically to slot and blade coating, and Q = 0.5 corresponds to drag flow. We have also included the shear stress curves that correspond to the CP level for each flow rate as asymptotes, which roughly help locate the inception of the slip layer IV. For slot coating, Figure 13a indicates that the shear stress drops sharply from its CP level just before the exit, and can exhibit a minimum followed by a maximum before the sharp drop for a thin coating film as illustrated for Q = 0.35. For any Q < 0.5, Figure 13a suggests that the flow separates; the vanishing of the shear stress is accompanied by an adverse pressure (Figure 8a). It seems that the separation always occurs at $x < 0$, at a location further upstream for the thinner coating film. In contrast, for blade coating, the shear stress plots in Figure 13b show that the stress experiences a rise near the exit. Although the stress can vanish for a thick coating film ($Q > 0.6$), there is no possibility of a separation given the favorable pressure gradient (see Figure 8b).

The rich dynamics and steep gradients exhibited by the shear stress for slot coating in Figure 13a, especially upstream of the exit, contrast sharply the smooth and mild behavior exhibited by the pressure in Figure 8a. This may seem at first physically unrealistic given the intimate coupling between stress and pressure gradients in the conservation equations. As a check, we examine the validity of the streamwise momentum equation along the wall for $x < 0$, which reduces to $-\mathrm{Re}p_x + \tau_{xz,z} = 0$. Clearly, upon noting from (104) that $\tau_{xz,z}(x < 0, z = 1) = 2G - \sum_{n=1}^{\infty} A_n e^{\beta_n x} V_n'(1)$, we deduce that $\mathrm{Re}p_x(x < 0, z = 1) = 2G - \sum_{n=1}^{\infty} A_n e^{\beta_n x} V_n'(1)$, which is the same result that is achieved upon evaluating the pressure gradient from (65) at $z = 1$. As to the transverse momentum equation, it reduces to $-\mathrm{Re}p_z + \tau_{zx,x} + \tau_{zz,z} = 0$ at $z = 1$. Noting that $\tau_{zx,x} + \tau_{zz,z} = u_{zx} + w_{zz} = -w_{zz} + w_{zz} = 0$ at the moving substrate, this leaves $p_z = 0$, which is the case to the current order as per (96). We thus confirm that momentum is conserved in the streamwise and transverse directions.

7. Discussion and Concluding Remarks

The planar laminar free-surface coating flow of a Newtonian fluid is investigated in the current study. The flow near and far from the channel exit is examined at moderate Reynolds and capillary numbers, subject to the substrate translation, and an adverse or a favorable constant pressure gradient applied far upstream of the channel exit as encountered in slot and blade coating flows. Although the flow far downstream is relatively simple to analyze as it becomes of the boundary-layer or the thin-film type, the treatment remains challenging in the vicinity of the exit. The method of matched asymptotic expansion is adopted to examine the influence of inertia and the applied pressure gradient on the shape of the free surface and the flow field. At the channel exit, a stress singularity occurs where the boundary condition changes from no-slip at the lower wall, to slip at the free surface, leading to the development of a boundary layer along the free surface. As a result, the flow domain consists of four different regions (Figure 1): the core layer I, the free-surface layer II, the slot or blade slip layer III, and the slip layer IV near the moving substrate. The layers II, III and IV are shear dominated, and the flow is obtained using a boundary-layer approach, but not all layers allow a similarity solution. In contrast, the core layer I is inviscid rotational where both shear and extensional flows are in balance. The small parameter in the problem turns out to be $\varepsilon = \mathrm{Re}^{-1/3}$, based on the balance of inertia and viscous effects in the free-surface layer II, allowing the asymptotic development of the flow by expanding the flow field.

As the fluid emerges from the channel in the form of the jet, it experiences a drastic drop in the shear stress and rise in normal stress at the channel exit as it slips along the free surface. This type of singularity constitutes a major hurdle in a theoretical methodology, particularly in a computational approach. Moreover, there is a wealth of physical mechanisms and phenomena that cannot be easily captured by a numerical approach. The boundary-layer structures in layers II and III are investigated in detail. The similarity solutions obtained in the two layers do not require matching at the exit. Consequently, the presence of the singularity is circumvented, constituting a major advantage of the present formulation. The shape of the free surface is determined by matching the free-surface layer flow with the core flow outside the channel exit.

The inviscid character of the core layer I can be perplexing given the parabolic Couette-Poiseuille flow as the leading-order velocity profile. This is a pivotal point which embeds the major premise on which the present theory is built, and can be explained as follows. The Couette-Poiseuille flow is only recovered in the limit of infinite Reynolds number ($\varepsilon \to 0$). In this limit, there is no viscous mechanism for the flow to change as it nears and traverses the exit. The fully developed Couette-Poiseuille velocity profile imposed far upstream retains its shape, and the free surface remains flat (horizontal). We therefore interpret the Couette-Poiseuille profile in the present context to correspond to inviscid rotational flow. Mathematically, any fully-developed profile: u = u(z), w = 0, satisfies Euler's equations, which are the inviscid limit of the Navier–Stokes' equations for infinite Reynolds number. In the present problem, although the Couette-Poiseuille profile is derived by integrating the (viscous) Navier–Stokes' equations, this profile does indeed satisfy Euler's equations as well as all the boundary conditions at the solid walls and flat free surface in the inviscid limit. Physically, one expects that if the Reynolds number is large enough (excluding turbulence), one should observe a profile close to the parabolic profile, even downstream (such as the case of water out of a garden hose). This hypothesis, and the present approach, have been amply validated for an axisymmetric jet (see, for instance, Philippe and Dumargue [33]). Thus, in this inviscid limit, the parabolic (or any) profile imposed far upstream remains unchanged as the fluid emerges at the exit. There is simply no viscous mechanism for the shear stress to relax or (equivalently) for a boundary layer to form along the free surface. Consequently, the free surface remains horizontal. In fact, the shear stress vanishes everywhere in the inviscid limit (zero viscosity) regardless of the value of the velocity gradient. In this limit, the dynamic conditions (6) and (7) are identically satisfied.

The core layer of the slot or blade coating flow should, apparently, be similar to the channel flow with a fine constriction considered by Smith [51]. Both flows comprise a leading-order fully developed contribution and an inviscid rotational first-order correction. However, there is an important difference as a result of the presence of a free surface in coating flow. In coating, the fully developed flow prevails everywhere, including the region downstream of the die or blade exit, *only in the limit of infinite Reynolds* number. In this limit, the free surface remains flat. The situation is different in Smith's case. The Poiseuille flow is recovered in the limit of zero slope of the indentation *at any Reynolds number*, and therefore exists in that limit. In the absence of the constriction, the flow physically exists and is certainly viscous. Therefore, we tend to consider the Couette-Poiseuille flow in coating as inviscid as it is reached only in the limit of infinite Re, satisfying Euler's equations without any viscous mechanism capable of altering the Couette-Poiseuille profile as the flow traverses the exit. This is an unphysical limit flow. It is important to emphasize that whether the Couette-Poiseuille flow is labelled as inviscid or viscous has no consequence on the present development and results.

The free surface is found to always contract near the channel exit regardless of the level of inertia and direction of the applied pressure gradient. The film tends to slightly expand further downstream for a thick blade coating film (Figure 2). For Ca = O(1), we show that the shape of the meniscus is not affected by surface tension. However, surface tension appears to alter significantly the dynamics in the normal stress, especially for blade coating where it causes a maximum to emerge near the exit (Figure 3). Further downstream where the flow becomes of the boundary-layer type, we illustrate how the near-exit solution can be matched to a thin-film formulation (Figure 4). Alternatively, the near-exit solution can be used as initial condition for a computational approach, thus avoiding the incorporation of the singularity.

Given the inviscid character of the core layer I for any finite Reynolds number, the core solution does not satisfy adherence at the walls where slip layers III and IV emerge. The detailed flow structure is obtained in the slip layers. In particular, we find as a result of the vanishing wall shear stress and the presence of adverse pressure (Figure 10), that a separation may eventually occur for a thin film in slot coating. Separation appears to be even more unavoidable near the moving substrate for slot coating (Figure 13). Finally, the undershoot exhibited by the pressure at the blade is captured in Figure 12. Both the current and existing numerical results from the literature appear to suggest that the pressure drop below atmospheric grows like $Re^{-1/7}$. However, unlike existing numerical predictions that show a linear decay sustained all the way to the exit, the current predictions show a further nonlinear drop, thus reflecting an upstream influence not predicted when using a numerical approach.

We would like to conclude this section by discussing two important issues that are not addressed in the present study and are worth elaborating on for future consideration. The first being the case of small-flow rate (Q < 1/3) in slot coating. The present formulation cannot handle such a situation, which involves bending of the free surface as it invades the region between the channel walls upstream of the exit. Consequently, the free-surface height experiences a significant departure from the infinite-Re limit of a flat surface, which violates the validity of expansion (19). We suspect that this would be an entirely different flow regime where the pressure gradient may no longer be assumed small relative to the Reynolds number. We also suspect that the flow in the immediate vicinity of the singularity with (approximate) solution to the full Navier–Stokes equations must be considered, which leads us to the second issue.

In fact, the second important issue concerns the flow very near the singularity. Much of twentieth-century boundary-layer theory was concerned with two problems, one the external flow over a plate (either infinite or semi-infinite length), the second flow in a slightly deformed channel by the presence of a mild constriction placed on the wall(s). The problem considered in this paper combines elements of both channel flow (through an upstream oncoming Couette–Poiseuille flow) with aspects of flow past a finite plate (through having the blade or die wall of the channel terminate with a free surface flow downstream of that point, so having elements of a wake-like flow). The present solution remains incomplete since the flow structure very near the blade or die exit is essentially left out. Only an approximation for the flow near the free surface downstream of the blade or die is derived. Since this boundary-layer solution (derived in Section 3) is only an approximate solution of the full Navier–Stokes equations, higher-order corrections remain envisageable.

The present development has a close parallel with the solution near the trailing point of a finite plate as proposed by Goldstein [47], who was able to develop approximate solutions to the wake, with the approximations including a singularity at the trailing edge. As in the present coating flow, the solution was developed in powers of $x^{1/3}$, where x was the distance from the trailing edge, and the singularity was in effect an infinite transverse velocity (order $x^{-2/3}$ as $x \to 0$) and a singularity in the pressure and stress from $x < 0$ to $x > 0$. The singularity was circumvented, and Goldstein was able to analyze the leading order far wake flow. It took three decades before the asymptotic structure was properly explained, when Stewartson [48] and Messiter [49] determined the triple-deck structure about the trailing edge point. Hence, the present paper puts the coating flow problem in a similar position to that for the trailing edge problem after Goldstein's development but before that of Stewartson and Messiter: a much more complex asymptotic structure near the blade or die is needed if the analysis is to be complete or extendable to higher order. Extension to either higher-order terms (in the present expansions) or larger parameter values may need a much more complicated interaction region. The idea is then is to seek a uniformly valid solution across the singularity, and thus provide the flow details very close to the edge. In analogy to the flow near the trailing edge, we anticipate the existence of a very small region near the edge of the blade or the die where derivatives of the flow variables are of the same order in both directions. In other words, both shearing and elongation are dominant mechanisms. In this case, the correct approximate (and not just the boundary-layer) solution of the full Navier–Stokes equations must be sought. Different attempts were made early on to achieve uniform validity but the approach developed by Stewartson [48] and Messiter [49], or the now well-established triple-deck approach, looks the most promising for future development.

Author Contributions: This article is based on preliminary work done by M.T.H. for his MASc thesis under the supervision of R.E.K. The work was then completely rewritten and calculations carried out by R.E.K. Conceptualization, R.E.K.; methodology, R.E.K.; software, R.E.K.; validation, M.T.H.; formal analysis, R.E.K. and M.T.H.; writing—original draft preparation, M.T.H.; writing—review and editing, R.E.K.; supervision, R.E.K.; project administration, R.E.K.; funding acquisition, R.E.K. All authors have read and agreed to the published version of the manuscript.

Funding: We would like to acknowledge the funding provided by the Natural Sciences and Engineering Council of Canada.

Conflicts of Interest: The authors declare no conflict of interest.

Appendix A. Matching between the Free-Surface and Core Layers

The matching rule employed by Van Dyke [57] is adopted here for the streamfunction and the pressure, but the pressure will turn out to be uniformly valid across the core and the free-surface layer. Thus, only the matching of the streamfunction is detailed here. In this case, Van Dyke's rule states that

$$E_n H_m \psi = H_m E_n \psi, \tag{A1}$$

where m and n are integers. Here, E_n is the core-expansion operator, which truncates immediately after the term of order ε^n where the expansion is expressed in terms of core variables. H_m is the corresponding boundary-layer expansion operator. The left- and right-hand sides of expressions (A1) must be exactly the same for all the values of m and n. Depending on the values of m and n, we need different levels of matching to obtain the boundary conditions for the inner and core solutions, and simultaneously to determine the free surface height to each order in ε.

We recall from (1) that, to leading order, the stream function in the core region is $\psi_0(z)$, which can be expressed in terms of $y = z - \zeta(x)$. In this case, the stream function expression (52) for the core flow must be cast in terms of y, and takes the following form (Khayat 2014, Tillett 1968):

$$\psi(x,z) = \left(\frac{1-G}{2}\right)(y+\varepsilon h)^2 + \frac{G}{3}(y+\varepsilon h)^3 + \varepsilon^3 \overline{\psi}(x, y+\varepsilon h) + O\left(\varepsilon^3\right). \tag{A2}$$

A useful expression is obtained upon expanding $\overline{\psi}(x,z) = \overline{\psi}(x, y+\varepsilon h)$ about $y = 0$:

$$\begin{aligned}\overline{\psi}(x,y+\varepsilon h) = \; & \overline{\psi}(x,0) + y\overline{\psi}_y(x,0) + \tfrac{y^2}{2}\overline{\psi}_{yy}(x,0) + \ldots.. \\ & +\varepsilon h\overline{\psi}_y(x,0) + \varepsilon h y\overline{\psi}_{yy}(x,0) + \ldots... \\ & +\tfrac{(\varepsilon h)^2}{2}\overline{\psi}_{yy}(x,0) + \ldots...\end{aligned} \tag{A3}$$

We first consider the matching between the free-surface and core layers to $O(\varepsilon)$. The asymptotic form (27)–(29) for Ψ_2 is determined by considering the application of (A1) for m = 2 and n = 0. Applying E_0 on (A2) gives

$$E_0\psi = \left(\frac{1-G}{2}\right)y^2 - \frac{G}{3}y^3. \tag{A4}$$

As this expression must be cast in inner variables when the operator H_2 is applied, it is rewritten in the following form:

$$E_0\psi = (1-G)\frac{\varepsilon^2\eta^2}{2} + G\frac{\varepsilon^3\eta^3}{3}. \tag{A5}$$

Therefore,

$$H_2E_0\psi = \left(\frac{1-G}{2}\right)\varepsilon^2\eta^2 = \left(\frac{1-G}{2}\right)y^2. \tag{A6}$$

On the other hand, to leading order, the expansion for the stream function in the free-surface layer is obtained from (20) as $\psi = \varepsilon^2\Psi_2$. Thus, $E_0H_2\psi = \varepsilon^2\Psi_2$, which, when matched with (A5), leads to $\Psi_2 \sim (1-G)\frac{\eta^2}{2}$ for large η or condition (27)–(29).

Next, we determine $h_0(x)$ in (19) by considering the application of (A1) for m = 2 and n = 1. Applying E_1 on (A2), we have

$$E_1\psi = \left(\frac{1-G}{2}\right)\left(y^2 + 2\varepsilon h_0 y\right) + \frac{G}{3}\left(y^3 + 3\varepsilon h_0 y^2\right). \tag{A7}$$

The terms surviving to $O\left(\varepsilon^2\right)$ are identified by expressing $E_1\psi$ in terms of $\eta = \frac{y}{\varepsilon}$ and using (A2) for m = 1 to yield:

$$H_2E_1\psi = \left(\frac{1-G}{2}\right)y^2 + \varepsilon(1-G)h_0 y. \tag{A8}$$

On the other hand, we note from (29) that $H_2\psi$ may be written in terms of the core variables as

$$H_2\psi = \varepsilon^2\xi^{2/3}f_2 \sim \left(\frac{1-G}{2}\right)\varepsilon^2\xi^{2/3}(\theta + c_1)^2 = \left(\frac{1-G}{2}\right)\varepsilon^2\xi^{2/3}\left(\eta\xi^{-1/3} + c_1\right)^2, \tag{A9}$$

which yields

$$E_1H_2\psi = \left(\frac{1-G}{2}\right)y^2 + \varepsilon(1-G)c_1x^{1/3}y. \tag{A10}$$

Equating (A8) and (A10) leads to

$$h_0(x) = c_1x^{1/3} = \frac{d_1}{(1-G)^{1/3}}x^{1/3}. \tag{A11}$$

Even to this order, we begin to sense the difference in behavior of the film height (thickness) between slot and blade coating for $G > 0$ and $G < 0$, respectively. Clearly, (A11) indicates that in both cases the height grows like $x^{1/3}$ but at a higher rate, leading ultimately to a thinner film for slot than for blade coating. A more accurate prediction of the film thickness will be achieved by considering the next order.

In order to determine the asymptotic behavior of $\Psi_3(\xi, \eta \to 0)$ in the free-surface layer, we set $n = 0$ and $m = 3$, and match the expressions:

$$H_3E_0\psi = 2y^2 - \frac{4}{3}y^3, \tag{A12}$$

$$E_0H_3\psi = E_0\left(\varepsilon^2\Psi_2 + \varepsilon^3\Psi_3\right), \tag{A13}$$

yielding $\Psi_3 \sim \frac{G}{3}\eta^3$ or condition (35).

Finally, to obtain $h_1(x)$ and $\overline{\psi}(x, 0)$, (A1) is applied for $m = n = 3$. This step is algebraically much more involved, and is only summarized here. Noting that $H_3\psi = \varepsilon^2\Psi_2 + \varepsilon^3\Psi_3 = \varepsilon^2\xi^{2/3}f_2 + \varepsilon^3\xi f_3$ and using the asymptotic forms (29) and (32) give

$$\begin{aligned} H_3\psi &= \varepsilon^2\xi^{2/3}\left\{(1-G)\frac{\left(\eta\xi^{-1/3}+c_1\right)^2}{2}\right\} \\ &+\varepsilon^3\xi\left[\frac{G}{3}\left\{\left(\eta\xi^{-1/3} + c_1\right)^3 - \frac{6}{1-G}\right\} + c_2\left(\eta\xi^{-1/3} + c_1\right)\right]. \end{aligned} \tag{A14}$$

Thus,

$$\begin{aligned} E_3H_3\psi &= (1-G)\frac{y^2}{2} + \frac{G}{3}y^3 + \varepsilon x^{1/3}yc_1[(1-G) + Gy] \\ &+\varepsilon^2x^{2/3}\left[\frac{(1-G)c_1^2}{2} + Gyc_1^2 + c_2y\right] + \varepsilon^3x\left[c_2c_1 + \frac{1}{3}\left(Gc_1^3 - \frac{6G}{1-G}\right)\right]. \end{aligned} \tag{A15}$$

Recalling (A2) and (A3), we observe that

$$\begin{aligned} E_3\psi &= \left(\frac{1-G}{2}\right)\left[y^2 + 2\varepsilon\left(h_0 + \varepsilon h_1 + \varepsilon^2h_2\right)y + \varepsilon^2\left(h_0^2 + 2\varepsilon h_0h_1\right)\right] \\ &+\frac{G}{3}\left(y^3 + 3\varepsilon\left(h_0 + \varepsilon h_1 + \varepsilon^2h_2\right)y^2 + 3\varepsilon^2\left(h_0^2 + 2\varepsilon h_0h_1\right)y + \varepsilon^3h_0^3\right) + \varepsilon^3\overline{\psi}(x,0). \end{aligned} \tag{A16}$$

Consequently,

$$\begin{aligned} H_3E_3\psi &= (1-G)\frac{y^2}{2} + G\frac{y^3}{3} + \varepsilon y(1-G+Gy)h_0 \\ &+\varepsilon^2\left[\left(\frac{1-G}{2} + Gy\right)h_0^2 + (1-G)yh_1\right] + \varepsilon^3\left[(1-G)h_0h_1 + G\frac{h_0^3}{3} + \overline{\psi}(x,0)\right]. \end{aligned} \tag{A17}$$

Equating (A15) and (A17), and recalling that $h_0 = c_1x^{1/3}$, the correction for the free surface height to the next order is obtained as

$$h_1(x) = \frac{c_2}{(1-G)}x^{2/3}. \tag{A18}$$

In addition, one has

$$\overline{\psi}(x,0) = -\frac{2G}{1-G}x \quad \text{(A19)}$$

Condition (A19) yields the third boundary condition in (59)–(61).

Appendix B. Values of Eigenvalues and Coefficients for the First Six Modes

n	β_n	A_n	$V_n'(1)$	Q	G
1	4.2800	−0.4202	−6.4042	0.3500	0.9000
2	7.3840	−1.0934	3.9830	0.3500	0.9000
3	10.4550	−1.9273	−3.0082	0.3500	0.9000
4	13.5250	−2.8259	2.4900	0.3500	0.9000
5	16.6000	−3.7257	−2.1727	0.3500	0.9000
6	19.6840	−4.5943	1.9603	0.3500	0.9000
1	3.7694	−0.8589	−1.7827	0.4000	0.6000
2	6.7715	−1.4428	1.4067	0.4000	0.6000
3	9.8205	−1.8071	−1.2695	0.4000	0.6000
4	12.9010	−2.0410	1.2002	0.4000	0.6000
5	15.9972	−2.2018	−1.1588	0.4000	0.6000
6	19.1061	−2.3235	1.1314	0.4000	0.6000
1	3.4080	−0.5525	−1.2056	0.4500	0.3000
2	6.4685	−0.6824	1.1067	0.4500	0.3000
3	9.5680	−0.7372	−1.0716	0.4500	0.3000
4	12.6800	−0.7652	1.0538	0.4500	0.3000
5	15.8100	−0.7835	−1.0430	0.4500	0.3000
6	18.9400	−0.7978	1.0358	0.4500	0.3000
1	2.7680	1.2325	−0.8453	0.6000	−0.6000
2	6.0642	0.9314	0.9162	0.6000	−0.6000
3	9.2630	0.8624	−0.9423	0.6000	−0.6000
4	12.4370	0.8303	0.9560	0.6000	−0.6000
5	15.5980	0.8126	−0.9644	0.6000	−0.6000
6	18.7550	0.8037	0.9701	0.6000	−0.6000
1	2.5130	2.4107	−0.7898	0.7000	−1.2000
2	5.9320	1.5087	0.8832	0.7000	−1.2000
3	9.1700	1.3402	−0.9186	0.7000	−1.2000
4	12.3650	1.2668	0.9375	0.7000	−1.2000
5	15.5400	1.2272	−0.9492	0.7000	−1.2000
6	18.7000	1.2053	0.9572	0.7000	−1.2000
1	2.3240	3.5018	−0.7660	0.8000	−1.8000
2	5.8400	1.8988	0.8672	0.8000	−1.8000
3	9.1100	1.6433	−0.9065	0.8000	−1.8000
4	12.3200	1.5358	0.9278	0.8000	−1.8000
5	15.5000	1.4783	−0.9411	0.8000	−1.8000
6	18.6700	1.4464	0.9503	0.8000	−1.8000

References

1. Ruschak, K.J. Coating flows. *Ann. Rev. Fluid Mech.* **1985**, *17*, 65–89. [CrossRef]
2. Weinstein, S.J.; Ruschak, K.J. Coating flows. *Ann. Rev. Fluid Mech.* **2004**, *36*, 29–53. [CrossRef]
3. Aidun, C.K.; Triantafillopoulos, N.G. High-speed blade coating. In *Liquid Film Coating*; Schweizer & Kistler—Springer: Dordrecht, The Netherlands, 1997; pp. 637–672.
4. de Ryck, A.; Quere, D. Inertial coating of a fibre. *J. Fluid Mech.* **1996**, *331*, 219–237. [CrossRef]
5. Landau, D.; Levich, B. Dragging of a liquid by a moving plate. *Acta Physicochim. USSR* **1942**, *17*, 42–54.
6. Lee, K.Y.; Liu, L.D.; Ta-Jo, L. Minimum wet thickness in extrusion slot coating. *Chem. Eng. Sci.* **1992**, *47*, 1703–1713. [CrossRef]
7. Becker, Y.; Wang, Y. Prediction of the minimum wet thickness of slot coating at small capillary number. In *IS & T's 50th Annual Conference*; The Society for Imaging Science and Technology: Springfield, VA, USA, 1997; pp. 470–474.

8. Chang, Y.R.; Chang, H.M.; Lin, C.F.; Liu, T.J.; Wu, P.Y. Three minimum wet thickness regions of slot die coating. *J. Colloid Interface Sci.* **2007**, *308*, 222–230. [CrossRef]
9. Carvalho, M.S.; Kheshgi, H.S. Low-flow limit in slot coating: Theory and experiments. *AIChE J.* **2000**, *46*, 1907–1917. [CrossRef]
10. Romero, O.J.; Suszynski, W.J.; Scriven, L.E.; Carvalho, M.S. Low-flow limit in slot coating of dilute solutions of high molecular weight polymer. *J. NonNewt. Fluid Mech.* **2004**, *118*, 137–156. [CrossRef]
11. Chin, C.P.; Wu, H.S.; Wang, S.S. Improved coating window for slot coating. *Ind. Eng. Chem. Res.* **2010**, *49*, 3802–3809. [CrossRef]
12. Yu, W.J.; Liu, T.J.; Yu, T.A. Reduction of the minimum wet thickness in extrusion slot coating. *Chem. Eng. Sci.* **1995**, *50*, 917–920. [CrossRef]
13. Tanner, R.I. *Engineering Rheology*; Oxford University Press: Oxford, UK, 2000.
14. Aidun, C.K. *An Overview of Blade Coating Systems*; Institute of Paper Science and Technology: Atlanta, GA, USA, 1993; pp. 1–30.
15. Saito, H.; Scriven, L.E. Study of coating flow by the finite element method. *J. Comput. Phys.* **1981**, *42*, 53–76. [CrossRef]
16. Jang, I.; Song, S. A model for prediction of minimum coating thickness in high speed slot coating. *Int. J. Heat Fluid Flow* **2013**, *40*, 180–185. [CrossRef]
17. Iliopoulos, L.; Scriven, L.E. A blade-coating study using a finite-element simulation. *Phys. Fluids* **2005**, *17*, 127101. [CrossRef]
18. Lin, C.F.; Hill Wong, D.S.; Liu, T.J.; Wu, P.Y. Operating windows of slot die coating: Comparison of theoretical predictions with experimental observations. *Adv. Polym. Techn.* **2010**, *29*, 31–44. [CrossRef]
19. Bajaj, M.; Prakash, J.R.; Pasquali, M. A computational study of the effect of viscoelasticity on slot coating flow of dilute polymer solutions. *J. NonNewt. Fluid Mech.* **2008**, *149*, 104–123. [CrossRef]
20. Taroni, M.; Breward, C.J.W.; Howell, P.D.; Oliver, J.M. Boundary conditions for free surface inlet and outlet problems. *J. Fluid Mech.* **2012**, *708*, 100–110. [CrossRef]
21. Lee, A.G.; Shaqfeh, E.S.; Khomami, B. A study of viscoelastic free surface flows by the finite element method: Hele–Shaw and slot coating flows. *J. NonNewt. Fluid Mech.* **2002**, *108*, 327–362. [CrossRef]
22. Pasquali, M.; Scriven, L.E. Free surface flows of polymer solutions with models based on the conformation tensor: Computational method and benchmark problems. *J. NonNewt. Fluid Mech.* **2002**, *108*, 363–409. [CrossRef]
23. Ashmore, J.; Shen, A.Q.; Kavehpour, H.P.; Stone, H.A.; McKinley, G.H. Coating flows of non-Newtonian fluids: Weakly and strongly elastic limits. *J. Eng. Math.* **2007**, *60*, 17–41. [CrossRef]
24. Mitsoulis, E.V. Numerical simulation of viscoelastic effects in blade-over-roll coating forming flows. *Comput. Methods Mater. Sci.* **2010**, *10*, 156–166.
25. Silliman, W.J. Viscous Film Flows with Contact Lines: Finite Element Simulation, A Basis for Stability Assessment and Design Optimization. Ph.D. Thesis, University of Minnesota, Minneapolis, MN, USA, 1979.
26. Ruschak, K.J. Limiting flow in a pre-metered coating device. *Chem. Eng. Sci.* **1976**, *31*, 1057–1060. [CrossRef]
27. Goren, S.L.; Wronski, S. The shape of low-speed capillary jets of Newtonian liquids. *J. Fluid Mech.* **1966**, *25*, 185–198. [CrossRef]
28. Benilov, E.S.; Benilov, M.S.; Kopteva, N. Steady rimming flows with surface tension. *J. Fluid Mech.* **2008**, *597*, 91–118. [CrossRef]
29. Timoshin, S.N. Instabilities in a high-Reynolds-number boundary layer on a film-coated surface. *J. Fluid Mech.* **1997**, *353*, 163–195. [CrossRef]
30. Tsang, J.M.F.; Dalziel, S.B.; Vriend, N.M. Interaction between the Blasius boundary layer and a free surface. *J. Fluid Mech.* **2018**, *839*, R1–R12. [CrossRef]
31. Tillett, J.P.K. On the laminar flow in a free jet of liquid at high Reynolds numbers. *J. Fluid Mech.* **1968**, *32*, 273–292. [CrossRef]
32. Miyake, Y.; Mukai, E.; Iemoto, Y. On a two-dimensional laminar liquid jet. *Bull. JSME* **1979**, *22*, 1382–1389. [CrossRef]
33. Philippe, C.; Dumargue, P. Étude de l'établissement d'un jet liquide laminaire émergeant d'une conduite cylindrique verticale semi-infinie et soumis à l'influence de la gravité. *Zeit. Ang. Math. Phys.* **1991**, *42*, 227–242. [CrossRef]

34. Wilson, D.E. A similarity solution for the axisymmetric viscous-gravity jet. *Phys. Fluids* **1986**, *29*, 632–639. [CrossRef]
35. Khayat, R.E. Slipping free jet flow near channel exit at moderate Reynolds number for large slip length. *J. Fluid Mech.* **2016**, *793*, 667–708. [CrossRef]
36. Khayat, R.E. Initial development of a free-surface wall jet at moderate Reynolds number. *J. Fluid Mech.* **2017**, *826*, 235–269. [CrossRef]
37. Khayat, R.E. Free-surface jet flow of a shear-thinning power-law fluid near the channel exit. *J. Fluid Mech.* **2014**, *748*, 580–617. [CrossRef]
38. Saffari, A.; Khayat, R.E. Flow of viscoelastic jet with moderate inertia near channel exit. *J. Fluid Mech.* **2009**, *639*, 65–100. [CrossRef]
39. Sobey, I.J. *Introduction to Interactive Boundary-Layer Theory*; Oxford University Press: Oxford, UK, 2005.
40. Higgins, B.G.; Scriven, L.E. Capillary pressure and viscous pressure drop set bounds on coating bead operability. *Chem. Eng. Sci.* **1980**, *35*, 673–682. [CrossRef]
41. Christodoulou, K.N.; Scriven, L.E. The fluid mechanics of slide coating. *J. Fluid Mech.* **1988**, *208*, 321–354. [CrossRef]
42. Ruschak, K.J.; Scriven, L.E. Developing flow on a vertical wall. *J. Fluid Mech.* **1977**, *81*, 305–316. [CrossRef]
43. Savage, M.D. Mathematical models for coating processes. *J. Fluid Mech.* **1982**, *117*, 443–455. [CrossRef]
44. Gaskell, H.; Savage, M.D.; Summers, J.L.; Thompson, H.M. The wall jet. *J. Fluid Mech.* **1995**, *298*, 113–137. [CrossRef]
45. Kelmanson, M. On inertial effects in the Moffatt–Pukhnachov coating-flow problem. *J. Fluid Mech.* **2009**, *633*, 327–353. [CrossRef]
46. Blythe, P.A.; Simpkins, P.G. Fibre coating: Non-unique solutions at small capillary numbers. *J. Fluid Mech.* **2004**, *515*, 353–370. [CrossRef]
47. Goldstein, S. Concerning some solutions of the boundary layer equations in hydrodynamics. *Proc. Camb. Philos. Soc.* **1939**, *26*, 1–30. [CrossRef]
48. Stewartson, K. On the flow near the trailing edge of a fiat plate. II. *Mathematika* **1969**, *16*, 106–121. [CrossRef]
49. Messiter, A.F. Boundary-layer flow near the trailing edge of a flat plate. *SIAM J. Appl. Math.* **1970**, *18*, 241–257. [CrossRef]
50. Mitsoulis, E.V.; Athanasopoulos, G. Numerical simulation of Blade-over-roll coating forming flows. *Comput. Methods Mater. Sci.* **2010**, *10*, 214–224.
51. Smith, F.T. Flow through constricted or dilated pipes and channels: Part 1. *Q. J. Mech. Appl. Math.* **1976**, *29*, 343–364. [CrossRef]
52. Smith, F.T. Flow through constricted or dilated pipes and channels: Part 2. *Q. J. Mech. Appl. Math.* **1976**, *29*, 365–376. [CrossRef]
53. McKinley, G.H. Dimensionless groups for understanding free surface flow of complex fluids. *SOR Rheol. Bull.* **2005**, 74–81.
54. Khayat, R.E.; Kim, K. Thin-film flow of a viscoelastic fluid on an axisymmetric substrate of arbitrary shape. *J. Fluid Mech.* **2006**, *552*, 37–71. [CrossRef]
55. Rojas, N.; Argentina, M.; Tirapegui, E. Inertial lubrication theory. *Phys. Rev. Lett.* **2010**, *104*, 187801–187804. [CrossRef]
56. Smith, F.T. The separating flow through a severely constricted symmetric tube. *J. Fluid Mech.* **1979**, *90*, 725–754. [CrossRef]
57. Van Dyke, M.D. *Perturbation Methods in Fluid Mechanics*; Academic Press: New York, NY, USA, 1964.

Publisher's Note: MDPI stays neutral with regard to jurisdictional claims in published maps and institutional affiliations.

© 2020 by the authors. Licensee MDPI, Basel, Switzerland. This article is an open access article distributed under the terms and conditions of the Creative Commons Attribution (CC BY) license (http://creativecommons.org/licenses/by/4.0/).

fluids

MDPI

Article

The Zoo of Modes of Convection in Liquids Vibrated along the Direction of the Temperature Gradient

Georgie Crewdson and Marcello Lappa *

Department of Mechanical and Aerospace Engineering, University of Strathclyde, James Weir Building, 75 Montrose Street, Glasgow G1 1XJ, UK; georgie.crewdson@strath.ac.uk
* Correspondence: marcello.lappa@strath.ac.uk

Abstract: Thermovibrational flow can be seen as a variant of standard thermogravitational convection where steady gravity is replaced by a time-periodic acceleration. As in the parent phenomena, this type of thermal flow is extremely sensitive to the relative directions of the acceleration and the prevailing temperature gradient. Starting from the realization that the overwhelming majority of research has focused on circumstances where the directions of vibrations and of the imposed temperature difference are perpendicular, we concentrate on the companion case in which they are parallel. The increased complexity of this situation essentially stems from the properties that are inherited from the corresponding case with steady gravity, i.e., the standard Rayleigh–Bénard convection. The need to overcome a threshold to induce convection from an initial quiescent state, together with the opposite tendency of acceleration to damp fluid motion when its sign is reversed, causes a variety of possible solutions that can display synchronous, non-synchronous, time-periodic, and multi-frequency responses. Assuming a square cavity as a reference case and a fluid with Pr = 15, we tackle the problem in a numerical framework based on the solution of the governing time-dependent and non-linear equations considering different amplitudes and frequencies of the applied vibrations. The corresponding vibrational Rayleigh number spans the interval from $Ra_\omega = 10^4$ to $Ra_\omega = 10^6$. It is shown that a kaleidoscope of possible variants exist whose nature and variety calls for the simultaneous analysis of their temporal and spatial behavior, thermofluid-dynamic (TFD) distortions, and the Nusselt number, in synergy with existing theories on the effect of periodic accelerations on fluid systems.

Keywords: thermovibrational convection; gravity modulation; thermofluid-dynamic distortions; patterning behavior

check for updates

Citation: Crewdson, G.; Lappa, M. The Zoo of Modes of Convection in Liquids Vibrated along the Direction of the Temperature Gradient. *Fluids* **2021**, *6*, 30. https://doi.org/10.3390/fluids6010030

Received: 9 December 2020
Accepted: 5 January 2021
Published: 8 January 2021

Publisher's Note: MDPI stays neutral with regard to jurisdictional claims in published maps and institutional affiliations.

Copyright: © 2021 by the authors. Licensee MDPI, Basel, Switzerland. This article is an open access article distributed under the terms and conditions of the Creative Commons Attribution (CC BY) license (https://creativecommons.org/licenses/by/4.0/).

1. Introduction

The direction of the prevailing temperature gradient is known to be an essential factor determining the properties of many distinct forms of thermal convection. This concept applies to both the stability scenario (i.e., the hierarchy of bifurcations undergone by the considered system when the applied temperature difference is increased) and the related patterning behavior. Archetypal situations for which the response of fluid systems to the application of an imposed temperature difference has been widely explored in the past are the cases where the thermal gradient is parallel or perpendicular to a fundamental reference direction typical of the considered problem.

For standard thermogravitational flow, i.e., fluid convection induced by buoyancy effects in non-thermally homogeneous fluids, this "reference direction" is obviously represented by gravity. Two fundamental conditions have enjoyed widespread attention over the years, leading to a true dichotomy in the literature, i.e., the cases where the angle Θ between the temperature gradient and gravity is either $0°$ or $90°$. The former circumstance corresponds to the paradigm of Rayleigh–Bénard (RB) convection (namely fluid motion in a domain uniformly heated from below and cooled from above), the latter (originally studied by Hadley [1]) can be technically realized considering systems laterally bounded by vertical walls kept at different constant temperatures.

This rigid classification can also be found in companion problems where gravity is replaced by surface-tension effects (that is the case of fluid systems with a free surface, i.e., an interface where a liquid is in contact with another immiscible fluid). Fluid flow takes a different nomenclature according to the angle formed by the prevailing temperature gradient and the free interface. In particular, $\Theta = 0^\circ$ and 90° correspond to the so-called thermocapillary and Marangoni–Bénard (MB) convection types, respectively.

Interestingly, buoyancy and surface-tension driven convection share more than a simple similarity in the related nomenclature. For both cases, a certain temperature difference (ΔT) must be exceeded before convection is produced when the temperature gradient is vertical (i.e., a "critical" ΔT exists), whereas fluid starts to move immediately (regardless of the effective magnitude of ΔT) when this gradient is horizontal.

However, while Marangoni–Bénard convection has been investigated essentially in infinite layers or domains uniformly heated from below with large horizontal extension, for Rayleigh–Bénard convection, a significant number of studies have been appearing where insights have been sought from consideration of the simplified setting corresponding to a square cavity. This configuration has enjoyed a widespread attention over the years owing to its simplicity and the concurrent possibility to produce flows that display or break the various intrinsic symmetries of this geometry (reflectional symmetries with respect to the vertical, horizontal, and diagonal directions). Moreover, it is known that if the Rayleigh number is increased beyond a certain critical threshold, the flow established inside the cavity can develop a complex sequence of instabilities (steady$\rightarrow$time-periodic$\rightarrow$low-dimensional chaos$\rightarrow$fully turbulent states) even if fluid motion is forced to maintain a two-dimensional (2D) structure. Owing to space limitations, we do not strive to review the incredible amount of work existing on this subject. Mizushima [2] provided a complete characterization of the critical Rayleigh number for the onset of convection from the diffusive state considering convective disturbances with different possible symmetries. Mizushima and Adachi [3] concentrated on the ensuing non-linear regime, illustrating that the combination of fundamental modes that become critical for similar values of the Rayleigh number can result in different patterns. Relevant examples of typical dynamics can also be found in the numerical study by Goldhirsch et al. [4], where several complicated flow structures and textural transitions were observed together with multi-stability effects, i.e., the possibility to obtain different solutions (for a fixed geometry and boundary conditions) depending on the considered initial conditions.

For the turbulent regime, the reader is referred to the interesting theoretical works by Villermaux [5] and Kadanoff [6] and the more recent review (specifically devoted to the square cavity) by Lappa [7] (and all references therein).

Although the field of RB convection has reached a level of maturity, in the sense that various techniques are in position to address almost any question, we may ask regarding the various possible states or regimes for this type of flow—unfortunately, the companion problem where steady gravity is replaced by an acceleration changing periodically in time has not received considerable (similar) attention.

This flow is generally referred to as 'thermovibrational convection' because the most obvious (simplest) way to create an acceleration that varies sinusoidally in time is to apply vibrations to a fluid container. The average value of the acceleration produced in this way is zero, which explains why this form of convection is extremely relevant to the microgravity environment (conditions established onboard the International Space Station and other orbiting platforms), and it has recently enjoyed a remarkable resurgence in interest.

Similar to all the other types of thermal convection discussed before, also in this case, the angle between the imposed temperature gradient and the direction of the acceleration can have a remarkable impact on the emerging flow. In addition to the acceleration amplitude, the imposed ΔT, and the aforementioned angle Θ, an additional degree of freedom influencing the dynamics is represented by the 'frequency' of vibrations. Nevertheless, only a limited number of studies are available in the literature where this variant of buoyancy flow was considered (Monti et al. [8]; Alexander [9]; Alexander et al. [10,11]; Feonychev and Dolgikh [12]; Gershuni and Lyubimov [13]; Monti et al. [14]; Naumann [15]; Mialdun et al. [16]; Melnikov et al. [17]; Lyubimova et al. [18]; Bouarab et al. [19]; Shevtsova et al. [20,21]; Maryshev et al. [22]; Vorobev and Lyubimova [23]; Lappa, [24–28]; Lappa and Burel [29]). Still fewer articles have been devoted to the case where vibrations and temperature gradient are parallel. Moreover, most of them examined cases where vibrations were combined with steady gravity (modulated gravity), leading to results (see, e.g., Lappa [30] for an exhaustive review), which have limited translational relevance to pure thermovibrational convection.

To the best of our knowledge, only Hirata et al. [31] considered the pure thermovibrational flow in a square cavity assuming zero gravity and no inclination between vibrations and the temperature gradient. In the present study, an attempt is made to extend that earlier investigation to larger values of the Prandtl number and the Rayleigh number and elaborate a unified picture of the related hierarchy of instabilities and patterning behavior.

2. Mathematical Model

2.1. The Geometry

In keeping with a large portion of the work outlined in the introduction for standard RB convection, a simple 2D square cavity is considered. Microgravity conditions are assumed (no steady gravity). The direction of the temperature gradient and vibrations as well as the wall boundary conditions can be seen in Figure 1.

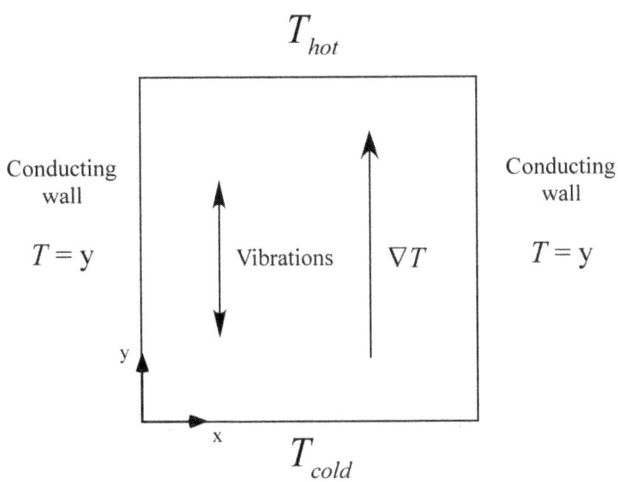

Figure 1. Square cavity with characteristic size L, delimited by solid walls (one at y = 0 cooled, the other at y = 1 heated, perfectly conducting conditions on the remaining sidewalls: T = y for x = 0 and x = 1).

The buoyancy force responsible for fluid motion is produced by a sinusoidal displacement of the cavity with respect to time in addition to a temperature difference imposed along the same direction of shaking (along the y axis in Figure 1). The time periodic displacement (i.e., the vibrations) can be modeled mathematically as:

$$\underline{s}^{lab}(t) = \text{bsin}(\omega\,\text{t})\underline{\hat{n}} \tag{1}$$

where b is the amplitude, $\omega = 2\pi f$ is the angular frequency of the displacement, and $\underline{\hat{n}}$ is the unit vector along the direction of vibrations. The ensuing time-varying acceleration can formally be obtained by taking the second derivative of Equation (1), which reads:

$$\underline{g}(t) = g_{\omega} \sin(\omega \mathrm{t})\underline{\hat{n}} \text{ where } g_{\omega} = \mathrm{b}\,\omega^2 \tag{2}$$

and satisfies the condition:

$$\frac{\omega}{2\pi}\int_0^{2\pi/\omega} \underline{g}(t)dt = \frac{\omega}{2\pi} g_{\omega}\underline{\hat{n}} \int_0^{2\pi/\omega} \sin(\omega \mathrm{t})dt = -\frac{1}{2\pi} g_{\omega}\underline{\hat{n}}[\cos(\omega \mathrm{t})]_0^{2\pi/\omega} = 0 \tag{3}$$

which shows that its time-averaged value over one period of oscillation $2\pi/\omega$ is zero.

2.2. Balace equations and boundary conditions

Using the Boussinesq approximation, and scaling the Cartesian coordinates (x,y), time (t), velocity ($\underline{V}$), pressure (p), and temperature (T) by the reference quantities L, L^2/α, α/L, $\rho\alpha^2/L^2$,and ΔT, respectively (where α is the fluid thermal diffusivity and ρ is the fluid density), the balance equations for mass, momentum, and energy can be cast in compact form as:

$$\underline{\nabla} \cdot \underline{V} = 0 \tag{4}$$

$$\frac{\partial \underline{V}}{\partial t} = -\underline{\nabla} p - \underline{\nabla} \cdot [\underline{V}\underline{V}] + \mathrm{Pr}\nabla^2\underline{V} + \mathrm{Pr}Ra_{\omega}T\sin(\Omega t)\underline{\hat{n}} \tag{5}$$

$$\frac{\partial T}{\partial t} + \underline{\nabla} \cdot [\underline{V}T] = \nabla^2 T \tag{6}$$

where:

$$\mathrm{Pr} = \frac{\nu}{\alpha} \tag{7}$$

$$\Omega = \frac{\omega L^2}{\alpha} \tag{8}$$

$$Ra_{\omega} = \frac{b\omega^2\beta_T\Delta T L^3}{\nu\alpha} \tag{9}$$

and β_T is the thermal expansion coefficient, ν is the fluid kinematic viscosity, and ΔT is the imposed temperature difference. As the reader will easily realize, Ω is the non-dimensional frequency of vibrations and Ra_{ω} might be seen as a variant of the classical Rayleigh number (based on the amplitude of the vibrations-induced acceleration $\mathrm{b}\omega^2$ in place of the standard gravity g).

Since in the present work, the vibrations and the imposed temperature gradient are both directed along the y axis, the projection of the momentum equation on the coordinate axes shown in Figure 1 reads:

$$\frac{\partial u}{\partial t} = -\frac{\partial p}{\partial x} - (\underline{V} \cdot \underline{\nabla V})_u + \mathrm{Pr}(\Delta \underline{V})_u \tag{10}$$

$$\frac{\partial v}{\partial t} = -\frac{\partial p}{\partial y} - (\underline{V} \cdot \underline{\nabla V})_v + \mathrm{Pr}(\Delta \underline{V})_v + \mathrm{Pr}Ra_{\omega}T\sin(\Omega t) \tag{11}$$

where:

$$(\underline{V} \cdot \underline{\nabla V})_u = \left(u\frac{\partial u}{\partial x} + v\frac{\partial u}{\partial y}\right),\ (\Delta \underline{V})_u = \left(\frac{\partial^2 u}{\partial x^2} + \frac{\partial^2 u}{\partial y^2}\right) \tag{12}$$

$$(\underline{V} \cdot \underline{\nabla V})_v = \left(u\frac{\partial v}{\partial x} + v\frac{\partial v}{\partial y}\right),\ (\Delta \underline{V})_v = \left(\frac{\partial^2 v}{\partial x^2} + \frac{\partial^2 v}{\partial y^2}\right). \tag{13}$$

These equations must be integrated with the relevant initial and boundary conditions, which in the present work are set as follows:

$$T = y \text{ and } u = v = 0 \text{ for } 0 \leq x \leq 1, 0 \leq y \leq 1 \text{ and } t = 0 \quad (14)$$

i.e., quiescent conditions and linear temperature distribution along y at the initial instant (t = 0) and

$$T = y \text{ and } u = v = 0 \text{ for } x = 0, x = 1, 0 \leq y \leq 1 \text{ and } t > 0 \quad (15)$$

$$T = 0 \text{ and } u = v = 0 \text{ for } y = 0, 0 \leq x \leq 1 \text{ and } t > 0 \quad (16)$$

$$T = 1 \text{ and } u = v = 0 \text{ for } y = 1, 0 \leq x \leq 1 \text{ and } t > 0 \quad (17)$$

i.e., no-slip conditions for all the solid walls, conducting thermal conditions for the walls parallel to imposed vibrations, and different constant temperatures on the perpendicular boundaries.

2.3. Embedded Symmetries

Notably, following Mizushima [2], assembled in this way, these equations and the related boundary conditions allow meaningful "a priori" reflections about the possible symmetries of the emerging convective modes in a square cavity, which can be categorized as follows:

- (ss): The symmetric–symmetric mode. This mode is characterized by an even number of rolls along the two coordinate axes. It reduces to a configuration with the central symmetry if the same number (m) of rolls affects both the x and y directions, i.e., $m_x = m_y$ (whereas a columnar arrangement is obtained if $m_y > m_x$).
- (sa): The symmetric–antisymmetric mode. This mode displays symmetry only with respect to the y-axis; accordingly, the flow typically features an odd number of rolls along y and an even number of rolls along the other axis.
- (as): The antisymmetric–symmetric mode. This mode displays symmetry only with respect to the x-axis; accordingly, the flow typically features an odd number of rolls along x and an even number of rolls along the other axis.
- (aa): The antisymmetric–antisymmetric mode. No symmetry is retained in this case, as the number of rolls is odd along both axes (a single column being obtained for $m_y > m_x = 1$).

This classification, largely used in the past (see, e.g., Mizushima [2] and Mizushima and Adachi [3]) to interpret the typical patterning behavior of standard buoyancy convection in square cavities heated from below and cooled from above (Rayleigh–Bénard convection), will be applied in the present work to characterize the spatial properties of the emerging thermovibrational flow.

3. Numerical Method

The integration of Equations (4)–(6) in addition to the initial and boundary conditions allows for the unknown pressure (p), velocity ($\underline{V}$), and temperature (T) fields to be found. The related procedure (time-marching algorithm) is described in the present section.

Along these lines, it is worth starting from the simple observation that, as implicitly made evident by the aforementioned set of equations, these three fundamental physical quantities ("primitive variables") display a varying degree of interrelation, depending on the specific couple considered. As an example, while $\underline{V}$ and p are intimately linked through the momentum equation, the temperature field (T) can be determined once the velocity field is known through the energy equation (Equation (6)).

In particular, the link between the first two unknowns is at the root of the so-called class of projection or fractional methods (Harlow and Welch [32]; Chorin [33]; Temam [34]; Gresho and Sani [35]; Gresho [36]; Guermond and Quartapelle [37]; Guermond et al. [38]). These techniques rely on the so-called Hodge decomposition theorem, which states that any vector field can be decomposed into a divergence-free contribution and the gradient of a scalar potential (a curl-free part). Stripped to its essentials, the related computational scheme can synthetically be described as follows. Initially, the pressure is artificially neglected in the balance of momentum in order to obtain an equation (Equation (18)) where only the velocity field requires solving for:

$$\frac{\partial \underline{V}^*}{\partial t} = -\underline{\nabla} \cdot [\underline{V}\underline{V}] + \mathrm{Pr}\nabla^2 \underline{V} + \mathrm{Pr}Ra_\omega T \sin(\Omega t)\underline{\hat{n}}. \tag{18}$$

In this way, even though the pressure is initially unknown, a time-marching procedure can be started. However, the field $\underline{V}^*$ obtained through integration of this equation is called 'provisional' because, obviously, it does not account for the impact of pressure on fluid flow; moreover, it does not satisfy the incompressibility constraint (represented by the separate equation for the balance of mass). Nevertheless, using the aforementioned Hodge decomposition theorem, $\underline{V}^*$ can formally be split into two contributions as follows:

$$\underline{V}^* = \underline{V} + C\underline{\nabla}p \tag{19}$$

where $\underline{V}$ and $\underline{\nabla}p$ play the role of divergence-free vector and the gradient of a scalar potential, respectively. This step is purely formal, as p is one of the unknowns. The next conceptual ingredient needed to obtain a complete time-marching procedure that consists of forcing $\underline{V} = \underline{V}^* - C\underline{\nabla}p$ into Equation (4). In this way, indeed, a 'working' equation for the effective determination of the pressure is obtained:

$$\nabla^2 p = \frac{1}{\Delta t}\underline{\nabla} \cdot \underline{V}^*. \tag{20}$$

This equation represents the 'core' of all variants pertaining to the aforementioned class of projection (or fractional) methods. After Equation (18) has been integrated, the pressure can be determined solving Equation (20); finally, the sought divergence-free velocity field can be computed from Equation (19) as $\underline{V} = \underline{V}^* - C\underline{\nabla}p$ (assuming $C = \Delta t$ where Δt is the time integration step).

The well-posedness of this approach is guaranteed by another important fundamental theorem—that is, the so-called theorem of the inverse calculus (see, e.g., Ladyzhenskaya [39]); it states that a vector field is uniquely determined when its divergence and curl are assigned; in the present case, these are $\underline{\nabla} \cdot \underline{V} = 0$ and $\underline{\nabla} \wedge \underline{V} = \underline{\nabla} \wedge \underline{V}^*$, where the latter equality follows from the well-known mathematical property of the curl operator to annihilate the gradient of a scalar function, i.e., $\underline{\nabla} \wedge \underline{\nabla}p = 0$ (Gresho [36]; Lappa [40]; Lappa and Boaro [41]).

In the present work, we have used the specific variant of this class of methods available in the OpenFOAM computational platform (OpenCFD Ltd 2019, London, UK), i.e., the so-called PISO (Pressure Implicit Split Operator) approach (originally elaborated by Issa [42]). The OpenFOAM implementation of this method relies on a collocated grid approach, which means that the unknowns are defined in the center of the computational cells. Moreover, in order to improve the coupling of velocity and pressure, a special interpolation of the velocity is applied on the cell faces (Rhie and Chow [43]), while the third unknown T is determined in a segregated manner after the computation of $\underline{V}$ and p. Finally, we wish to remark that we have used backward differencing in time and upwind differencing schemes in space for both the convective term and diffusion term, while the solution of Equation (20) has been based on a Generalized Geometric-Algebraic Multi-Grid (GAMG) strategy.

4. Validation

Before numerical results can be interpreted, it is imperative that the strategy for the solution of the governing equations is validated against available relevant benchmarks. Given our specific target, the earlier study by Hirata et al. [31] is specifically considered for such a purpose. In particular, three cases are chosen: (a) $\Omega = 200$, $Ra_\omega = 7 \times 10^4$, (b) $\Omega = 500$, $Ra_\omega = 10^5$, and (c) $\Omega = 200$, $Ra_\omega = 10^5$ (the velocity signals for these cases are readily available to the reader in the original study, making the comparison with the current results straightforward). As the reader will realize by inspecting Figure 2, the present results are in excellent agreement with the original signals reported by Hirata et al. [31], both in shape and periodicity.

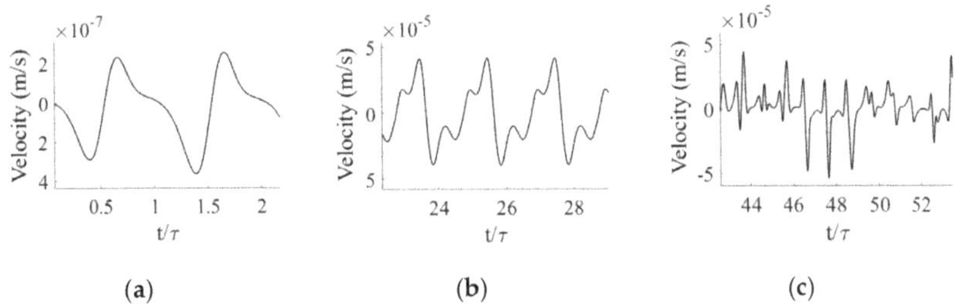

Figure 2. Time evolution of velocity at (**a**) Ω = 200, Ra_ω = 7 × 10^4, (**b**) Ω = 500, Ra_ω = 10^5, and (**c**) Ω = 200, Ra_ω = 10^5, respectively. Corresponding to cases (b), (c), and (d) of Figure 4 in Hirata et al. [31].

For the sake of completeness, additional validation has been obtained considering quantitative comparison with other well-known benchmarks in the literature concerned with classical RB convection in a square cavity ([44,45]). These comparisons have already been published in Lappa and Inam [46] and are not duplicated here for the sake of brevity.

5. Grid Refinement Study

Due to the potential complexity of the flow considered in this study, close attention must also be given to the grid adopted for the numerical simulations. Although mesh refinement criteria for thermovibrational flow are not available, meaningful indications in this regard can be obtained through 'analogies' with parent forms of convection. As an example, relevant similarities with standard gravitational convection in high-Pr fluids include the development of thermal boundary layers for relatively high values of the Rayleigh number. For the case of standard buoyancy flow, as an example, Russo and Napolitano [47] showed that a working correlation for the thickness of the thermal boundary layer can be introduced as follows:

$$\delta_{BL} \cong Ra^{-1/4}. \quad (21)$$

The presence of such boundary layers cannot be ignored when designing an adequate mesh and can be translated into precise numerical requirements. As an example, relevant information along these lines can be found in the study on RB convection by Shishkina et al. [48], where the number of cells required in the thermal boundary layer has been specified directly as a function of *Ra*, i.e.,

$$N_{BL} \cong 0.35Ra^{0.15}. \quad (22)$$

Another important influential factor to be taken into account (especially when one targets high-*Ra* regimes) in the preliminary definition of a computational grid is the so-called Kolmogorov length scale, i.e., the need to keep the size of the computational cell sufficiently small to capture the 'eddies' that are produced when the flow assumes a turbulent behavior. It is known that for standard RB convection (De et al. [49]), this characteristic (non-dimensional) length scales as

$$\zeta_{Ra} = 1.336(Ra)^{-0.32}. \quad (23)$$

All these criteria should be regarded as a set of multiple requirements finally leading to a relevant mesh. Assuming the worst conditions considered in the present work, i.e., the highest possible value of Ra_ω, all these constraints taken together would return a uniform mesh 102 × 102.

To verify the consistency of this way of thinking with the standard approach generally used to define a suitable mesh (i.e., a 'classical grid refinement study'), we have increased progressively the number of computational nodes until convergence has been obtained. The outcomes of such a study are presented in Figure 3, where the maximum of the difference between the instantaneous temperature and the corresponding time-averaged (over the period of vibrations) value has been plotted (the so-called thermofluid-dynamic distortion). It can be seen that the results for a grid of size 82 × 82 are extremely close to that of the grid size 102 × 102, which implicitly indicates that (toward the end to save computational time) one may limit to considering the former coarser mesh. However, in order to meet all the possible criteria described before (see Table 1), we have decided to use the 102 × 102 mesh for all the cases considered in the present work.

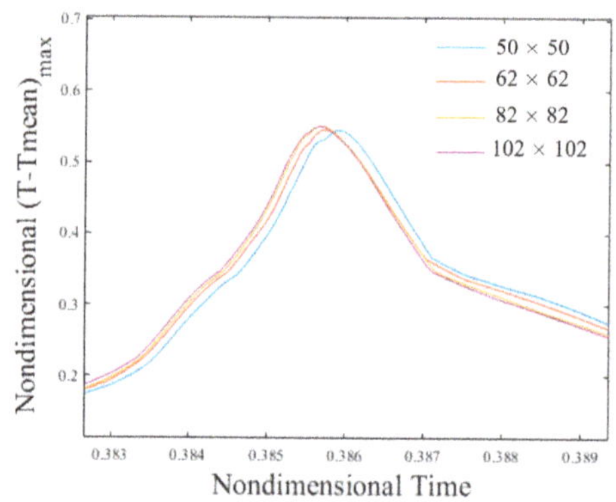

Figure 3. Convergence of the thermofluid-dynamic disturbances as a result of grid refinement for the case Pr = 7, $Ra_\omega = 10^6$, $\Omega = 100$.

Table 1. 102 × 102 grid results versus theoretical requirements.

Criteria	Theoretical Values	Values Provided by the Numerical Simulation
Cell size determined by Kolmogorov length scale	1.61×10^{-2}	9.80×10^{-3}
Boundary layer thickness (Russo and Napolitano)	3.16×10^{-4}	2.98×10^{-2}
Number of cells required in boundary layer (Shishkina et al.)	3	3

As an additional layer of validation, comparisons with an in-house explicit code have also been made (this is presented a posteriori in Section 6.2, where the same mesh resolution of 102 × 102 is used).

6. Results

With the numerical approach robustly tested by (1) verifying its ability to capture available results in the literature and (2) checking its convergence under mesh refinement (forcing the used grids to also satisfy existing practical and theoretical criteria about mesh design), we have then simulated a vast range of cases by varying parametrically the vibrational Rayleigh number and the frequency of vibrations over three orders of magnitude. In particular, the same value of the Prandtl number originally considered by Lappa [7] has been examined (Pr = 15) with the two-fold purpose of (1) extending that earlier study conducted for classical RB convection in a square cavity to the case of thermovibrational flow and (2) expanding the space of parameters originally examined by Hirata et al. [31]. In order to consider conditions for which the flow can still be considered laminar or weakly chaotic (the investigation of the fully turbulent regime being beyond the scope of the present work), the maximum value of the Rayleigh number has been limited to 10^6.

The work/study progresses with the aid and support of both global parameters and detailed velocity fields for a better representation of the emerging dynamics. While a coarse-grained macroscopic perspective is used at the beginning by providing results in terms of maps (Sections 6.1 and 6.2) and general trends in terms of 'distortions' (Section 6.3) and Nusselt number (Section 6.4), the problem is considered from a fine-grained micromechanical level in Section 6.5 (in terms of flow temporal behavior and related symmetries). We finally link (Section 7) the resulting statistics to the evolution of global parameters to provide useful information about the underlying cause-and-effect relationships.

6.1. Regime Classification

As the present study has been expressly conceived as an extension of the numerical investigation originally conducted by Hirata et al. [31], the simplest way to place the present results in an adequate context is to start from an overview of that work (see Figure 4).

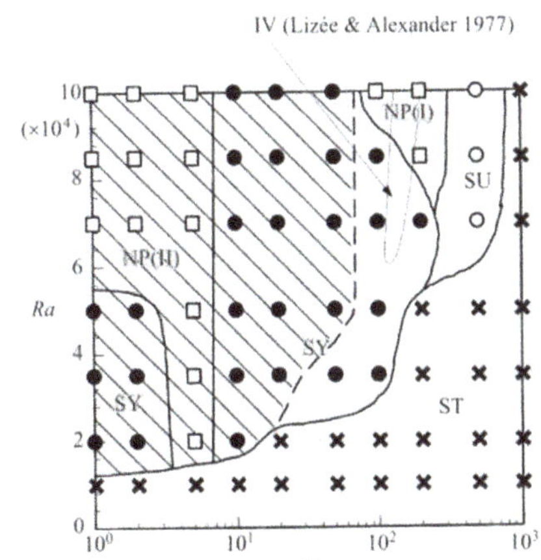

Figure 4. Periodicity map for Pr = 7: ● Syncronous case (SY); ○ 1/2 Subharmonic case (SU); □ Non-periodic case (NP), ✖ Stable case (ST). The shaded area represents the cases where the flow is stationary over a certain time interval. Figure after [31].

The diversity of flow regimes that exist for pure thermovibrational convection in a square cavity when the temperature gradient is imposed parallel to the vibration can be clearly seen in this figure. Hirata et al. [31] split these regimes into four possible categories, namely, Synchronous (SY), Subharmonic (SU), Non-periodic (NP), and Stable (ST) solutions. Unfortunately, these authors limited themselves to considering values of the vibrational Rayleigh number in the range ($10^4 \leq Ra_\omega \leq 10^5$, the constraint on the upper value being essentially an outcome of the limited computational resources available at that time). This figure is instructive also for another reason. It shows the well-known stabilization of thermovibrational flow when the frequency of vibrations is increased (Simonenko and Zen'kovskaja [50]; Simonenko [51]; Gershuni and Zhukhovitskii [52]; Gershuni et al. [53]; Gershuni and Zhukhovitskii [54]); we will come back to this fundamental concept later.

A total of 66 new simulations have been conducted in the present work. In line with Hirata et al. [31], Ω has been varied from $\Omega = 1$ to $\Omega = 10^3$, with the addition of a very high value of $\Omega = 10^4$. As explained before, a fluid with Pr = 15 has been considered in place of Pr = 7. Another distinguishing mark of the present analysis is the extension to $Ra_\omega = 10^6$.

Before starting to deal with the detailed discussion of these results, we wish to anticipate that although the regimes (SY, SU, NP, and ST) were sufficient to compartmentalize the flow in the range of parameters considered by Hirata et al. [31], additional classes of flow have been identified for the cases simulated here in the interval ($10^5 < Ra_\omega < 10^6$). These new findings have implicitly led to the need to introduce a distinction within the synchronous regime whereby a flow may be synchronous and periodic or synchronous and non-periodic (SY-P or SY-NP). Relevant examples of such new solutions are shown in Figure 5a,b, respectively. In particular, Figure 5a relates to the circumstances where the velocity signal is identical over each period for all periods (therefore, the signal can be considered synchronous and periodic). By contrast, although the signal shown in Figure 5b is still synchronous in time with the applied forcing (i.e., the vibrations), it also exhibits turbulent busts every period; moreover, the signal is not periodic, as the bursts display a more or less erratic evolution in time.

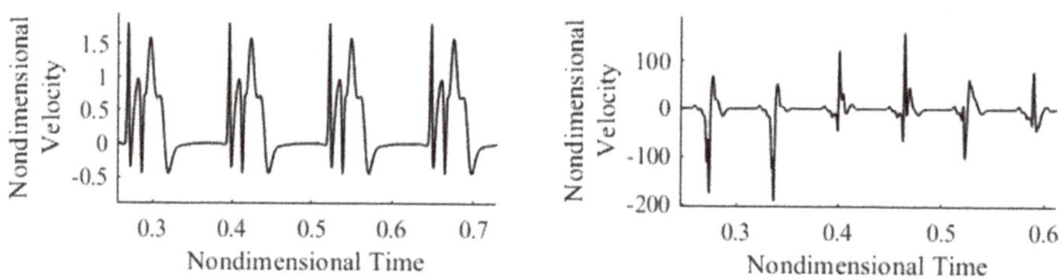

(a) SY-P (b) SY-NP

Figure 5. Cases $Ra_\omega = 2 \times 10^5$, $\Omega = 50$ and $Ra_\omega = 7 \times 10^5$, $\Omega = 100$ respectively (**a**) shows the case synchronous and periodic(SY-P) and (**b**) the case synchronous and non-periodic(SY-NP).

6.2. Map Extension

The main outcome of the present parametric investigation, i.e., the extended map, is presented in Figure 6. It is divided into regions where clusters of convective modes can be observed. Although stabilization is achieved at extremely high frequencies ($\Omega = 10^4$), the predominant flow regime apparent for high vibrational Rayleigh numbers ($Ra_\omega > 5 \times 10^5$) is the aforementioned SY-NP case denoted by ▲. The red sinusoidal signal included in the insets represents the forcing applied to the system; it is instrumental in making evident that the turbulent bursts for the SY-NP mode occur synchronously with the forcing but do not display the same behavior for each period. These solutions are less ordered than those found for the lower vibrational Rayleigh numbers whereby they appear in less organized clusters. The appearance of the SY-P regime is more sporadic. However, this mode of convection also exists at extremely high Rayleigh numbers ($Ra_\omega = 10^6$), which indicates that an increase in Ra_ω does not systematically lead to a more chaotic system (in some regions of parameters ($\Omega = 10^3$), the flow reverts from an NS-NP back to an SY-P state).

In agreement with Hirata et al. [31], in the range of low frequencies $\Omega < 200$, the fluid displays a stationary behavior over a certain sub-interval of the period τ ($= 2\pi/\Omega$) of the applied vibrations.

In addition to enlightening the reader on the complexity and unpredictability of the flow regime (with the exception of extremely high values of Ω for which the dynamics reduce to the emergence of a quiescent thermally diffusive state), this map provides a picturesque description of the velocity signals and the fluid behavior.

The *thermal response* of the system is described in the next section where some dedicated ('ad hoc') definitions and concepts are introduced.

However, before starting to deal with thermal effects, we wish to remark that an extra layer of validation (with respect to that illustrated in Section 4) has been implemented comparing the results obtained here for the new cases (using OpenFOAM) with those provided by the same code used by Lappa [25] and Lappa and Burel [29] (see Figure 7). Both OpenFOAM and this code pertain to the same class of pressure–velocity methods discussed before. However, while OpenFOAM is based on an implicit approach (for what concerns the time integration) and on a collocated distribution of unknowns, the computational platform used by Lappa [25] and Lappa and Burel [29] relies on an explicit approach and a 'staggered' arrangement of variables, respectively.

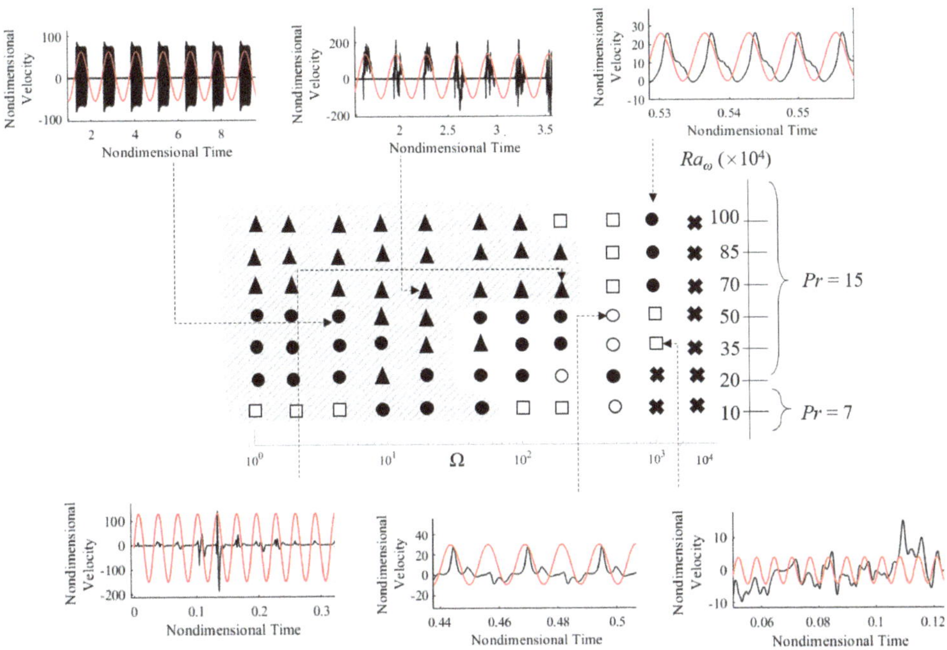

Figure 6. Response of the velocity field to the imposed periodic acceleration (Pr = 15): ● Synchronous and periodic case (SY-P); ○ 1/2 Subharmonic case (SU); ▲ Synchronous and non-periodic case (SY-NP); □ Non-periodic and non-synchronous case (NP-NS); Stable case (ST) ✖ (quiescent states). The shaded area represents the cases where the flow is stationary over a certain sub-interval of the period of the applied vibrations.

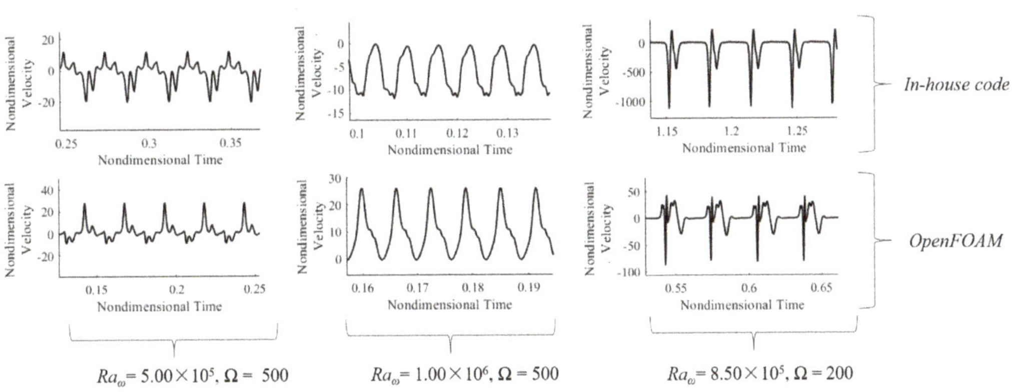

Figure 7. Validation of OpenFOAM solver against an in-house solver for various areas of the map.

It can be seen that despite the differences highlighted above, the shape and periodicity of the signals produced by both codes are in agreement (the discrepancy in signal amplitude can be expected as the exact location of the probes and the positive direction of the axes of the reference Cartesian system differ from solver to solver).

6.3. Themofluid-Dynamic Disturbances

A further understanding of the observed dynamics can be gained through the so-called thermofluid-dynamic (TFD) distortions. These characteristic quantities have enjoyed a widespread use in past studies concerned with the effect of vibrations on non-isothermal fluid systems (see, e.g., Monti et al. [14]). They can be used to characterize in a synthetic way the thermal response of the fluid to the application of a time-varying acceleration.

However, a proper introduction of these characteristic quantities requires a short excursus on the peculiar properties of thermovibrational flows. In particular, it is worth recalling that a non-isothermal fluid subjected to vibrations can develop a stationary response in addition to the oscillatory velocity field directly induced by the time-periodic acceleration. The latter can easily be explained assuming a straightforward cause-and-effect relationship between the time-varying buoyancy force and the induced fluid motion. The former requires a more involved interpretation. This stationary response (detectable through analysis of the time-averaged flow field) is an outcome of the non-linear nature of the balance equations (Savino and Lappa [55]). It becomes significant when the frequency of vibrations is sufficiently high and their amplitude is relatively small, i.e., in the so-called Gershuni regime (Simonenko and Zen'kovskaja [50]; Simonenko [51]; Gershuni and Zhukhovitskii [52]; Gershuni et al. [53]; Gershuni and Zhukhovitskii [54]; Savino and Lappa [55]; Lappa [30]). Indeed, for the opposite circumstances for which the frequency is small and the amplitude large, the linear response (direct proportionality between the oscillatory flow and the time-dependent acceleration) is dominant [55]. The time-averaged and fluctuating components of the velocity and temperature fields can formally be defined as:

$$\overline{\underline{V}} = \frac{\Omega}{2\pi}\int_0^{2\pi/\Omega} \underline{V}dt, \overline{T} = \frac{\Omega}{2\pi}\int_0^{2\pi/\Omega} Tdt \tag{24}$$

and

$$\underline{V}' = \underline{V} - \overline{\underline{V}}, T' = T - \overline{T}. \tag{25}$$

In the present work, these quantities have been determined "a posteriori" after evaluating $\underline{V}$ and T via direct numerical solution of the governing equations in their complete time-dependent and non-linear form, as illustrated in Section 3.

The above-mentioned distortions can be defined accordingly as follows:

$$\delta T(x,y,t) = T(x,y,t) - T_{diff}(x,y) \tag{26}$$

where T_{diff} represents the temperature field that would be established in the absence of convection (in other words, a purely diffusive temperature profile, which using the reference system indicated in Figure 1 would simply read T_{diff} = y).

Taking into account that (as illustrated above) the local temperature can be split into a time-averaged steady component plus a fluctuating part ($T_j = \overline{T}_j + T_j{}'$), Equation (26) can be further expanded as:

$$\delta T = T'(x,y,t) + \overline{T}(x,y,t) - T_{diff}(x,y) = \overline{\delta T} + T'(x,y,t) \tag{27}$$

where $\overline{\delta T}$ represents the companion *averaged distortion*, i.e.,

$$\overline{\delta T} = \overline{T}(x,y,t) - T_{diff}(x,y). \tag{28}$$

Global measures can be defined accordingly as:

$$\text{TFD} = \max(\delta T) \text{ for } 0 \le x \le 1, 0 \le y \le 1\ t_o \le t \le t_o + \tau \text{ (where } \tau = 2\pi/\Omega) \tag{29}$$

$$\text{TFD}_{\text{averaged}} = \max\left(\overline{\delta T}\right) \text{ for } 0 \leq x \leq 1,\ 0 \leq y \leq 1 \tag{30}$$

These quantities are reported in Figures 8 and 9 for different circumstances (δT and $\overline{\delta T}$ represented by dashed and solid lines, respectively).

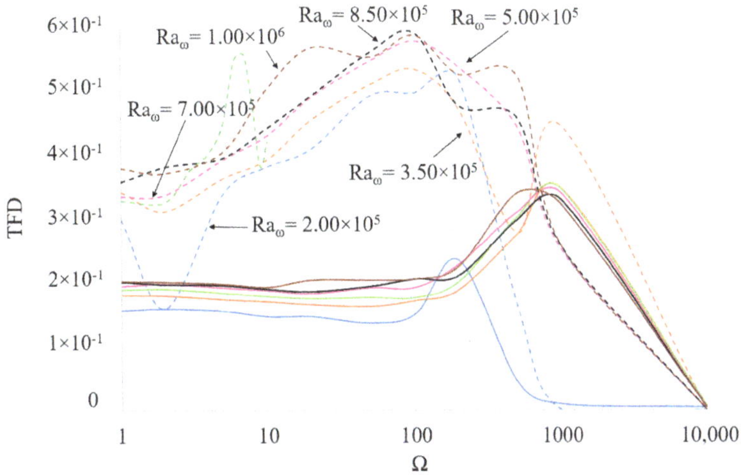

Figure 8. Influence of Ω on the global thermofluid-dynamic (TFD) disturbances (the dashed and solid lines indicating instantaneous and time-averaged variants, respectively).

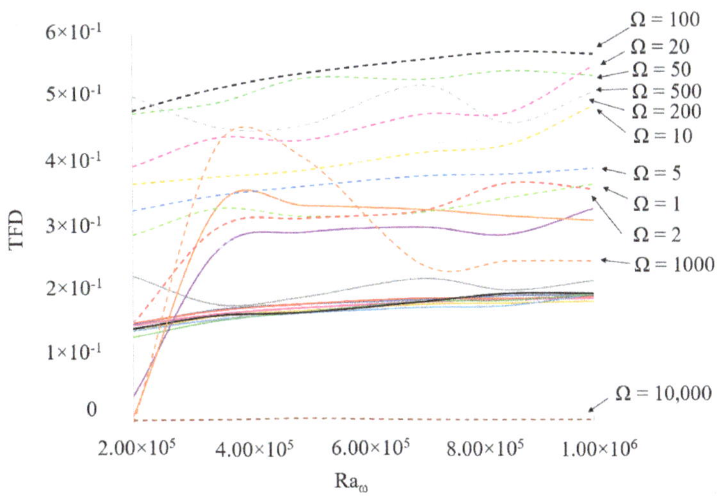

Figure 9. Influence of Ra_ω on the global thermofluid-dynamic (TFD) disturbances (the dashed and solid lines indicating instantaneous and time-averaged variants, respectively).

As quantitatively substantiated by these figures, the oscillatory thermofluid dynamic disturbance (TFD) is generally higher that the time-averaged one over the considered range of frequencies.

In particular, the time-averaged disturbances are approximately constant if their dependence on either Ra_ω or Ω is considered until the critical value of $\Omega = 500$ is attained, where these disturbances are seen to increase (with the exception of the case $Ra_\omega = 2 \times 10^5$ for which the $TFD_{averaged}$ tends to 0).

For what concerns the oscillatory disturbances, an increase in TFD occurs until the critical value of $\Omega = 100$, while for $\Omega > 100$, the opposite trend can be seen.

Remarkably, all the TFD distortions tend to zero as the frequency grows. A simple way to think about this scenario is to consider that it reflects the existence of the almost quiescent states already reported in Figure 6. However, from a physical point of view, this trend can be interpreted directly, taking into account a well-known property of thermovibrational flow for high frequencies (for $\Omega \geq 10^4$, i.e., when the Gershuni regime is approached (Savino and Lappa [55]). As originally argued by Birikh at al. [56], indeed, in the limit as the frequency tends to infinite, if temperature distortions with respect to the purely diffusive case are present, the major role of the mean vibration force is that of forcing isotherms to turn and become perpendicular to the vibration direction. To elucidate further the significance of this observation, one should keep in mind that in other words, this simply means that an intrinsic property of thermovibrational convection induced by vibrations parallel to the imposed temperature difference is to tend naturally to a quiescent thermally diffusive state as Ω is increased (which provides the sought physical justification for the ST states reported in the existence map).

Apart from showing that the oscillatory disturbances prevail over the time averaged ones, Figures 8 and 9 are also instrumental in revealing that the increase of the vibrational amplitude (Ra_ω) affects the two types of distortions differently: as already explained to a certain extent before, the time-averaged disturbances seem quasi-independent of the increase in Ra_ω (or value of Ω in fact), whereas the instantaneous (complete) TFD are appreciably affected by both Ra_ω and Ω.

6.4. Evaluation of the Nusselt Number

In keeping with the previous section, further analysis of the thermal behavior of the system may be carried out by looking at another global parameter, i.e., the classical Nusselt number, namely the ratio of heat transfer due to convection over the heat transfer due to conduction along a given boundary. In our case, this non-dimensional number can be defined as:

$$Nu = \int_0^1 \frac{\partial T}{\partial y} dx \tag{31}$$

and we introduce accordingly

$$Nu_{max} = \max (Nu) \text{ for } t_o \leq t \leq t_o + \tau. \tag{32}$$

In the case of high frequencies, it has been shown in the previous section that when parallel to the temperature gradient, the vibrations have a stabilizing effect on the flow. This trend can still be appreciated when cases with $Ra_\omega > 10^5$ and the SY-NP and NS-NP regimes are considered. As witnessed by Figure 10, a remarkable decrease in Nu_{max} occurs for $\Omega = 1000$ (Nu_{max} ideally tending to 1 in the limit as $\Omega \to \infty$). However, as still evident in this figure, a peak is located $\Omega = 100$. For all values of Ra_ω at low frequencies ($\Omega < 20$), Nu_{max} remains constant; then, it grows for intermediate frequencies ($50 < \Omega < 100$) and finally decreases for high values of Ω ($\Omega > 1000$).

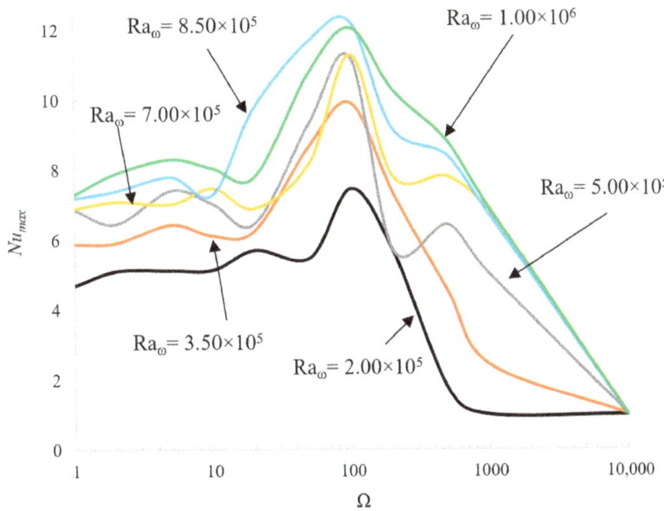

Figure 10. Influence of Ω on the maximum Nusselt number (Nu_{max}) across the heated wall of the cavity.

However, upon increasing Ra_ω (as expected), Nu_{max} tends to become higher for all values of Ω (Figure 11).

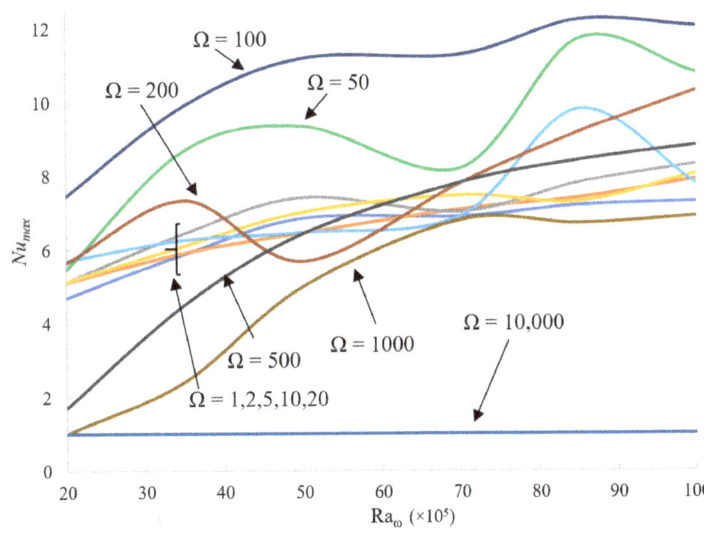

Figure 11. Influence of Ra_ω on the maximum Nusselt number (Nu_{max}) across the heated wall of the cavity.

Notably, the peak located at Ω ≅ 100 is consistent with the maximum taken by the TFD (see again Figure 8), and the key to understanding this finding lies in considering that, given the dominance of instantaneous effects on time-averaged ones, the effective configuration of the temperature field (in terms of topology of the isotherms and ensuing heat exchange at the boundaries) must essentially be ascribed to the fluctuating components of velocity and temperature. As already outlined above, the tendency of the Nusselt number toward the unit value as Ω is increased is indirect evidence of the fact that fluid motion tends to be suppressed in those conditions.

6.5. Streamlines and Patterning Behaviors

In this section, we finally concentrate on the effective patterning behavior of the flow for the different regimes reported in Figure 6. Emphasis is put on the interval $10^5 < Ra_\omega \leq 10^6$, as these circumstances were not covered in the earlier study by Hirata et al. [31].

Along these lines, we begin from the case $Ra_\omega = 10^6$, $\Omega = 10^4$, i.e., a condition for which the flow is almost negligible ("stable state"). As shown by Figure 12, it manifests itself as an (ss) convective mode characterized by two rolls along each coordinate axis. This extremely weak flow starts as a four-roll configuration. From there, small rolls nucleate at the corners of the cavity and, as time passes, they tend to merge with their respective neighbors until the original quadrupolar arrangement is recovered. This nucleation occurs twice in the space of a period rapidly regaining the four-roll configuration. The periodicity of this evolutionary scenario is consistent with the velocity signal (which is sinusoidal and synchronous with the forcing period).

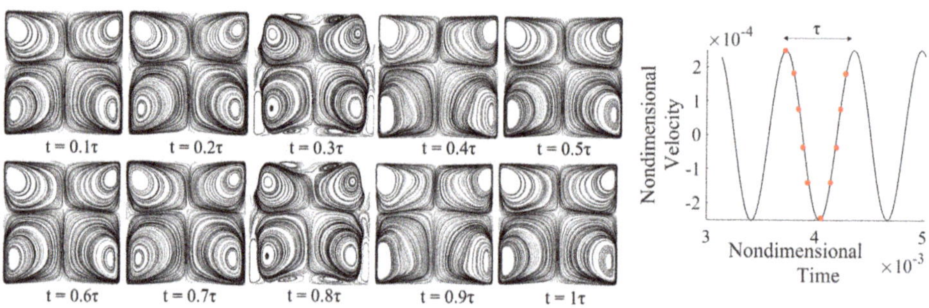

Figure 12. Instantaneous patterning behavior for the case ST, where it is shown that the nucleation of the external rolls occurs at approximately 0.3τ and 0.8τ and that the four-roll configuration is re-established fully when the acceleration tends to zero.

Another characteristic type of solution present in the map (Figure 6) is the synchronous and periodic state (SY-P). This mode of convection can be found mainly at the center and at the left side of the map for $Ra_\omega < 5 \times 10^5$. As illustrated in Figure 13, this regime presents periodically identical instantaneous velocity fields and streamlines. In this case, the quadrupolar (four-roll) configuration is interrupted at each period by the genesis of two small rolls in the center of the lower part of the cavity, which are eventually flattened, hence allowing the flow to return to the original pattern.

The next figure of the sequence (Figure 14) illustrates a Subharmonic case (SU). In this figure, the typical behavior of a subharmonic mode of convection can be recognized in both the velocity field and streamlines snapshots. The period of the flow is double with respect to that of the forcing. However, the signature of the forcing period τ can still be recognized if one considers the two spikes (one large and one small) visible in the signal (then, these spikes are repeated with a slightly lower amplitude in the second forcing period). This is also quantitatively substantiated by the panels (a) and (b), where the magnitude of the first velocity field is (slightly) higher than that of the second.

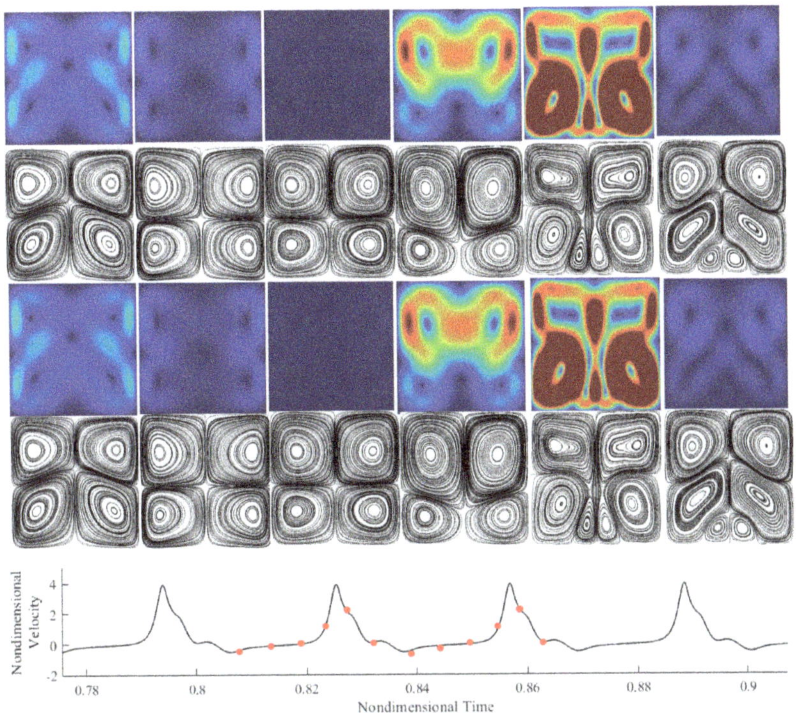

Figure 13. Instantaneous streamlines and velocity magnitude over two periods for the case $Ra_\omega = 3.5 \times 10^5$, $\Omega = 200$ (SY-P) accompanied by the velocity signal. The 12 red dots represent the time at which the snapshots are taken (six snapshots for each period).

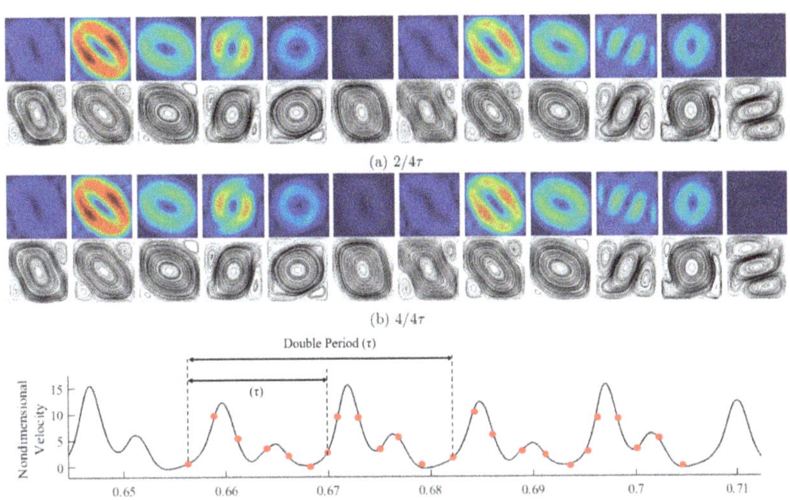

Figure 14. Instantaneous streamlines and velocity magnitude over two periods of forcing (panel (**a**): first period, panel (**b**): second period) for the case $Ra_\omega = 3.5 \times 10^5$, $\Omega = 500$ (SU), accompanied by the velocity signal. The 24 red dots represent the time at which the snapshots are taken (six snapshots for each period).

In terms of spatial symmetry, the (aa) type is dominant (one single roll). However, during one period of flow oscillation, modes with the (ss) symmetry are excited, which combined with the main roll give rise to one diagonal clockwise-oriented vortex with two small counter-rotating eddies located in opposite corners of the cavity or a columnar arrangement of three superposed rolls slightly inclined to the left.

The next case serves to reveal the intrinsic features of the synchronous and non-periodic regime (SY-NP) found at higher vibrational Rayleigh numbers and at low and intermediate frequencies. As already explained in Section 6.1, a distinguishing mark of this type of solutions is the existence of bursts in the velocity signal, which display a more or less random nature.

In particular, here, the case of $Ra_\omega = 8.5 \times 10^5$, $\Omega = 100$ is taken as a representative example (Figure 15). As a fleeting glimpse into this figure would immediately confirm, the fluid becomes almost quiescent over a fixed sub-interval of each period.

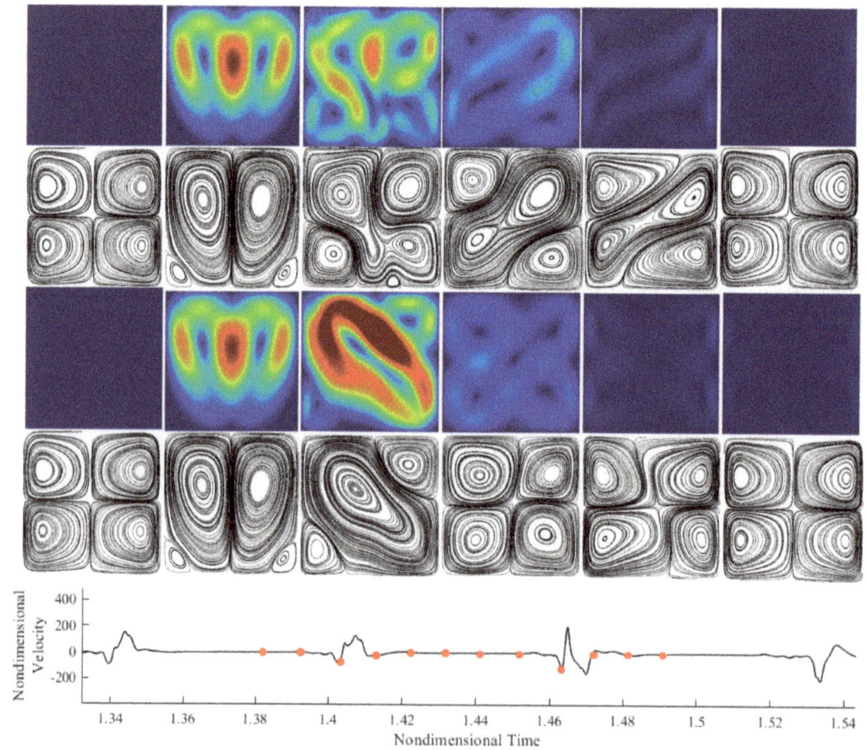

Figure 15. Instantaneous streamlines and velocity magnitude over two periods for the case $Ra_\omega = 8.5 \times 10^5$, $\Omega = 100$, accompanied by the velocity signal. The 12 red dots represent the time at which the snapshots are taken (six snapshots for each period).

Although the velocity profile shows a large difference in behavior over time, the instantaneous velocity field and streamlines witness that the actual pattern is different at each turbulent burst. This is evident in the third, fourth, and fifth snapshots of the sequence for each period. Although the bust occurs at the same point in time, the flow structure changes considerably.

When the velocity magnitude is close to zero, the streamlines present again the four-roll configuration; when a burst occurs, as shown in snapshot 3, existing vortices merge, and new rolls appear randomly. In terms of spatial symmetry, although in certain sub-intervals of the period solutions with the (sa) symmetry also appear (two vertically extended rolls in a side-by-side configuration), the (ss) mode with four rolls is generally dominant.

Figure 16 can be finally used to get insights into the non-synchronous and non-periodic regime (NS-NP) apparent in the region of high vibrational frequencies. Easily identifiable, these solutions exhibit no adherence to the imposed vibrational forcing. As the reader will easily realize by inspecting Figure 16, the behavior of the flow changes randomly and presents a number of interesting and unpredictable topological (in terms of streamlines) features, which essentially result from the excitation and superposition of convective modes with different symmetries.

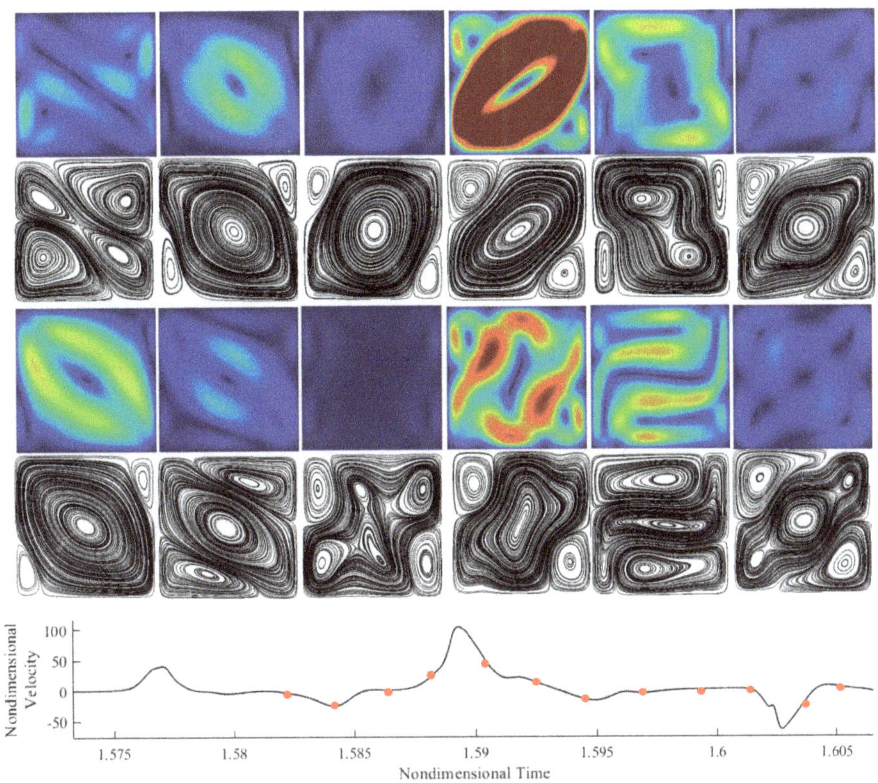

Figure 16. Instantaneous streamlines and velocity magnitude over two periods for the case $Ra_\omega = 10^6$, $\Omega = 500$ (NS-NP), which are accompanied by the velocity signal. The 12 red dots represent the time at which the snapshots are taken (six snapshots for each period).

7. Discussion and Conclusions

Originally conceived as an extension of other works in the literature, the present study has confirmed that a kaleidoscope of solutions can be obtained in an apparently innocuous configuration such as a square cavity subjected to vibrations parallel to the applied temperature difference. Considering relatively high values (heretofore unexplored) of the vibrational Rayleigh number and non-dimensional angular frequency, two new states have been identified in the space of parameters, namely, periodic and non-periodic synchronous modes of convection. The peculiarity of the latter resides in its ability to produce turbulent bursts, which occur synchronously with the forcing but do not display the same behavior in each period of oscillation. Although the appearance of the synchronous periodic solution is more sporadic, this mode of convection can manifest itself also for high Rayleigh numbers, which indicates that an increase in Ra_{ω} does not systematically lead to more chaotic phenomena.

Comparison with equivalent studies conducted for the same value of the Prandtl number, same geometry, and same range of values of the Rayleigh number for classical Rayleigh–Bénard convection indicates that the set of potentially excitable modes with different symmetries is greatly expanded when the steady gravity is replaced by a time-periodic acceleration. From the limited series of snapshots included in the present work, the predominantly occurring symmetries are represented by the symmetric–symmetric mode (ss) (generally appearing as a quadrupolar pattern) and the diagonal mode characterized by a predominant central vortex ornated with two smaller outer corner rolls. However, occasional manifestations of other patterns are also possible, including (but not limited to) the vertical two-roll configuration (sa), the columnar arrangement of three horizontally stretched rolls, as well as other three-roll configurations. In line with earlier studies on the companion problem of standard RB convection, we argue that an explanation for this variety of multicellular states can be rooted in the existence of multiple solutions, i.e., different possible modes of convection that coexist in the space of parameters and can be excited for comparable values of the driving force (Mizushima [2]; Hof et al. [57]; Leong [58]; Lappa [59]). These modes are not mutually exclusive, nor are they truly progressive, which means that they can be excited at different times or at the same time, resulting in new patterns due to their non-linear combination.

A comparison of instantaneous and time-averaged effects also leads to meaningful conclusions. The former is generally dominant over the entire range of values of and Ω considered. The dependence on the problem parameters also displays notable differences. While time-averaged quantities are almost independent from the vibrational Rayleigh number, instantaneous ones grow (as expected) with this parameter. However, as the angular frequency of the imposed vibrations is increased, both fluctuating and time-averaged (stationary) effects (as properly quantified through the so-called TFD distortions) tend to be damped until a completely motionless state is attained (for a cut-off value of the frequency that grows with the considered value of Ra_{ω}). This scenario is consistent with that revealed by the Nusselt number (which tends to 1 as this cut-off value is exceeded, thereby indicating that purely thermally diffusive conditions are established). It is also congruent with the so-called Birikh's law, i.e., the expected tendency of the time-averaged vibration force to create isotherms perpendicular to the direction of vibrations when the frequency becomes relatively high (thereby causing a strong increase in the value of the Rayleigh number needed to produce convection, which ideally tends to infinite in the limit as $\Omega \rightarrow \infty$, [60]).

The peak visible in both the instantaneous TFD and Nu plot at $\Omega \cong 100$ for relatively high values of the vibrational Rayleigh number calls for a complementary explanation. This can be further elaborated in its simplest form on the basis of the argument that the fluctuating components of velocity and temperature are dominant and therefore play a crucial role in determining the intensity of the heat exchange at the solid boundaries. Moreover, we argue that an explanation for the non-monotone behavior must be sought in the spatial symmetries of the dominant flow. Indeed, the specific value of the angular frequency for which the maximum is attained ($\Omega \cong 100$) in the range of high values of the vibrational Rayleigh number corresponds to conditions where the (ss) symmetry (multicellular state) is dominant (its reverberation on the heat exchange being an increase in magnitude).

Given the lack of studies specifically conceived to investigate the properties of pure thermovibrational flow in conditions for which the temperature gradient is parallel to the direction of shaking, this work has been conducted under the optimistic idea that it may provide a common point of origin from which many studies in the community may depart (including meaningful extensions to three-dimensional configurations). These future works might be based on the same variegated approach used in the present work, which has proven instrumental in unraveling processes that are interwoven or overshadowed and successful at illuminating the dynamical mechanisms at play on these systems, thereby making research results easier to compare and providing researchers with reasonable values to assume for areas outside their experience.

Author Contributions: Data curation, G.C.; Formal analysis, M.L.; Investigation, G.C. and M.L.; Methodology, M.L.; Software, G.C. and M.L.; Supervision, M.L.; Visualization, G.C.; Writing—original draft, G.C. and M.L.; Writing—review and editing, M.L. All authors have read and agreed to the published version of the manuscript.

Funding: This work has been supported by the UK Space Agency (grants ST/S006354/1 and ST/V005588/1) in the framework of the "particle vibration" (T-PAOLA) project.

Institutional Review Board Statement: Not applicable.

Informed Consent Statement: Informed consent was obtained from all subjects involved in the study.

Data Availability Statement: Publicly available datasets were analyzed in this study. These data can be found in the pure repository of the University of Strathclyde.

Conflicts of Interest: The authors declare no conflict of interest.

Nomenclature

Nomenclature	
b	Vibration amplitude
Nu	Nusselt number
p	Pressure
Pr	Prandtl number
Ra	Rayleigh number
s	Displacement
T	Temperature
t	Time
u	Velocity component along x
V	Velocity
v	Velocity component along y
x	Horizontal coordinate
y	Vertical coordinate
Greek Symbols	
α	Thermal diffusivity
β_T	Thermal expansion coefficient
ν	Kinematic viscosity
ρ	Fluid density
ω	Dimensional angular frequency
Ω	Non-dimensional angular frequency
ΔT	Temperature difference
τ	Non-dimensional period of vibrations
δ	Thickness
ζ	Kolmogorov length scale
Subscripts	
BL	Boundary layer
Cold	Cold
diff	Diffusive
Hot	Hot
max	Maximum
Superscripts	
Lab	Laboratory

References

1. Hadley, G. Concerning the cause of the general trade winds. *Philos. Trans. R. Soc. Lond.* **1735**, *29*, 58–62.
2. Mizushima, J. Onset of thermal convection in a finite two–dimensional box. *J. Phys. Soc. Jpn.* **1995**, *64*, 2420–2432. [CrossRef]
3. Mizushima, J.; Adachi, T. Sequential Transitions of the Thermal Convection in a Square Cavity. *J. Phys. Soc. Jpn.* **1997**, *66*, 79–90. [CrossRef]
4. Goldhirsch, I.; Pelz, R.B.; Orszag, S.A. Numerical simulation of thermal convection in a two-dimensional finite box. *J. Fluid Mech.* **1989**, *199*, 1–28. [CrossRef]
5. Villermaux. Memory-induced low frequency oscillations in closed convection boxes. *Phys. Rev. Lett.* **1995**, *75*, 4618–4621. [CrossRef]
6. Kadanoff, L.P. Turbulent Heat Flow: Structures and Scaling. *Physics Today* **2001**, *54*, 34–39. [CrossRef]
7. Lappa, M. Some considerations about the symmetry and evolution of chaotic Rayleigh–Bénard convection: The flywheel mechanism and the "wind" of turbulence. *C. R. Mécanique* **2011**, *339*, 563–572. [CrossRef]
8. Monti, R.; Langbein, D.; Favier, J.J. Influence of residual accelerations on fluid physics and material science experiments. In *Fluid and material Science in Space: A European Perspective*; Walter, H.U., Ed.; Springer: Berlin, Germany, 1987; Chapter XVIII; pp. 637–680.
9. Alexander, J.I.D. Low gravity experiment sensitivity to residual acceleration: A review. *Microgravity Sci. Technol.* **1990**, *3*, 52–68.
10. Alexander, J.I.D.; Ouazzani, J.; Rosenberger, F. Analysis of the low gravity tolerance of Bridgman-Stockbarger crystal growth, II. Transient and periodic accelerations. *J. Cryst. Growth* **1991**, *113*, 21–38. [CrossRef]
11. Alexander, J.I.D.; Garandet, J.-P.; Favier, J.J.; Lizee, A. g-jitter effects on segregation during directional solidification of tin-bismuth in the MEPHISTO furnace facility. *J. Cryst. Growth* **1997**, *178*, 657–661.
12. Feonychev, A.I.; Dolgikh, G.A. Influence of vibration on heat and mass transfer in microgravity conditions. *Microgravity Q.* **1994**, *4*, 233–240.
13. Gershuni, G.Z.; Lyubimov, D.V. *Thermal Vibrational Convection*; Wiley: Chichester, UK, 1998.
14. Monti, R.; Savino, R.; Lappa, M. Microgravity sensitivity of typical fluid physics experiment. In Proceedings of the 17th Microgravity Measurements Group Meeting, Cleveland, OH, USA, 24–26 March 1998; Published in the Meeting Proceedings in NASA CP-1998-208414. pp. 1–15.
15. Naumann, R.J. An analytical model for transport from quasi-steady and periodic accelerations on spacecraft. *Int. J. Heat Mass Transf.* **2000**, *43*, 2917–2930. [CrossRef]
16. Mialdun, A.; Ryzhkov, I.I.; Melnikov, D.E.; Shevtsova, V. Experimental Evidence of Thermal Vibrational Convection in a Nonuniformly Heated Fluid in a Reduced Gravity Environment. *Phys. Rev. Lett.* **2008**, *101*, 084501. [CrossRef] [PubMed]
17. Melnikov, D.; Ryzhkov, I.; Mialdun, A.; Shevtsova, V. Thermovibrational Convection in Microgravity: Preparation of a Parabolic Flight Experiment. *Microgravity Sci. Technol.* **2008**, *20*, 29–39. [CrossRef]
18. Lyubimova, T.P.; Perminov, A.V.; Kazimardanov, M.G. Stability of quasi-equilibrium states and supercritical regimes of thermal vibrational convection of a Williamson fluid in zero gravity conditions. *Int. J. Heat Mass Transf.* **2019**, *129*, 406–414. [CrossRef]
19. Bouarab, S.; Mokhtari, F.; Kaddeche, S.; Henry, D.; Botton, V.; Medelfef1, A. Theoretical and numerical study on high frequency vibrational convection: Influence of the vibration direction on the flow structure. *Phys. Fluids* **2019**, *31*, 043605. [CrossRef]
20. Shevtsova, V.; Lyubimova, T.; Saghir, Z.; Melnikov, D.; Gaponenko, Y.; Sechenyh, V.; Legros, J.C.; Mialdun, A. IVIDIL: On-board g-jitters and diffusion controlled phenomena. *J. Phys. Conf. Ser.* **2011**, *327*, 012031. [CrossRef]
21. Shevtsova, V.; Mialdun, A.; Melnikov, D.; Ryzhkov, I.; Gaponenko, Y.; Saghir, Z.; Lyubimova, T.; Legros, J.C. The IVIDIL experiment onboard the ISS: Thermodiffusion in the presence of controlled vibrations. *C. R. Mécaniq* **2011**, *339*, 310–317. [CrossRef]
22. Maryshev, B.; Lyubimova, T.; Lyubimov, D. Two-dimensional thermal convection in porous enclosure subjected to the horizontal seepage and gravity modulation. *Phys. Fluids* **2013**, *25*, 084105. [CrossRef]
23. Vorobev, A.; Lyubimova, T. Vibrational convection in a heterogeneous binary mixture. Part I. Time-averaged equations. *J. Fluid Mech.* **2019**, *870*, 543–562. [CrossRef]
24. Lappa, M. Control of convection patterning and intensity in shallow cavities by harmonic vibrations. *Microgravity Sci. Technol.* **2016**, *28*, 29–39. [CrossRef]
25. Lappa, M. The patterning behavior and accumulation of spherical particles in a vibrated non-isothermal liquid. *Phys. Fluids* **2014**, *26*, 093301. [CrossRef]
26. Lappa, M. Numerical study into the morphology and formation mechanisms of threedimensional particle structures in vibrated cylindrical cavities with various heating conditions. *Phys. Rev. Fluids* **2016**, *1*, 25. [CrossRef]
27. Lappa, M. On the multiplicity and symmetry of particle attractors in confined non-isothermal fluids subjected to inclined vibrations. *Int. J. Multiphase Flow* **2017**, *93*, 71–83. [CrossRef]
28. Lappa, M. On the formation and morphology of coherent particulate structures in non-isothermal enclosures subjected to rotating g-jitters. *Phys. Fluids* **2019**, *31*, 11. [CrossRef]
29. Lappa, M.; Burel, T. Symmetry Breaking Phenomena in Thermovibrationally Driven Particle Accumulation Structures. *Phys. Fluids* **2020**, *32*, 23. [CrossRef]
30. Lappa, M. *Thermal Convection: Patterns, Evolution and Stability*; John Wiley & Sons, Ltd.: Chichester, UK, 2009.

31. Hirata, K.; Sasaki, T.; Tanigawa, H. Vibrational effects on convection in a square cavity at zero gravity. *J. Fluid Mech.* **2001**, *445*, 327–344. [CrossRef]
32. Harlow, F.H.; Welch, J.E. Numerical calculation of time-dependent viscous incompressible flow with free surface. *Phys. Fluids* **1965**, *8*, 2182–2189. [CrossRef]
33. Chorin, A.J. Numerical solutions of the Navier-Stokes equations. *Math. Comput.* **1968**, *22*, 745–762. [CrossRef]
34. Temam, R. Sur l'approximation de la solution des èquations de Navier-Stokes par la mèthode des pas fractionnaires (I). *Arch. Rat. Mech. Anal.* **1969**, *33*, 377–385. [CrossRef]
35. Gresho, P.M.; Sani, R.T. On pressure boundary conditions for the incompressible Navier-Stokes equations. *Int. J. Numer. Meth. Fluids* **1987**, *7*, 1111–1145. [CrossRef]
36. Gresho, P.M. Incompressible fluid dynamics: Some fundamental formulation issues. *Ann. Rev. Fluid Mech.* **1991**, *23*, 413–453. [CrossRef]
37. Guermond, J.-L.; Quartapelle, L. On stability and convergence of projection methods based on pressure Poisson equation. *Int. J. Numer. Meth. Fluids* **1998**, *26*, 1039–1053. [CrossRef]
38. Guermond, J.-L.; Minev, P.; Shen, J. An Overview of Projection Methods for Incompressible Flows. Comput. Methods. *Comput. Methods Appl. Mech. Eng.* **2006**, *195*, 6011–6045. [CrossRef]
39. Ladyzhenskaya, O.A. *The Mathematical Theory of Viscous Incompressible Flow*, 2nd ed.; Gordon and Breach: New York, NY, USA; London, UK,, 1969.
40. Lappa, M. Strategies for parallelizing the three-dimensional Navier-Stokes equations on the Cray T3E. In *Science and Supercomputing at CINECA*; Voli, M., Ed.; CINECA: Bologna, Italy, 1997; Volume 11, pp. 326–340. ISBN 10:88-86037-03-1.
41. Lappa, M.; Boaro, A. Rayleigh-Bénard convection in viscoelastic liquid bridges. *J. Fluid Mech.* **2020**, *904*. [CrossRef]
42. Issa, R.I. Solution of the implicitly discretized fluid flow equations by operator-splitting. *J. Comp. Phys.* **1986**, *62*, 40–65. [CrossRef]
43. Rhie, C.M.; Chow, W.L. Numerical study of the turbulent flow past an airfoil with trailing edge separation. *AIAA J.* **1983**, *21*, 1525–1532. [CrossRef]
44. Lappa, M.; Inam, S. Thermogravitational and hybrid convection in an obstructed compact cavity. *Int. J. Thermal Sci.* **2020**, *156*, 21. [CrossRef]
45. Ouertatani, N.; Cheikh, N.B.; Beya, B.B.; Lili, T. Numerical simulation of two-dimensional Rayleigh-Benard convection in an enclosure. *C. R. Mecanique* **2008**, *336*, 464–470. [CrossRef]
46. Soong, C.Y.; Tzeng, P.Y.; Chiang, D.C.; Sheu, T.S. Numerical study on mode-transition of natural convection in differentially heated inclined enclosures. *Int. J. Heat Mass Transf.* **1995**, *39*, 2869–2882. [CrossRef]
47. Russo, G.; Napolitano, L.G. Order of Magnitude Analysis of unsteady Marangoni and Buoyancy free convection. In Proceedings of the 35th Congress of the International Astronautical Federation, Lausanne, Switzerland, 7–13 October 1984.
48. Shishkina, O.; Stevens, R.J.A.M. Grossmann, S.; Lohse, D. Boundary layer structure in turbulent thermal convection and its consequences for the required numerical resolution. *New J. Phys.* **2010**, *12*, 17. [CrossRef]
49. De, A.K.; Eswaran, V.; Mishra, P.K. Scalings of heat transport and energy spectra of turbulent Rayleigh-Bénard convection in a large-aspect-ratio box. *Int. J. Heat Fluid Flow* **2017**, *67*, 111–124. [CrossRef]
50. Simonenko, I.B.; Zen'kovskaja, S.M. On the effect of highfrequency vibrations on the origin of convection. *Izv. Akad. Nauk SSSR. Ser. Meh. Zhidk. Gaza* **1966**, *5*, 51–55.
51. Simonenko, I.B. A justification of the averaging method for a problem of convection in a field rapidly oscillating forces and other parabolic equations. *Mat. Sb.* **1972**, *129*, 245–263. [CrossRef]
52. Gershuni, G.Z.; Zhukhovitskii, E.M. Free thermal convection in a vibrational field under conditions of weightlessness. *Sov. Phys. Dokl.* **1979**, *24*, 894–896.
53. Gershuni, G.Z.; Zhukhovitskii, E.M.; Yurkov Yu, S. Vibrational thermal convection in a rectangular cavity. *Izv. Akad. Nauk SSSR Mekh. Zhidk. Gaza* **1982**, *4*, 94–99. [CrossRef]
54. Gershuni, G.Z.; Zhukhovitskii, E.M. Vibrational thermal convection in zero gravity. *Fluid Mech. Sov. Res.* **1986**, *15*, 63–84.
55. Savino, R.; Lappa, M. Assessment of the thermovibrational theory: Application to g-jitter on the Space-station. *J. Spacecraft Rockets* **2003**, *40*, 201–210. [CrossRef]
56. Birikh, R.V.; Briskman, V.A.; Chernatynski, V.I. Roux, B. Control of thermocapillary convection in a liquid bridge by high frequency vibrations. *Microgravity Q.* **1993**, *3*, 23–28.
57. Hof, B.; Lucas, G.J.; Mullin, T. Flow state multiplicity in convection. *Phys. Fluids* **1999**, *11*, 2815–2817. [CrossRef]
58. Leong, S.S. Numerical study of Rayleigh-Bénard convection in a cylinder. *Numer. Heat Transf. Part A* **2002**, *41*, 673–683. [CrossRef]
59. Lappa, M. On the Nature of Fluid-dynamics. In *Understanding the Nature of Science*; Lindholm, P., Ed.; Science, Evolution and Creationism, BISAC: SCI034000; Nova Science Publishers Inc.: Hauppauge, NY, USA, 2019; Chapter 1; pp. 1–64. ISBN 978-1-53616-016-1. Available online: https://novapublishers.com/shop/understanding-the-nature-of-science/ (accessed on 10 October 2020).
60. Gershuni, G.Z.; Zhukhovitskii, E.M. Convective instability of a fluid in a vibration field under conditions of weightlessness. *Fluid Dyn.* **1981**, *16*, 498–504. [CrossRef]

fluids

MDPI

Article

Vibroconvective Patterns in a Layer under Translational Vibrations of Circular Polarization

Victor Kozlov *, Kirill Rysin and Aleksei Vjatkin

Laboratory of Vibrational Hydromechanics, Perm State Humanitarian Pedagogical University, Sibirskaya av. 24, 614990 Perm, Russia; rysin@pspu.ru (K.R.); a.a.vjatkin@mail.ru (A.V.)
* Correspondence: kozlov@pspu.ru

Abstract: This article experimentally investigates thermal vibrational convection in horizontal layers, subject to circular translational oscillations in the horizontal plane. The definite direction of translational vibrations lacks investigation, and the case of a layer heated from above is considered. At large negative values of the gravitational Rayleigh number, the thermovibrational convection appears in a threshold manner with an increase in the vibration intensity. Our results show that in the case of strong gravitational stabilization, thermovibrational convection develops in the form of patterns with strong anisotropy of spatial periods in orthogonal directions. The vibroconvective patterns have the form of parallel rolls divided along their length into relatively short segments. The layer thickness determines the distance between the rolls, and the longitudinal wavelength, depends on the Rayleigh number. Convective cells are studied using the noninvasive thermohromic methodic. It is found that when using the tracers for flow visualization, the concentration and type of the visualizer particles have a serious impact on the shape of the observed vibroconvective structures. In particular, the presence of even a small number of tracers (used in the study of velocity fields by the PIV method) generates flows and intensifies the heat transfer below the threshold of thermovibrational convection excitation.

Keywords: plane layer; circular translational vibrations; thermal vibrational convection; convective patterns

check for updates

Citation: Kozlov, V.; Rysin, K.; Vjatkin, A. Vibroconvective Patterns in a Layer under Translational Vibrations of Circular Polarization. *Fluids* **2021**, *6*, 108. https://doi.org/10.3390/fluids6030108

Academic Editors: D. Andrew Rees and Marcello Lappa

Received: 1 February 2021
Accepted: 2 March 2021
Published: 6 March 2021

Publisher's Note: MDPI stays neutral with regard to jurisdictional claims in published maps and institutional affiliations.

Copyright: © 2021 by the authors. Licensee MDPI, Basel, Switzerland. This article is an open access article distributed under the terms and conditions of the Creative Commons Attribution (CC BY) license (https://creativecommons.org/licenses/by/4.0/).

1. Introduction

The study of thermal vibrational convection (TVC) (Gershuni and Lyubimov [1]) is the object of high interest nowadays, in connection with its potential use in effectively controlling the convection and heat transfer in various technological devices, particularly at zero gravity (Gershuni and Zhukhovitskii [2], Mialdun et al. [3]). According to Gershuni and Lyubimov [1], TVC is the averaged convection, excited in a non-isothermal fluid by the oscillating force fields. This is done by the oscillating inertial forces arising in the non-inertial frame of reference associated with the cavity, as a result of the cavity vibrations. The non-isothermal liquid oscillations lead to the appearance of averaged mass forces as a result of nonlinear effects, which generates the averaged convective flows. The last has been well-studied, both theoretically and experimentally, in the case of linear translational vibrations of a cavity with a non-uniform in density (for example, non-isothermal or multi-component) liquid. Wide theoretical studies are carried out both in the limiting case of high vibration frequencies (using the equations of averaged convection) and in the region of arbitrary frequencies (Simonenko and Zen'kovskaja [4], Gershuni et al. [5], Smorodin et al. [6], Vorobev and Lyubimova [7], Lappa [8]], Crewdson and Lappa [9]). The thermal vibrational convection equations obtained by the averaging method in the approximation of high dimensionless frequencies $\omega \equiv \Omega h^2/\nu \gg 1$ (Simonenko and Zen'kovskaja [4]) are often used. They correspond to the case when the characteristic size of the cavity h is much greater than the thickness of the viscous boundary layers $\delta \equiv \sqrt{2\nu/\Omega}$, and the oscillations of the fluid are inviscid. In this case, the thermovibrational mechanism is determined by the vibrational parameter $R_v = (b\Omega\beta\Theta h)^2/2\nu\chi$, where b is the vibration amplitude, $\Omega = 2\pi f$—radian frequency of vibration, Θ—characteristic temperature difference, β, χ and ν—coefficients of volumetric expansion, thermal conductivity, and kinematic viscosity of the fluid.

Linear translational vibrations, which have the preferred direction of oscillations, determines the shape of convective structures. For instance, in plane layers, the linear vibrations lead to two-dimensional convective structures, elongated in the general case across the vibration axis. Exceptions are the linear vibrations directed along the temperature gradient (Crewdson and Lappa [9]).

A more general case is the one, when the cavity performs translational oscillations of elliptical polarization, considered theoretically by Kozlov [10]. The case of circular translational vibrations, when the cavity moves along a circular path, is of special interest. This type of vibration has no preferred direction—it leads to a rotating inertial force field in a non-inertial reference frame associated with the cavity. A theoretical description of TVC (equations of thermal vibrational convection obtained by the averaging method in the high-frequency approximation) for this case, as well as the problem of vibroconvective stability of liquid in a horizontal layer with isothermal boundaries of different temperatures, are considered in Reference [10]. It is shown that in the horizontal layers, the vibrations of circular polarization in the horizontal plane act similarly to vibrations of linear polarization. The fundamental difference lies in the absence of a preferred direction. According to the results of the linear stability theory, at the threshold, the most dangerous disturbances are characterized only by the square of the wavenumber, while the shape of convective structures (two-dimensional rolls, hexagonal or rectangular convective cells) remains undefined. Determining the type of disturbances developing in the supercritical region requires a nonlinear analysis. In this regard, a theoretical study by Pesch et al. [11] is interesting, where thermal convection in a horizontal layer performing unidirectional and circular translational oscillations in its own plane is studied. The analysis is based on the Galerkin method, as well as on direct numerical simulations of the underlying Boussinesq equations, in contrast to Reference [10], where the equations of vibrational convection obtained by the averaging method in the high frequency approximation were used. The studies [11] were carried out in the region of moderate dimensionless frequencies and have demonstrated that in the case of circular shaking, the onset of convection corresponds to the transverse roll, like in the case of unidirectional vibrations, with the roll axis selected spontaneously.

Experimental study of thermovibrational convection in a horizontal layer performing circular translational oscillations is performed by Kozlov et al. in Reference [12]. Since the experiments were carried out in a gravity field, in addition to the vibrational mechanism, thermal convection was determined by the thermogravitational mechanism, which is characterized by the Rayleigh number $Ra = g\beta\Theta h^3/\nu\chi$. To exclude the destabilizing effect of the gravitational convective mechanism, the case of a fluid stably stratified in the gravity field, when the layer is heated from above, is considered. The studies are carried out in the area of large negative values of the Rayleigh number, at $Ra \sim -10^5$. According to the experimental observations, the thermovibrational convection, in this case, develops in the form of two-dimensional rolls; the wavelength of convective structures is determined by the layer thickness. The visualization of convective structures was carried out with the help of calliroscope particles, which is a small amount were located at the lower (cold) boundary of the layer. The convection excitation threshold found experimentally is in good agreement with the results of the linear stability theory (Ivanova and Kozlov [13]). It is interesting, that the wavenumber of two-dimensional convective rolls, observed in the experiment, is significantly lower than one predicted by the linear theory for large negative values of the gravitational Rayleigh number; at the same time, it is in good agreement with the results of the theory in the absence of gravity. In accordance with (Ivanova and Kozlov [13]), an increase in the negative value of the Rayleigh number, along with an increase in the threshold value of the vibrational parameter R_v, is accompanied by a sharp increase in the critical wavenumber.

Let us mention an interesting case of translational-rotational or pendulum vibrations of the cavity. Such combined vibrations result in an additional thermovibrational mechanism (Kozlov [14], Kozlov and Selin [15]) which is linear in density inhomogeneity, and thus, analogous to the Rayleigh mechanism of gravitational convection. It is important that in thin layers, due to the specific form of the isothermal component of the pulsation velocity, excited by the rotational component of vibrations, the combined translational-rotational vibrations result in the renormalization of the external static force field. One can see that the equations of thermal vibrational convection under pendulum vibrations in a complete nonlinear formulation have the same form as in the case of translational vibrations, both in the case of linear and circular ones, Kozlov [10]. The only difference is in the renormalization of the static force field. This allows to compensate the effect of gravity in ground-based experiments and to study thermovibrational convection in layers under conditions that simulate the microgravity. This was confirmed in an experimental study of thermovibrational convection in a layer oscillating as a plane pendulum (Ivanova and Kozlov [13]) and spherical pendulum (Kozlov and Selin [15]). By the way, the experiments have shown that the use of different types of visualizing particles has a significant effect on the structure of convective flows, up to the fact that convective cells appear to significantly lower the excitation threshold of thermovibrational convection. In the case of circular translational vibrations, the presence of even an insignificant number of particles of practically neutral buoyancy can change the heat transfer in the subthreshold region, without, however, affecting the excitation threshold of thermovibrational convection.

The aim of this work is the experimental study of the structure of vibroconvective flows in a horizontal layer heated from above and subjected to circularly polarized translational vibrations, as well as the study of the effect of different types of tracers on the structure of observed convective flows and convective heat transfer, in particular, below the excitation threshold of thermal vibrational convection. Along with the visualization of convective flows with the help of different types of tracers, the structure of vibroconvective cells was studied using a thermochromic film, which allows visualization of temperature fields in a liquid in the absence of tracers, that is, without disturbing the liquid.

2. Formulation and Methods

2.1. Problem Formulation

Convection in a horizontal plane layer with boundaries of different temperatures T_1 and T_2 is considered (Figure 1). The upper boundary temperature is higher than that of the lower one ($T_1 > T_2$), so the gravity field plays a stabilizing role. The layer performs translational circular vibrations in the horizontal plane with an amplitude $b = 4.4$ cm and frequency Ω, which varies in the interval $\Omega = 0 - 60\ \mathrm{s}^{-1}$. The oscillations are set using a mechanical vibrator, a description available in Kozlov et al. [12]. The radius vector **r**, drawn from the origin of a laboratory coordinate system xyz to the origin of the moving system $x'y'z'$ associated with the cavity, rotates with a frequency $\Omega \equiv 2\pi f$ around the vertical axis z. Thus, the cavity performs translational displacement along the circle path (translational vibrations of circular polarization). At this, a spatially uniform inertial force field $\Omega^2\mathbf{r}$ rotates in the non-inertial frame of reference associated with the vibrating cavity. The direction of the inertial force field coincides with the vector **r**.

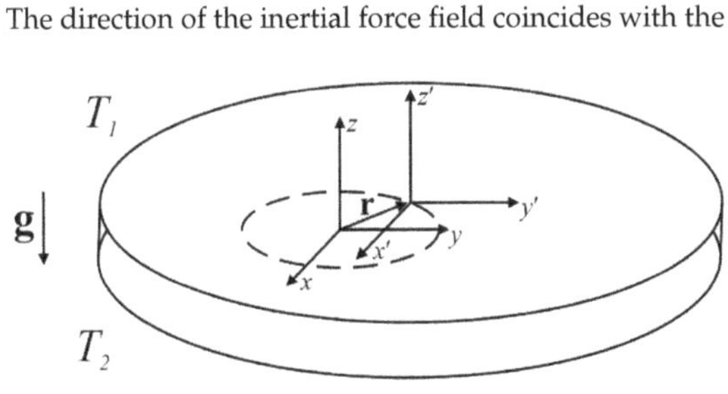

Figure 1. The problem statement.

The layer used in the experiments (Figure 2) is formed by two parallel flat heat exchangers. The bottom heat exchanger is an aluminum plate 1, with channels through which the liquid of constant temperature is pumped. The heat exchanger is equipped with a heat flux sensor, which consists of a thin plate with low thermal conductivity 2 between the heat exchanger and top-mounted aluminum plate 3 with a thickness of 3 mm. The heat flux is measured according to the temperature drop ΔT at this thermal resistance. The thickness of the liquid layer in the experiments is $h = 0.42$ cm or 0.32 cm, the lateral border diameter -13.8 cm. The upper heat exchanger 4 is transparent and serves to observe the convective structures in the layer. It is formed by two polished silicate glass plates 5 with a thickness of 4 mm, between which the fluid of a given temperature circulates. The temperature of the working surfaces of the liquid layer is measured by resistance thermometers with an accuracy 0.1 of °C. The temperature of the heat exchangers has constant values, due to the circulation of water at a given temperature coming from powerful jet thermostats. The working fluid is ethanol.

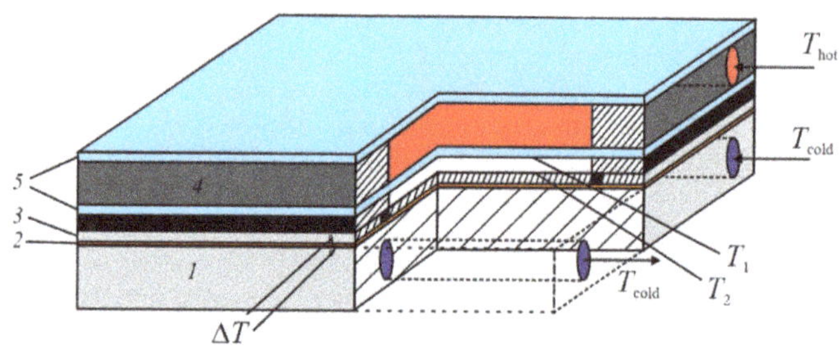

Figure 2. Scheme of the experimental cavity.

2.2. Experimental Technique and Procedure

In the course of the experiments, at given values of the temperature of the upper and lower heat exchangers, circular oscillations of a certain amplitude and frequency are reported to the cavity. The amplitude b is measured accurately to the closest 0.1, mm using the marker at the cuvette. For this purpose, the moving cavity is photographed by high-resolution photo camera, and the radius of the marker trajectory is measured. The value of the amplitude is controlled in all the experiments. The radian frequency of vibration varies in the interval $\Omega = 0 - 60\ \mathrm{s}^{-1}$ and is kept constant at each step of the experiment with an accuracy of $0.1\ \mathrm{s}^{-1}$.

At a monotonous stepwise increase of the vibration frequency (while controlling vibration amplitude), the temperature difference between the layer boundaries $\Theta \equiv T_2 - T_1$ and temperature drop ΔT at thermal resistance are measured. The measurements at each step are performed after the system reaches the stationary convection regime, which takes place five minutes after the vibration frequency change. The use of the transparent upper heat exchanger and the visualizing particles in the working fluid allows for the observation, and registration, of the convective patterns that form after the thermal convection excitation. A stroboscopic lamp, or a flash lamp, was used for the layer illumination.

To visualize the vibroconvective structures, along with traditional methods (using the traces—calliroscope or particles of Rezin-Amberlite), a thermochromic film was used in the experiments. This film is the microencapsulated liquid-crystal ink protected on both sides by a transparent polymer base. The ink changes its color when the temperature changes in a relatively narrow range, so the film can be used as a color temperature indicator. For the film used in the experiment, the activation range is from 30 °C to 35 °C. Outside this area, the film is black. At 30 °C, the film turns yellow-green and above 33 °C—blue and dark blue. The film with a thickness of 0.15 mm adheres to the lower boundary of the liquid layer over the heat flux sensor. To exclude the contact with the working fluid, the thermochromic film is additionally covered with a transparent film with a thickness equal to 0.10 mm. For the visualization of the vibroconvective patterns, the temperature at the cold lower boundary of the liquid layer is set slightly lower than 30 °C. In this case, under the conditions of convective equilibrium, the film at the surface has a uniform dark color. In the supercritical area, when the convective cells appear, the flows of the hot liquid descending from the upper boundary cause an increase in temperature near the surface of the lower layer boundary. As a result, the film changes its color in the areas of hot stream hitting. The experiments show that this technique resolves the convective cells even of a relatively small size. An important feature of the described technique is that it is not associated with the visualizing particles, which are added to the liquid and play the role of perturbations.

3. Results

The liquid layer is heated from the top. At definite temperatures of heat exchangers and definite amplitude of vibrations, the frequency of vibrations is increased in a stepwise manner. In the absence of the visualizer in the range of low frequencies of vibrations, the temperature difference between the layer boundaries Θ does not change with frequency; the liquid is in a quasi-equilibrium state (Figure 3, points 3). In this case, the heat flux, which is characterized by a temperature drop on ΔT, remains unchanged (Figure 3, points 1). At reaching a certain critical frequency f_c there is a threshold increase in heat flux through the layer ΔT (points 1), accompanied by a decrease in the temperature difference between the layer boundaries (points 3).

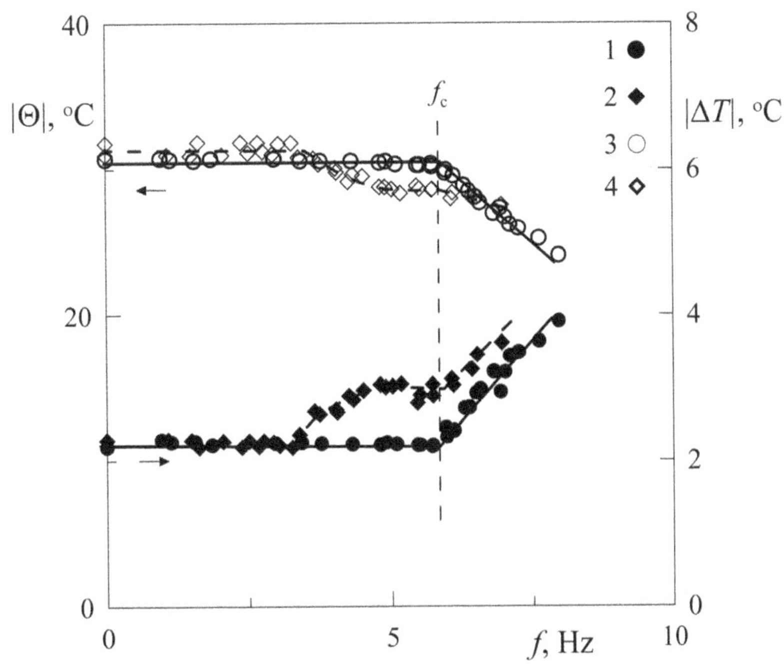

Figure 3. The dependence of the temperature difference Θ at the layer boundaries and the temperature drop ΔT on heat flux sensor on the vibration frequency f when using the different types of visualizer (points 1, 3 and 2, 4). The layer thickness $h = 4.2$ mm.

To visualize the vibroconvective flows, the calliroscope particles with the density slightly exceeding the density of the liquid were used, the characteristic particle size equals 40 ± 5 μm. Experiments with and without particles added [12] showed that this visualizer does not affect the excitation threshold of vibrational convection, as well as the heat flux.

Below the threshold of the thermovibrational convection excitation, at $f < f_c$, the particles were randomly distributed at the bottom of the cavity (Figure 4a). At the threshold, their redistribution in the form of periodically arranged rows is observed (Figure 4b). The photos of structures in Figure 4 are obtained in the conditions corresponding to the experiments presented in Figure 3 (points 1, 3). The photos show the top view of the working layer. At the top, a channel is visible through which the isothermal liquid is supplied to the central part of the upper transparent heat exchanger. In the top left part of fragment *a*, one can see the resistance thermometer sensor, which is a thin copper wire glued in the form of a snake to the inner boundary of the liquid layer. The sensor covers a relatively large area, which, unlike a point temperature sensor, makes it possible to measure the area-average boundary temperature even after developing cellular convection.

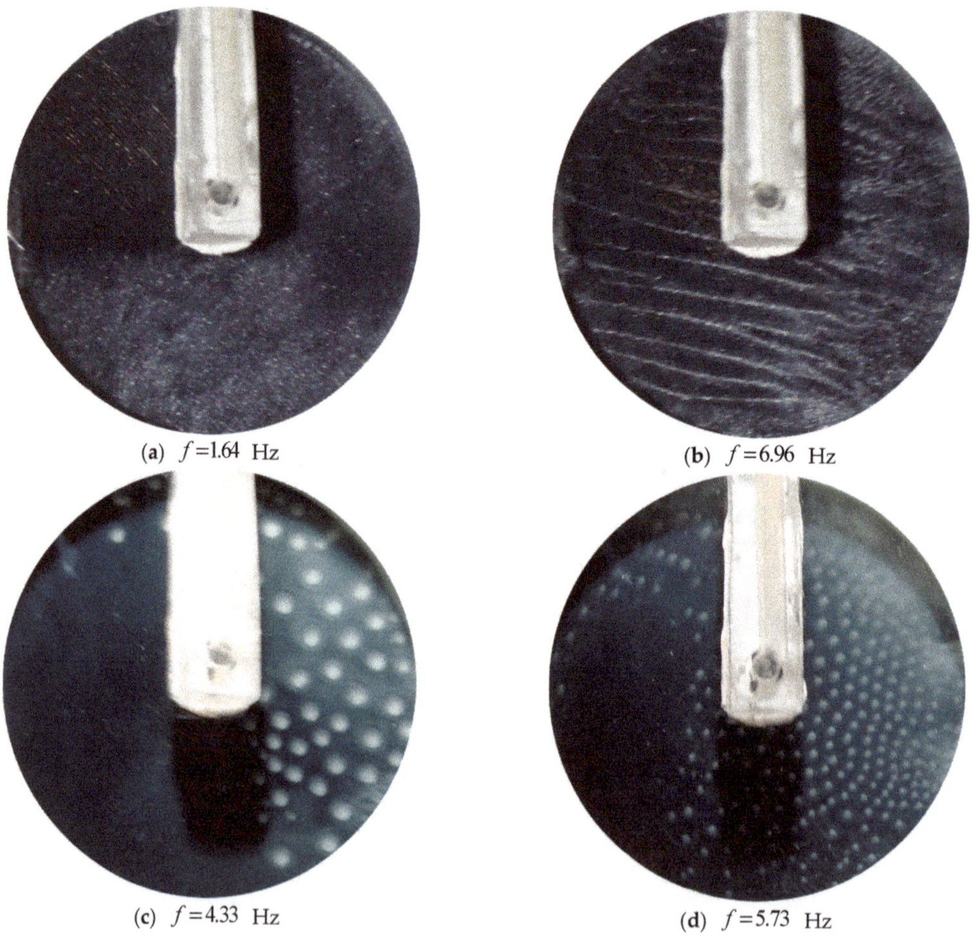

Figure 4. *Cont.*

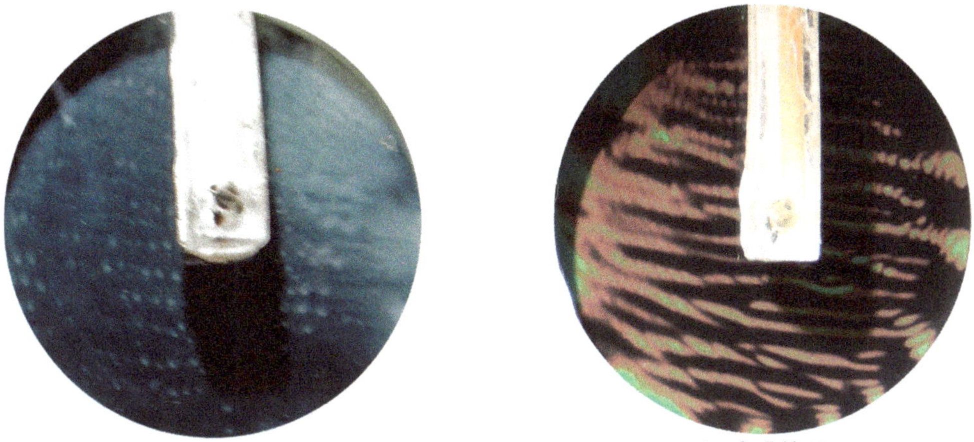

(**e**) $f = 6.52$ Hz (**f**) $f = 7.44$ Hz

Figure 4. The distribution of the calliroscope particles at the bottom of the layer below (**a**) and above (**b**) the threshold of the thermal vibrational convection excitation. The distribution of the Resin Amberlite particles in the layer with a thickness of $h = 0.42$ cm below (**c**,**d**) and above (**e**) the excitation of the thermal vibrational convection. Convective structures in the supercritical region, visualized using the thermochromic film (**f**), $h = 0.32$ cm.

The results obtained in the experiments with the visualization by the Resin Amberlite particles (mass fraction less than 0.14%) with an average size 30 ± 10 μm are shown by points 2, 4. At the vibration frequency $f < 4$ Hz, the presence of the particles has practically no effect on the heat transfer. At the same time, already at these frequencies, the particles begin to gather in heaps located at a significant distance from each other (Figure 4c).

The presence of visualizer particles in the liquid leads to an increase in the heat flux through the layer, and as a result, the lowering of the temperature difference between the layer boundaries (Figure 3) in the frequency range ($f = 4 \div 6$ Hz) up to the excitation threshold of thermovibrational convection in the absence of tracers. The particles of the visualizer form spatial periodic structures in the layer in the form of hills spaced at a considerable distance from each other (Figure 4c). With an increase in the intensity of vibrations, the distance between the hills decreases (Figure 4d). It can be assumed that, in this case, the intensification of heat transfer is caused by steady streaming flows [16], which are generated by hillocks as a result of fluid oscillations. In this case, the mounds themselves are formed by these averaged flows. This phenomenon, which was previously observed in Kozlov, Selin (Reference [15]), is of independent interest, but its description is beyond the scope of this work. It should be noted that a large number of studies have been devoted to the influence of vibrations on phase inclusions and their accumulation against the background of oscillations; a literature review can be found in Lappa (Reference [17]).

When the frequency exceeds the critical value ($f > f_c$), the structure of the hills and the distribution of the visualizer particles qualitative change. At $f = 6.52$ Hz (Figure 4e), the visualizer particles form the elongated parallel rows similar to the case with the heavier particles of the calliroscope (Figure 4b). However, the structures formed by the lighter particles have a clearly visible periodic distribution of the particles along the rows length. It should be noted that in the supercritical area (at $f > f_c$), the change of ΔT and Θ, with the vibration frequency in the absence of tracers and in their presence are in qualitative agreement (Figure 3), which indicates the transition to the thermal vibrational convection mode.

The comparison of the fragments b and e in Figure 4 shows that in the supercritical area, the PIV particles visualize vibroconvective structures in the form of two-dimensional rolls, the distance between which is consistent with the results obtained using a calliroscope (Figure 4b). The question of the nature of the PIV particles accumulation into periodic heaps along the rows length remains open, if this distribution is associated with the interaction between the particles aggregations similar to ones observed in the subthreshold region (Figure 4d). The thermochromic visualization of the temperature distribution near the cold layer boundary answers this question. This technique excludes the presence of any particles that disturb the convective flows. Below the threshold of thermal vibrational convection excitation, at $f < 6$ Hz in the considered case, the thermochromic film at the boundary of the layer has one color, which indicates the constancy of its temperature and the absence of convective structures. The photo of the film in the supercritical region (Figure 4f) demonstrates the presence of the spatial convective cells in the layer, and these cells are characterized by different spatial periods in perpendicular directions. The distance between the parallel rolls, which is characterized by the wavelength λ_1, corresponds to the thickness of the working layer, $\lambda_1 \sim 2h$. Such structures were previously observed in Reference [12]; it was shown that the distance between the parallel rolls is completely determined by the layer thickness. The visualization of convective flows with calliroscope in Reference [12] allowed an assumption about the two-dimensional form of the convective structures developing at the excitation threshold of the thermal vibrational convection. As one can see in Figure 4f, the convective cells are characterized not only by λ_1 but also by a spatial period λ_2. The reference directions of this spatial periods are perpendicular and the value of λ_2 is much less than that of λ_1.

The characteristic linear dimensions of the structures (wavelength λ) depending on the vibration frequency are shown in Figure 5; the dashed lines f_c indicates the threshold of convection excitation in layers $h = 0.42$ cm. The visualizer is the Resin Amberlite particles, the layer is heated from above, $\Theta = -30.7$. At a low frequency, the particles in the form of hillocks are distributed fairly uniformly in space, but without pronounced regularity. In Figure 5 points 1 correspond to the average distance between the hills tops. With increasing the vibration frequency, the average distance between the centers of adjacent hills λ_0 decreases. At frequency f_c there is a redistribution of the particles and a qualitative change in the structures shape: The particles form the regular parallel rows. The average distance between the rows λ_1 is shown in Figure 5 by the points 2, and the average distance between the hills in a row λ_2 (at $f > f_c$)—by the points 3. The points 4 and 5 show the results of λ_1 and λ_2 measuring using the thermochromic technique of visualization of the convective flows in a layer $h = 0.32$ cm (f_c'—the threshold frequency). One can see, the thinner the layer, the shorter is the wavelength λ_1 (compare points 2 and 4).

It should be noted that under vibrations, the particles of the visualizer accumulate in the heaps relatively regularly located at the bottom of the cavity (Figure 4c,d) in the isothermal case too. The distance between the heaps is smaller than the non-isothermal case. With increasing the frequency, the distance between the heaps decreases monotonically. There is no redistribution of the particles at any frequency in an explored range; the distribution of the hillocks remains random, close to the hexagonal one.

In the experiments with using the calliroscope particles as a visualizer (concentration equals 0.02%) in the absence of heating, the regular patterns are not observed at any vibration parameters. The visualizer particles with relatively high density are randomly located at the bottom of the layer. Only when thermovibrational convection is excited, the visualizer distributes in the form of two-dimensional rows. This indicates that the presence of a small amount of the calliroscope particles in the liquid does not affect the excitation threshold of thermal vibrational convection. At the same time, the calliroscope particles do not visualize the flows with a relatively short wavelength.

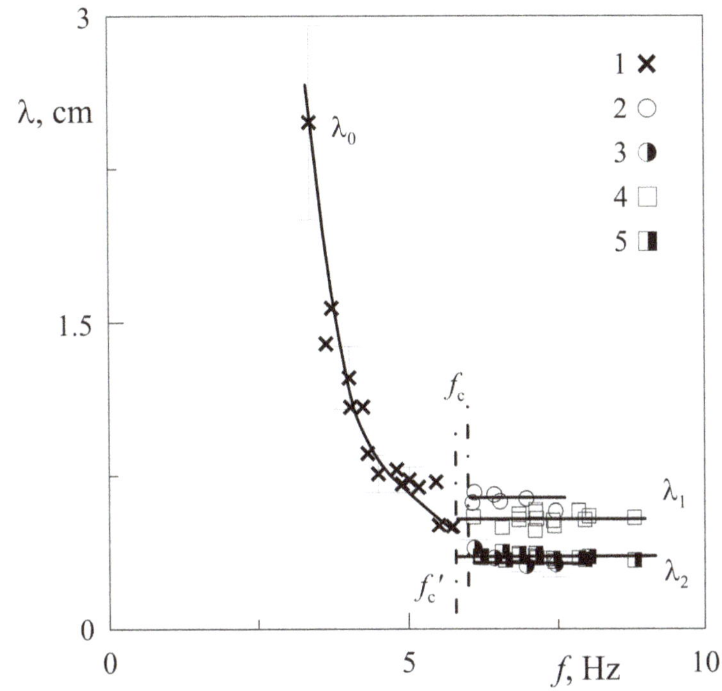

Figure 5. The wavelength of the structures (PIV particles) depending on the vibration frequency in the ethanol layer with a thickness $h = 0.42$ cm (points 1–3); thermochromic visualization in layer $h = 0.32$ cm (points 4 and 5).

4. Discussion

Figure 6 shows the dimensionless wavenumber k of the spatial structures depending on the vibrational parameter R_v. The experimental conditions and designations correspond to Figure 5. The wavenumber k_0 for the points corresponds to the case of spatial structures in the form of heaps located in an order similar to the hexagonal one (points 1 in Figure 5). To calculate the wavenumber, the equation for the hexagonal structures is used, $k_0 = 4\pi h/\sqrt{3}\lambda_0$ [18].

To calculate the dimensionless wavenumber k in the case of convective structures characterized by a periodic change in mutually perpendicular directions (points 2 and 3 in Figure 5), the term $k = \sqrt{k_1^2 + k_2^2}$ is used. Here the wavenumber $k_1 = 2\pi h/\lambda_1$ corresponds to the periodic distribution with a bigger wavelength (Figure 5, points 2 and 4), and $k_2 = 2\pi h/\lambda_2$ characterizes the distribution of the hills along each row.

It should be noted that the wavenumber k_1 in the supercritical region turns out to be constant and plactically the same in the layers of different thicknesses, $h = 0.42$ and 0.32 cm (Figure 6, transparent circles and squares). Moreover, its value is close to the theoretical value $k = 3.23$ [1] of the most dangerous disturbances in the problem of vibrational convection in plane layer in zero gravity. It is important to note that this wavenumber corresponds to the optimal size of two-dimensional convective rolls, consistent with the thickness of the liquid layer; the wavelength is close to twice the thickness of the layer. At the same time, the total wavenumber of the convective cells k under the conditions of the performed experiment turns out to be much higher and is in good agreement with the theoretical value predicted by the linear theory of stability (shown by the dashed line). This is achieved due to the short-wave modulation of the structure of convective rolls along their length with the wavenumber $k_2 = 2\pi h/\lambda_2$. The results indicate that in the absence of a selected direction of translational vibrations (at the vibrations of circular polarization) under the conditions of a strong stabilizing effect of a static gravity field, the structure of vibroconvective cells is characterized by the strong inhomogeneity in perpendicular directions.

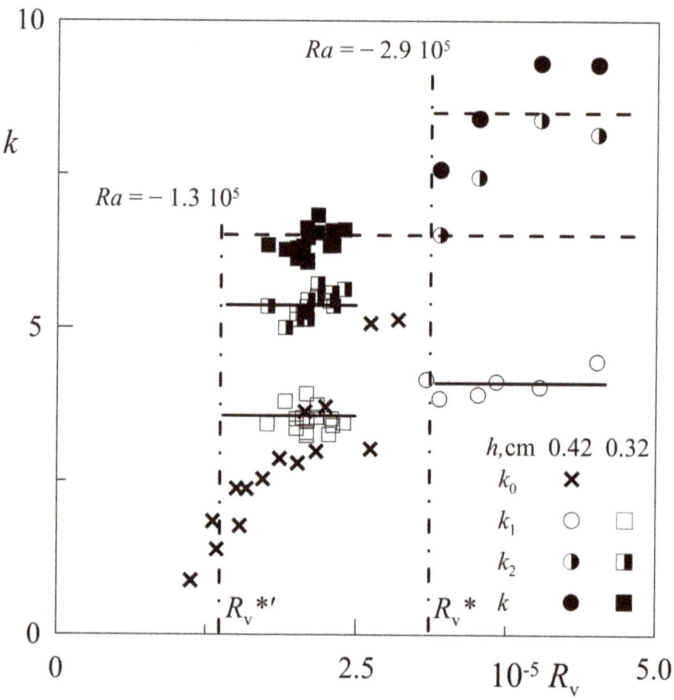

Figure 6. The dimensionless wavenumbers of different structures depending on the vibrational parameter R_V. Crosses and circles correspond to the experiments with a layer of thickness $h = 0.42$ cm at $Ra \approx -2.9 \cdot 10^5$, squares—to experiments with a layer of thickness $h = 0.32$ cm at $Ra \approx -1.3 \cdot 10^5$ (thermochromic methodic of convective patterns visualization). The horizontal dashed line shows the theoretical values of the wavenumber k in the threshold at corresponding values of Ra [13]. The dash-dotted vertical lines show the critical values of vibrational parameter R_V at corresponding values of Ra [13].

One can see that with increase of negative value of Ra the threshold value of corresponding vibrational parameter R_V^* grows in good agreement with the theory [13]. At this, the critical value of total wavenumber k also significantly grows; but the wavenumber k_1 practically does not change, it keeps the value, which corresponds to the vortexes occupying the whole layer thickness, and thus, providing the optimal heat transfer. The growth of k with $|Ra|$, which also takes place in good agreement with the linear theory, happens due to rapid growth of k_2. Such a phenomenon is found for the first time and could be named "the beads instability", it takes place at high negative Rayleigh numbers ($|Ra| > 10^5$), due to strong interaction of two mechanisms, thermogravitational (playing the stabilizing role) and the destabilizing thermovibrational one. The dominant role of vibroconvective structures in the form of parallel rolls with a wavenumber $k_1 \sim 3$ discovered in the experiments is consistent with the results of theoretical analysis [11], where it was shown that in the case of circular shaking, when the system is isotropic in the time average sense, the onset of convection corresponds to the 2D system of rolls with the roll axis selected spontaneously, in the case of positive and negative values of Ra. Reference [11], however, corresponds to the smaller negative values of $|Ra|$ and moderate dimensionless frequencies ($\omega \sim 50$), while in the present study the dimensionless frequency of vibrations is an order higher ($\omega \sim 500$). The experiments suggest that the increase of heat transfer in the subthreshold (in terms of vibrational thermal convection) area in the experiments with visualizer particles (PIV particles) located near the bottom of the cavity, is associated with the excitation of the averaged flows generated as a result of the system of heaps formation. As it has been mentioned, the appearance of the heaps is not of thermal origin. It should be noted that a similar phenomenon of the formation of a spatially ordered system of hills was observed on the surface of quartz sand in a cavity filled with liquid, performing high-frequency circular oscillations in the horizontal plane [19]. In that case, the emergence of a quasi-stationary "frozen" relief at the interface (liquefied granular medium and pure liquid), performing tangential circular oscillations, was associated with developing the Kelvin-Helmholtz instability. This fact was confirmed by comparison with a similar "frozen" relief at the boundary of liquids at circular vibrations of the cavity [20].

A similar distribution of the particles of the visualizer suspended near the bottom of the cavity was previously observed in a flat layer performing the oscillations of a spherical pendulum [15]. In this case, in agreement with the results of the present experiment, the wavenumber of structures increases with a dimensionless frequency and practically does not depend on the temperature conditions at the layer boundaries.

5. Conclusions

This article experimentally instigated the structure of the vibroconvective flows excited in the horizontal plane layer by the translational vibrations of circular polarization under conditions of high negative Rayleigh numbers. An important feature of the problem is the absence of a preferred direction of translational vibration. It was found that, in this case, the vibroconvective cells, developing in a threshold manner upon reaching the threshold value of the vibration parameter R_v, are characterized by a strong heterogeneity of sizes in different directions. The wavenumbers, characterizing the spatial periods in perpendicular directions, differ several times; the convective structures have the form of a system of parallel rolls, with a regular change in their dimensions along the length. The transverse size of the rolls is consistent with the thickness of the liquid layer; at the same time, the total wavenumber of the vibroconvective cells is much higher and is consistent with the theoretical value predicted by the linear analysis of vibroconvective stability. Such a selection of the vibroconvective patterns (in the form of asymmetric cells) in the absence of a definite direction of vibrations is observed for the first time. It will be interesting to check if such an overcritical behavior (when the size of vibroconvective vortexes is near to the layer thickness) will remains at much higher negative values of Ra, when predicted by the linear theory wavenumber k will be extremely high.

It was found that the presence of the relatively dense calliroscope particles in the liquid, which are located directly at the bottom of the layer, does not affect the excitation threshold of the thermal vibrational convection. At the same time, the particles of the visualizer close in density to the liquid, which are suspended near its boundary and under vibrations gather into the hillocks, cause the increase in heat transport in the subthreshold region (below the threshold of thermovibrational convection excitation). This is because the particles change the bottom surface profile, as a result of the redistribution, which leads to the steady flows and an increase in the heat transfer. With an increase in the dimensionless frequency, the wavenumber characterizing the spatial arrangement of the hills increases.

Author Contributions: Conceptualization, V.K.; methodology, A.V. and V.K.; formal analysis, A.V.; investigation, K.R. and V.K.; data curation, A.V.; writing—original draft preparation, K.R. and V.K.; writing—review and editing, V.K.; visualization, A.V.; supervision, V.K.; project administration, V.K. All authors have read and agreed to the published version of the manuscript.

Funding: This research was funded by the Russian Science Foundation (project 18-71-10053).

Conflicts of Interest: The authors declare no conflict of interest.

References

1. Gershuni, G.Z.; Lyubimov, D.V. *Thermal Vibrational Convection*; Wiley: Chichester, UK, 1998.
2. Gershuni, G.Z.; Zhukhovitskii, E.M. Free thermal convection in a vibrational field under conditions of weightlessness. *Akademiia Nauk SSSR Dokl.* **1979**, *24*, 894–896.
3. Mialdun, A.; Ryzhkov, I.I.; Melnikov, D.E.; Shevtsova, V. Experimental evidence of thermal vibrational convection in a nonuniformly heated fluid in a reduced gravity environment. *Phys. Rev. Lett.* **2008**, *101*, 084501. [CrossRef] [PubMed]
4. Simonenko, I.B. On the effect of high frequency vibrations on the origin of convection. *Izvestiya Akademii Nauk SSSR* **1966**, *5*, 51–55.
5. Gershuni, G.Z.; Keller, I.O.; Smorodin, B.L. Vibrational and convective instability of a plane horizontal fluid layer at finite vibration frequencies. *Fluid Dyn.* **1996**, *31*, 666–671. [CrossRef]
6. Smorodin, B.L.; Ishutov, S.M.; Myznikova, B.I. On the convection of a binary mixture in a horizontal layer under high-frequency vibrations. *Microgravity Sci. Technol.* **2018**, *30*, 95–102. [CrossRef]
7. Vorobev, A.; Lyubimova, T. Vibrational convection in a heterogeneous binary mixture. Part I. Time-averaged equations. *J. Fluid Mech.* **2019**, *870*, 543–562. [CrossRef]
8. Lappa, M. *Thermal Convection: Patterns, Evolution and Stability*; John Wiley & Sons: Chichester, UK, 2009.
9. Crewdson, G.; Lappa, M. The zoo of modes of convection in liquids vibrated along the direction of the temperature gradient. *Fluids* **2021**, *6*, 30. [CrossRef]
10. Kozlov, V.G. Vibrational convection in a cavity undergoing spatial pendulum oscillations. *Heat Transf. Sov. Res.* **1991**, *23*, 999–1008.
11. Pesch, W.; Palaniappan, D.; Tao, J.; Busse, F. Convection in heated fluid layers subjected to time-periodic horizontal accelerations. *J. Fluid Mech.* **2008**, *596*, 313–332. [CrossRef]
12. Kozlov, V.; Rysin, K.; Vjatkin, A. Vibroconvective stability of liquid in horizontal plane layer subject to circular translational vibrations. *Microgravity Sci. Technol.* **2019**, *31*, 759–765. [CrossRef]
13. Ivanova, A.A.; Kozlov, V.G. Thermal vibrational convection in a cavity under nontranslational oscillations. *Fluid Dyn.* **2003**, *38*, 372–386. [CrossRef]
14. Kozlov, V.G. Vibrational thermal convection in a cavity executing high-frequency rocking motions. *Fluid Dyn.* **1988**, *23*, 437–442. [CrossRef]
15. Kozlov, V.G.; Selin, N.V. Pendulum thermal vibrational convection in a liquid layer with internal heat generation. *Fluid Dyn. Mater. Process.* **2006**, *2*, 107–117.
16. Riley, N. Steady streaming. *Annu. Rev. Fluid Mech.* **2001**, *33*, 43–65. [CrossRef]
17. Lappa, M. The patterning behavior and accumulation of spherical particles in a vibrated non-isothermal liquid. *Phys. Fluids* **2014**, *26*, 093301. [CrossRef]
18. Gershuni, G.Z.; Zhukhovitskii, E.M. *Convective Stability of an Incompressible Fluid*; Nauka: Moscow, Russia, 1972. (In Russian)
19. Ivanova, A.A.; Kozlov, V.G. Sand-Fluid Interface under Vibration. *Fluid Dyn.* **2002**, *37*, 277–293. [CrossRef]
20. Ivanova, A.; Kozlov, V.; Legros, J.C. Two Liquids Interface Dynamics in Cavity Subject to Translational Vibration of Circular Polarization. In Proceedings of the 1st International Symposium on Microgravity Research and Applications in Physical Sciences and Biotechnology, Sorrento, Italy, 10–15 September 2000; pp. 823–828.

fluids

MDPI

Article

Determination of Critical Reynolds Number for the Flow Near a Rotating Disk on the Basis of the Theory of Stochastic Equations and Equivalence of Measures †

Artur V. Dmitrenko [1,2]

[1] Department of Thermal Physics, National Research Nuclear University MEPhI, Kashirskoyeshosse 31, 115409 Moscow, Russia; AVDmitrenko@mephi.ru

[2] Department of Thermal Engineering, Russian University of Transport MIIT, Obraztsova Street 9, 127994 Moscow, Russia

† This article is dedicated to the memory of Academician N. A. Anfimov.

Abstract: The determination of the flow regime of liquid and gas in power plants is the most important design task. Performing the calculations based on modern calculation methods requires a priori knowledge of the initial and boundary conditions, which significantly affect the final results. The purpose of the article is to present the solution for the critical Reynolds number for the flow near a rotating disk on the basis of the theory of stochastic equations of continuum laws and equivalence of measures between random and deterministic motions. The determination of the analytical dependence for the critical Reynolds number is essential for the study of flow regimes and the thermal state of disks and blades in the design of gas and steam turbines. The result of the calculation with using the new formula shows that for the flow near a wall of rotating disk, the critical Reynolds number is 325,000, when the turbulent Reynolds is 5 ÷ 10 and the degree of turbulence is 0.01 ÷ 0.02. Therefore, the result of solution shows a satisfactory correspondence of the obtained analytical dependence for the critical Reynolds number with the experimental data.

Keywords: stochastic equations; equivalence of measures; nature of turbulence; critical Reynolds number

check for updates

Citation: Dmitrenko, A.V. Determination of Critical Reynolds Number for the Flow Near a Rotating Disk on the Basis of the Theory of Stochastic Equations and Equivalence of Measures. *Fluids* **2021**, *6*, 5. https://doi.org/10.3390/fluids6010005

Received: 24 November 2020
Accepted: 22 December 2020
Published: 25 December 2020

Publisher's Note: MDPI stays neutral with regard to jurisdictional claims in published maps and institutional affiliations.

Copyright: © 2020 by the author. Licensee MDPI, Basel, Switzerland. This article is an open access article distributed under the terms and conditions of the Creative Commons Attribution (CC BY) license (https://creativecommons.org/licenses/by/4.0/).

1. Introduction

The development of new physical and mathematical theories for phenomena occurring in the nature and technical devices requires the constant application of the theory for observed varieties of this phenomenon. Therefore, each application of the theory to a specific process also requires the theoretical comparison between the existing fundamental methods instead of only the mathematical solutions. It makes possible for scientists and specialists in various fields to understand the evolution of theoretical ideas as well as the difference, the essence, and the advantage of the new scientific methodology.

The theories on the nature of the turbulence were formulated in [1–10]. The main principles of the theory of measures of random stationary processes are described in the publications of Kolmogorov and Khinchin. These works underlie the statistical theory of turbulence. On the basis of statistical theory, Obukhov and, later, Heisenberg proposed the statistical theory for the process of generation of the turbulent field. It is also worth to mention the publications of J. Taylor in which one first tried to determine the critical Reynolds number in the function of initial parameters of the fluctuation. However, J. Taylor determined this dependence semi-empirically for one type of flow. The linear theory of turbulence led to a certain success. This is especially true for the Orr–Sommerfeld equation, which allowed calculating only the critical Reynolds numbers. A special place is occupied by the Landau theory, which has a qualitative character when describing the turbulence as a quasi-periodic process. However, even in this case, it is impossible to calculate the flow characteristics despite the qualitative description of the turbulence. Klimontovich investigated the Leontovich and Sato equations and presented the mathematical formulation of the entropy change, when the turbulence occurs. However, this theory only enables us to represent the turbulence process qualitatively without calculating the characteristics of the phenomenon.

The development of the theory of strange attractors and mathematical methods for obtaining a strict solution of the Navier–Stokes equation are presented in [11–32]. It is known that the theory of strange attractors is based on the measure theory, which allows one deducing the Kolmogorov–Sinai entropy. Somewhat later, a more general formula for the Renyi entropy made it possible to extend the application of attractor theory. However, this theory allows determining only the increase in the number of degrees of freedom in time. This theory does not allow explaining the spatial change in increasing number of degrees of freedom.

The theory of solitons was also considered to be useful for certain time for explaining the origin of the turbulence [10–12]. However, the results of the investigation of solutions to the Korteweg–de Vries equation provide no basis for determining and calculating the characteristics of the turbulence phenomenon.

The statistical and stochastic equations and the numerical methods for investigating the turbulent processes are presented in [33–53]. The above fundamentals of statistical hydrodynamics and the development of computer technology and numerical methods enabled us to implement the solution of moment equations for second- and higher-order correlations using the RANS method [39–50], and later the LES method [39–43].

The most powerful numerical method called DNS is represented by three methodologies that are fundamentally different from each other. Therefore, the study of the origin of turbulence by each of the methods has several features that require explanation. However, all DNS methods are very sensitive to the initial and boundary conditions for each type of the hydrodynamic flow.

In certain cases, the instantaneous Navier–Stokes equations undergo the artificial "stochasticization" by adding the left-side additional term. In this case, we need to do the same in the continuity and energy equations. However, it is necessary to be aware that if it is done for the instantaneous equations, then an open thermodynamic system is obtained. However, these methods failed in determining the unified physics and cause of the turbulence process [33–35].

The special attention was focused on the theoretical solutions for the critical Reynolds number. It should be noted that the most well-known ratio based on the theory of dimension was, as is known, determined with using the experimental data [37–42,52,53]. Therefore, on the basis of these experimental formulas, it was impossible to obtain the new theory for determining analytically the dependences for the critical Reynolds number of turbulence in different flows.

At the same time, it is known that in an arbitrary hydrodynamic flow, there are initial disturbances generated by various causes, both technical and natural. Therefore, it is obvious that whether or not the particular disturbance, which arises, exists and develops, depends on the interaction of the main undisturbed motion with this initial disturbance.

The stochastic theory of turbulence based on the stochastic equations and the theory of equivalent measures make it possible deriving the analytical dependences for the first and second critical Reynolds numbers in the cases of the isothermal and non-isothermal flows on the smooth flat plate and in the round tube [54–59]. The progress of this theory gives a new method for determining the analytical dependences for the profiles of averaged velocity and the temperature fields [60,61], the friction and heat transfer coefficients [62–64], the second-order correlations [58,65,66], the correlation dimension of the attractor in the boundary layer [67–70], the theoretical solutions for the spectral function of the turbulent medium [71–73], and the formula for the Reynolds analogy [74–76].

It should also be noted that from the results obtained on the basis of the theory of stochastic equations and the theory of equivalent measures, it was possible for the first time to investigate the analytical relations for calculating the spatial distribution of the number of degrees of freedom of a strange attractor. Such distributions of the attractor correlation dimension were calculated for the flow in a tube, on a flat plate, and in the Earth's atmosphere.

Also, as a result, it was determined that the spectrum $E(k)_j$ depends on the wave numbers k for the interval of generation of turbulence in the form $E(k)_j \sim k^n$, $n = 1.2 \div 1.5$. This formula was named the ratio of uncertainty in turbulence generation [62].

The uncertainty relation derived analytically determines the fact that in the turbulence generation region, there is the family of perturbations—the vortices, which have a space–energy similarity $(E \cdot L^{-a})$ = constant, and each of the perturbations of this family can interact with the main flow, which leads to the origin and development of turbulence [62]. Moreover, for each type of flow, whether or not it is the flow in a pipe or along a flat plate, the spatial–energy similarity has its own value of the indicator "a". Therefore, there is an uncertainty in both the geometric and energy parameters of the perturbation when determining the interaction with the main motion.

In accordance with [49–51], the essence of the discovery of the theory of equivalence of measures in the stochastic process determines the beginning of the interaction between the deterministic and random field. It is found that this interaction begins when the mass shift, the momentum shift, and the energy shift of the main undisturbed flow is equal to the mass fluctuation, momentum, and energy of the random field in the space edge, which is commensurate with the linear measure of the perturbation at a fixed time. As a result, the equivalence of substantial time derivatives is observed in the interaction domain.

It should be noted that the main part of publications [54–76] is devoted to such types of fluid flows as the flow in a round tube and the flow along a smooth flat plate for which there is a considerable experimental material, which allows calculating the parameters with using new formulas. Therefore, it is interesting to consider other types of fluid flows, which are also important for both theory and practice. In this connection, we presented here the solution for the critical Reynolds number for the motion near a rotating disk.

2. Conservation Equations for Stochastic Process

The equations derived in [54–59] take the form:

The equation of mass (continuity)

$$\frac{d(\rho)_{col_{st}}}{d\tau} = -\frac{(\rho)_{st}}{\tau_{cor}} - \frac{d(\rho)_{st}}{d\tau}, \tag{1}$$

the momentum equation

$$\frac{d\left(\rho\vec{U}\right)_{col_{st}}}{d\tau} = div(\tau_{i,j})_{col_{st}} + div(\tau_{i,j})_{st} - \frac{(\rho\vec{U})_{st}}{\tau_{cor}} - \frac{d(\rho\vec{U})_{st}}{d\tau} + F_{col_{st}} + F_{st} \tag{2}$$

and the energy equation

$$\begin{aligned} \frac{dE_{col_{st}}}{d\tau} &= div(\lambda\frac{\partial T}{\partial x_j} + u_i\tau_{i,j})_{col_{st}} + div(\lambda\frac{\partial T}{\partial x_j} + u_i\tau_{i,j})_{st} \\ &- \left(\frac{E_{st}}{\tau_{cor}}\right) - \left(\frac{dE_{st}}{d\tau}\right) + (u_iF)_{col_{st}} + (u_iF)_{st} \end{aligned} \tag{3}$$

Here, E, ρ, $\vec{U}$, u_i, u_j, u_l, μ, τ, $\tau_{i,j}$ are the energy, the density, the velocity vector, and the velocity components in the directions x_i, x_j, x_l $(i, j, l = 1, 2, 3)$; the dynamic viscosity, the time, and the stress tensor $\tau_{i,j} = P + \sigma_{i,j}$, $\delta_{ij} = 1$ if $i = j$, $\delta_{ij} = 0$ for $i \neq j$. P is the pressure of liquid or gas; λ is the thermal conductivity; c_p and c_v are the specific heat at constant pressure and volume, respectively; F is the external force, and $\sigma_{i,j} = \mu\left(\frac{\partial u_i}{\partial x_j} + \frac{\partial u_j}{\partial x_i}\right) - \delta_{ij}(\xi - \frac{2}{3}\mu)\frac{\partial u_l}{\partial x_l}$, $(\tau_{cor}) = \frac{L}{\left((E_{st})_{U,P}/\rho\right)^{1\backslash 2}}$.

Furthermore, $L = L_{U,P} = L_U$ is the scale of turbulence. The subscripts $(_{U,P})$ and $(_U)$ refer to the velocity field and the subscript $(_T)$ refers to the temperature field. L_y on $x_2 = y$, or L_x,

$x_1 = x$. Here, x_1 and x_2 are the coordinates along the wall and normal to it. The subscript "col_{st}" refers to the components, which are actually the deterministic. The subscript "st" refers to the component, which are actually the stochastic. As a result, using the law of the equivalency of measures between the random and deterministic process at the critical point, we obtained the sets of stochastic equations of mass, momentum, and energy for the next space–time areas: (1) the onset of generation (subscript 1, 0, or 1); (2) the generation of turbulence (subscript 1,1); (3) the diffusion (1,1,1) or 1; (1,1), and (4) the dissipation of the turbulent fields.

The resulting set of equations can be described by a correlator, which can be written for each of the four space–time domains (N, M). This correlator also determines the probability of the fractal origin of this interaction instead of only determining the set of equations for the interaction of the random and deterministic fields. Therefore, in accordance with [54–59], this correlator in space–time is

$$\lim_{m_i \to m_c; r_i \to r_c; \Delta\tau_i \to \tau_c} (D_{N,M}(m_i; r_i; \Delta\tau_i)) = 0 \tag{4}$$

$$D_{N,M}(m_c; r_c; \tau_c) = \sum_i \lim_{m_i \to m_c} \lim_{r_i \to r_c} \lim_{\Delta\tau_i \to \tau_c} \left\{ m\left(T^M Z^* \cap T^N Y^*\right) - R_{1_{T^M Z^* T^N Y^*}} m\left(T^M Z^*\right) \right\} \tag{5}$$

The subscript j denotes the parameters m_{cj} (j = 3 means the mass, the momentum, and the energy). For the case of the binary intersections, it was written that $X = Y + Z + W$. Subsets Y, Z, and W are called extended in X as $\{Y^*, Z^*, W^*\}$ if the measures $m(Y)$, $m(Z)$, and $m(W)$ have the properties [54–59]:

$$\begin{aligned}
&m(Y) = m(Y^*) = m(T^n Y) + \bigcup_{k=0}^{k=n-1} m(T^k(G_1^{n-k}))\text{and wandering subsets} \bigcup_{k=0}^{k=n-1} m(T^k(G_1^{n-k})) \subset Y;\\
&m(Z) = m(Z^*) = m(T^n Z^*) + \bigcup_{k=0}^{k=n-1} m\left(T^k\left(G_2^{n-k}\right)\right)\text{and wandering subsets} \bigcup_{k=0}^{k=n-1} m(T^k(G_2^{n-k})) \subset Z;\\
&m(W) = m(W^*) = m(T^n W) + \bigcup_{k=0}^{k=n-1} m\left(T^k\left(G_3^{n-k}\right)\right)\text{and wandering subsets} \bigcup_{k=0}^{k=n-1} m(T^k(G_3^{n-k})) \subset W.
\end{aligned} \tag{6}$$

Here G_1^n is the wandering subset of the expanded subset $Y^* \subset X$, G_2^n is the wandering subset of the expanded subset $Z^* \subset X$, G_3^n is the wandering subset of the expanded subset W^*.

Here subscripts "cr" or "c" refer to the critical point $r(x_{cr}, \tau_{cr})$ or r_c: the space–time point of the onset of the interaction between the deterministic field and the random field, which leads to the turbulence. In addition, subsets Y, Z, W are called extended in X. For the transfer of the substantial quantity Φ (mass (density ρ), momentum ($\rho \boldsymbol{U}$), energy (E)) of the deterministic (laminar) motion into the random (turbulent) one, for domain 1 of the start of turbulence generation, the pair $(N, M) = (1, 0)$ with the equivalence of measures is written $(d\Phi_{col_{st}})_{1,0} = -R_{1,0}(\Phi_{st})$ and $\left(\frac{d(\Phi)_{col_{st}}}{d\tau}\right)_{1,0} = -R_{1,0}\left(\frac{\Phi_{st}}{\tau_{cor}}\right)$. Applying the correlation $D_{N,M}(m_c; r_c; \tau_c) = D_{1,1}(m_c; r_c; \tau_c)$ derived in [39–44], the equivalence relation for pair (N,M) = (1,1) was defined as $(d\Phi_{col_{st}})_{1,1} = -R_{1,1}(d\Phi_{st})$, $\left(\frac{d(\Phi)_{col_{st}}}{d\tau}\right)_{1,1} = -R_{1,1}\left(\frac{d\Phi_{st}}{d\tau}\right)$, where $R_{1,0}$ and $R_{1,1}$ are the fractal coefficients, $\Phi_{col_{st}}$ is the part of the field of Φ, exactly, its deterministic component (subscript col_{st}) is the stochastic component, the measure of which is zero; Φ_{st} is the part of Φ, exactly, the proper stochastic component (subscript st). It should be noted that the stochastic equations (1)–(3) derived in [54–59] include free terms of gradient and non-gradient structures on their right-hand side.

3. Sets of Stochastic Equations

The flow near a rotating disk, as well as the previous ones, has an important scientific and applied significance as the three-dimensional motion of a fluid on a solid surface, but the forced motion is caused here by the rotational motion of the body. Then, taking into account the previously presented set of stochastic equations of the considered theory of equivalent measures in the case of a continuous isothermal medium, we write set (1)–(3) of equations of mass, momentum, and energy in accordance with [54–59]. For the area (1) —the onset of generation (subscript 1, 0, or 1) referring the pair (N, M) = (1,0) is:

$$\left\{\begin{array}{l}\left\{\begin{array}{l}\left(\frac{d(\rho)_{col,st}}{d\tau}\right)_{1,0}=-\frac{\rho_{st}}{\tau_{cor}};\\ \left(\frac{d(\rho\vec{U})_{col,st}}{d\tau}\right)_{1,0}=-\left(\frac{(\rho\vec{U})_{st}}{\tau_{cor}}\right);\\ div(\tau_{i,j})_{col,st1}=\frac{(\rho\vec{U})_{st}}{\tau_{cor}}\end{array}\right.\\ \left(\frac{d(E)_{col,st}}{d\tau}\right)_{1,0}=-\left(\frac{(E)_{st}}{\tau_{cor}}\right)_{1,0};\\ div(\lambda\frac{\partial T}{\partial x_j}+u_i\tau_{i,j})_{col,st1}=\left(\frac{(E)_{st}}{\tau_{cor}}\right)_{1,0}\end{array}\right. \tag{7}$$

Set (1)–(3) of equations for the area (2)—the area of generation of turbulence (subscript 1,1) referring to the pair (N, M) = (1,1) is written as:

$$\left\{\begin{array}{l}\left\{\begin{array}{l}\left(\frac{d(\rho)_{col,st}}{d\tau}\right)_{1,1}=-\left(\frac{d\rho_{st}}{d\tau_{cor}}\right);\\ \left(\frac{d(\rho\vec{U})_{col,st}}{d\tau}\right)_{1,1}=-\left(\frac{d(\rho\vec{U})_{st}}{d\tau}\right);\\ div(\tau_{i,j})_{col,st2}=\frac{d(\rho\vec{U})_{st}}{d\tau}\end{array}\right.\\ \left(\frac{d(E)_{col,st}}{d\tau}\right)_{1,1}=-\left(\frac{d(E)_{st}}{d\tau}\right)_{1,1};\\ div(\lambda\frac{\partial T}{\partial x_j}+u_i\tau_{i,j})_{col,st2}=\left(\frac{d(E)_{st}}{d\tau}\right)_{1,1}.\end{array}\right. \tag{8}$$

Set (1)–(3) of equations for the area (3) is the diffusion of the turbulence referring to the pair (N = *p*, M = *k*, *l*) = (1,1,0) was written as

$$\begin{array}{c}\frac{d(\rho)_{st}}{d\tau}=-\frac{(\rho)_{st}}{\tau_{cor}};\ \frac{d\rho_{st}}{d\tau}=\left(\frac{d\rho_{st}}{d\tau}\right)_{1,0}+\left(\frac{d\rho_{st}}{d\tau}\right)_{1,1};\\ \frac{d(\rho\vec{U})_{st}}{d\tau}=-\frac{(\rho\vec{U})_{st}}{\tau_{cor}};\ \frac{d(\rho\vec{U})_{st}}{d\tau}=\left(\frac{d(\rho\vec{U})_{st}}{d\tau}\right)_{1,0}+\left(\frac{d(\rho\vec{U})_{st}}{d\tau}\right)_{1,1}\\ \left|\frac{(dE_{st})}{d\tau}\right|_{1;(1,0)}=(R_{zTz})_{1;(1,0)}\left|\frac{(E_{st})_j}{\delta\tau}\right|_{1,(1,0)};\ \left|\frac{(dE_{st})}{d\tau}\right|=\left|\frac{(dE_{st})_1}{d\tau}\right|+\left|\frac{(dE_{st})_2}{d\tau}\right|\end{array} \tag{9}$$

Therefore, for the area (3) of diffusion, we have two fractal equations. The first equation is written as

$$\frac{d(E_{st})_j}{d\tau}=-(R_{zTz})_{(1,1,1)}\frac{(E_{st})_j}{\tau_{cor1}} \tag{10}$$

Here, (E_{st}) is the field-energy component, which is actually the stochastic one (subscript '*st*'), the subscript j = 1 refers to the space–time area of the diffusion of turbulence 3).

4. Equations for Critical Reynolds Number

The solution for the velocity field (u, v, and w are the components of the velocity in the radial "r", circumferential "ϕ", and axial "z" directions) of the deterministic (laminar) motion is presented in [39–41]. According to this solution, the velocity components are expressed as the dependences in the radial, circumferential, and axial directions, respectively: $u=r\omega F(\xi)$, $v_\varphi=r\omega G(\xi)$, $w=\sqrt{v\omega}H(\xi)$. The values $F(\xi)$, $G(\xi)$, $H(\xi)$, and their derivatives in the function $\xi=z\sqrt{\omega/v}$, are also given in [41]. In this case, the flow-motion mode is represented as a function of the coefficient of the moment of resistance of the disk.

$$C_M=\frac{2M}{0.5\cdot\rho\cdot\omega^2\cdot R^5} \tag{11}$$

Here, $M=-2\pi\int\limits_0^R r^2\tau_{z\varphi}dr$, $Re=\frac{R^2\omega}{v}$ is the Reynolds number and ω, R, are the rotational velocity and the radius of the disc. Taking into account that the velocity $v_\varphi=r\omega G(\xi)$, the value of the stress $\tau_{z\varphi}$ is written as

$$\tau_{z\varphi}=\mu\frac{dv_\varphi}{dz}=\rho r v^{1/2}\omega^{3/2}G'(\xi) \tag{12}$$

Here $v_\varphi=r\omega G(\xi)$ is the velocity in the circumferential direction, for z = 0, $G(\xi)=1$, for the current value of z near the critical point $G(\xi)\approx 1-Kz\left(\frac{L_y}{z}\right)\sqrt{\frac{\omega}{v}}$ and δ is the thickness of the boundary layer, in accordance with [41], $\delta^{-1} \approx K\sqrt{\frac{\omega}{\nu}}$. L_x is the scale of the disturbance along the radial direction (along the current radius "r"), and L_y is the scale of the disturbance along the axial direction "z". Then, according to the equivalence of measures of deterministic and random motion, we write that

$$div(u_i\tau_{i,j})_{col,st1} = \left(\frac{(E)_{st}}{\tau_{cor}}\right)_{1,0} \tag{13}$$

In the first approximation, the left-hand side of the equation takes the form:

$$\begin{gathered} div(u_i\tau_{i,j}) \approx div(v_\phi\tau_{z\phi}) \approx G'(\xi)\rho\omega\nu^{1/2}\omega^{3/2}G(\xi)\tfrac{d}{dz}r^2 = 2G'(\xi)]\rho\omega\nu^{1/2}\omega^{3/2}G(\xi)r\tfrac{L_x}{L_y} \approx \\ 2G'(\xi)\rho\omega\nu^{1/2}\omega^{3/2}\tfrac{L_x}{L_y}r\left(1 - Kz\left(\tfrac{L_y}{z}\right)\sqrt{\tfrac{\omega}{\nu}}\right) = 2*0.616\rho\omega\nu^{1/2}\omega^{3/2}\tfrac{L_x}{L_y}r\left(1 - Kz\left(\tfrac{L_y}{z}\right)\sqrt{\tfrac{\omega}{\nu}}\right) \\ G'(\xi) = 0.616, \end{gathered} \tag{14}$$

see [41].

Then, we have the expression

$$1.232\,\rho\omega R\nu^{1/2}\omega^{1/2}\frac{L_x}{L_y}\frac{r}{R}\left(1 - Kz\left(\frac{L_y}{z}\right)\sqrt{\frac{\omega}{\nu}}\right) = \frac{E_{st}}{\tau^0_{cor}} \tag{15}$$

From the obtained expression, we can determine the dependence for a dimensionless number at which there is an equivalence of measures between deterministic and random motion called in the hydrodynamics the critical Reynolds number for the flow in the boundary layer near the disk surface corresponding to the values of the correlation times $(\tau^0_{cor})_1$, $(\tau^0_{cor})_2$, $(\tau^0_{cor})_3$, $\tau_{motion} = [\omega]^{-1}$. Thus, for the case of the correlation time $(\tau^0_{cor})_1 = \frac{L}{\sqrt{E_{st}/\rho}}$, we write

$$1.232\left(\tfrac{L_x}{R}\right)\left(\tfrac{r}{R}\right)\left(\tfrac{\omega^2R^2}{E_{st}/\rho}\right)\tfrac{\nu\omega}{E_{st}/\rho}\tfrac{R^2\omega}{R^2\omega}\sqrt{\tfrac{E_{st}/\rho}{\nu\omega}}\left(1 - Kz\left(\tfrac{L_y}{z}\right)\sqrt{\tfrac{\omega}{\nu}}\right) = 1 \tag{16}$$

$$1.232\left(\frac{L_x}{R}\right)\left(\frac{r}{R}\right)\left(\frac{\omega^2R^2}{E_{st}/\rho}\right)\frac{R^2\omega^2}{E_{st}/\rho}\frac{\nu}{R^2\omega}\sqrt{\frac{E_{st}/\rho}{\nu\omega}}\left(1 - Kz\left(\frac{L_y}{z}\right)\sqrt{\frac{\omega}{\nu}}\right) = 1 \tag{17}$$

$$1.232\left(\frac{L_x}{R}\right)\left(\frac{r}{R}\right)\left(\frac{\omega^2R^2}{E_{st}/\rho}\right)^2\frac{\nu}{R^2\omega}\sqrt{\frac{E_{st}/\rho}{\nu\omega}\frac{R^2\omega}{R^2\omega}}\left(1 - Kz\left(\frac{L_y}{z}\right)\sqrt{\frac{\omega}{\nu}}\right) = 1 \tag{18}$$

$$1.232\left(\frac{L_x}{R}\right)\left(\frac{r}{R}\right)\left(\frac{\omega^2R^2}{E_{st}/\rho}\right)^2\frac{\nu}{R^2\omega}\sqrt{\frac{E_{st}/\rho}{R^2\omega^2}\frac{R^2\omega}{\nu}}\left(1 - Kz\left(\frac{L_y}{z}\right)\sqrt{\frac{\omega}{\nu}}\right) = 1 \tag{19}$$

$$\sqrt{\frac{R^2\omega}{\nu}} = 1.232\left(\frac{L_x}{R}\right)\left(\frac{r}{R}\right)\left(\frac{\omega^2R^2}{E_{st}/\rho}\right)^{3/2}\left(1 - Kz\left(\frac{L_y}{z}\right)\sqrt{\frac{\omega}{\nu}}\right) \tag{20}$$

Finally, we obtain

$$Re \approx 1.5\left(\frac{L_x}{R}\right)^2\left(\frac{r}{R}\right)^2\left(\frac{\omega^2R^2}{E_{st}/\rho}\right)^3\left(1 - Kz\left(\frac{L_y}{z}\right)\sqrt{\frac{\omega}{\nu}}\right)^2, \tag{21}$$

$$Re \approx 1.5\left(\tfrac{L_x}{R}\right)^2\left(\tfrac{r}{R}\right)^2\left(\tfrac{\omega R}{\sqrt{E_{st}/\rho}}\right)^6\left(1 - Kz\left(\tfrac{L_y}{z}\right)\sqrt{\tfrac{\omega}{\nu}}\right)^2 \tag{22}$$

Correspondently, for the correlation time $(\tau^0_{cor})_2 = \frac{L^2}{\nu}$, we have

$$Re \approx \left\{1.5\left(\tfrac{L_x}{R}\right)^2\left(\tfrac{r}{R}\right)^2\left(\tfrac{\omega R}{\sqrt{E_{st}/\rho}}\right)^6\left(1 - Kz\left(\tfrac{L_y}{z}\right)\sqrt{\tfrac{\omega}{\nu}}\right)^2\right\}Re^2{}_{st} \tag{23}$$

or

$$Re \approx \left\{1.5\left(\tfrac{L_x}{R}\right)^2\left(\tfrac{r}{R}\right)^2\left(\tfrac{\omega R}{\sqrt{E_{st}/\rho}}\right)^6\left(1 - Kz\left(\tfrac{L_y}{z}\right)\sqrt{\tfrac{\omega}{\nu}}\right)^2\right\}Re^2{}_{st} \tag{24}$$

For the correlation time $(\tau^0_{cor})_3 = \frac{\nu}{E_{st}/\rho}$, we obtain the value

$$Re \approx \left\{1.5\left(\tfrac{L_x}{R}\right)^2\left(\tfrac{r}{R}\right)^2\left(\tfrac{\omega R}{\sqrt{E_{st}/\rho}}\right)^6\left(1 - Kz\left(\tfrac{L_y}{z}\right)\sqrt{\tfrac{\omega}{\nu}}\right)^2\right\}\frac{1}{Re^2{}_{st}} \tag{25}$$

or

$$\mathrm{Re} \approx \left\{ 1.5 \left(\frac{L_x}{R} \right)^2 \left(\frac{r}{R} \right)^2 \left(\frac{\omega R}{\sqrt{E_{st}/\rho}} \right)^6 \left(1 - Kz\left(\frac{L_y}{z} \right) \sqrt{\frac{\omega}{\nu}} \right)^2 \right\} \frac{1}{\mathrm{Re}^2{}_{st}} \quad (26)$$

5. The Equation for the Critical Point

Now let us determine the position of the critical point. The definition of the critical point is found from the equation as

$$\int\limits_{-\Delta V|2}^{+\Delta V|2} d(E_{col_{st}})_{1;0} = \int\limits_{X} dE_{st} \quad (27)$$

Here E_{st} is the random energy component in the space X with the measure $m(E_{st}) < \infty$

$$E_{st} = E_{st}(\vec{x_i}, \tau_i, m_i) < \infty \quad (28)$$

In accordance with the ergodic theory [39,40]

$$\int\limits_{X} dE_{st} = \frac{1}{\Delta V} \int\limits_{V} E_{st}\delta((\Delta V)_{critic} - \Delta V)dV = \frac{1}{\tau^0_{cor}} \int\limits_{\tau} E_{st}\delta(\tau^0_{cor} - \tau)d\tau = (E_{st})_{critic} \quad (29)$$

$(E_{st})_{critic}$ is the energy of the stochastic field in the critical point,

or

$$\int\limits_{X} dE_{st} = \frac{1}{L} \int\limits_{L} E_{st}\delta((x_i)_{critic} - x_i)dL = \frac{1}{\tau^0_{cor}} \int\limits_{\tau} E_{st}\delta(\tau^0_{cor} - \tau)d\tau = (E_{st})_{critic} \quad (30)$$

L is the scale of the disturbance.

Then, taking into account the values of functions $F(\xi)$, $G(\xi)$, $H(\xi)$, in the neighborhood of critical point and using the equations for Formulas (11) and (12), we find $\left(E_{col_{st}}\right)_{1;0} = 0.5\rho v_\phi^2$, $v_\phi = r\omega G(\xi)$ is the velocity in the circumferential direction, for z = 0, $G(\xi) = 1$, for the current value of z near the critical point $G(\xi) \approx 1 - Kz\left(\frac{L_y}{z}\right)\sqrt{\frac{\omega}{\nu}}$.

Then, we may write that

$$\int\limits_{-V|2}^{+V|2} d\left(E_{col_{st}}\right)_{1;0} \approx \int\limits_{-L|2}^{+L|2} d\left(E_{col_{st}}\right)_{1;0} \approx 0.5\rho\omega^2 \left\{ \left[(r + L_x/2)^2 \left(1 - K(z + L_y/2\right)^2 \frac{\omega}{\nu} \right] - \left[(r - L_x/2)^2 \left(1 - K(z - L_y/2\right)^2 \frac{\omega}{\nu} \right] \right\}$$
$$\approx 0.5\rho\omega^2 r L_x \left[1 - K^2 z^2 \left(\frac{L_y}{z} \right) \frac{\omega}{\nu} \right] \quad (31)$$

So, we obtain

$$\frac{L_x}{R}\frac{r}{R}\left[1 - \left(\frac{L_y}{z}\right)K^2 z^2 \frac{\omega}{\nu}\right] = \frac{E_{st}/\rho}{(R\omega)^2} \quad (32)$$

Then, for the Equation (32), we can write

$$\left(\frac{L_x}{R}\frac{r}{R}\right)^2 = \left(\frac{E_{st}/\rho}{(R\omega)^2}\right)^2 \frac{1}{\left[1 - K^2 z^2 \left(\frac{L_y}{z}\right)\frac{\omega}{\nu}\right]^2} \quad (33)$$

6. The Solution for the First Critical Reynolds Number

We substitute Equation (33) in expression (22) for the Reynolds number, then, for the correlation time $(\tau^0_{cor})_1 = \frac{L}{\sqrt{E_{st}/\rho}}$, the critical Reynolds number is

$$\mathrm{Re} \approx 1.5 \left(\frac{L_x}{R} \right)^2 \left(\frac{r}{R} \right)^2 \left(\frac{\omega R}{\sqrt{E_{st}/\rho}} \right)^6 \left(1 - Kz\left(\frac{L_y}{z} \right) \sqrt{\frac{\omega}{\nu}} \right)^2 \quad (34)$$

Then, we obtain

$$\mathrm{Re} \approx 1.5 \left(\frac{\omega^2 R^2}{E_{st}/\rho} \right) \frac{\left(1 - Kz\left(\frac{L_y}{z} \right) \sqrt{\frac{\omega}{\nu}} \right)^2}{\left[1 - K^2 z^2 \left(\frac{L_y}{z}\right)\frac{\omega}{\nu}\right]^2} \quad (35)$$

or

$$\mathrm{Re} \approx 1.5\left(\frac{\omega R}{\sqrt{E_{st}/\rho}}\right)^2 \frac{\left(1 - Kz\left(\frac{L_y}{z}\right)\sqrt{\frac{\omega}{\nu}}\right)^2}{\left[1 - K^2 z^2\left(\frac{L_y}{z}\right)\frac{\omega}{\nu}\right]^2} \tag{36}$$

For the correlation time $\left(\tau^0_{cor}\right)_2 = \frac{L^2}{\nu}$, we have

$$\mathrm{Re} \approx \left(1.5\left(\frac{\omega R}{\sqrt{E_{st}/\rho}}\right)^2 \frac{\left(1 - Kz\left(\frac{L_y}{z}\right)\sqrt{\frac{\omega}{\nu}}\right)^2}{\left[1 - K^2 z^2\left(\frac{L_y}{z}\right)\frac{\omega}{\nu}\right]^2}\right)\mathrm{Re}^2{}_{st} \tag{37}$$

Then, for the correlation time $\left(\tau^0_{cor}\right)_3 = \frac{\nu}{E_{st}/\rho}$, we obtain

$$\mathrm{Re} \approx \left(1.5\left(\frac{\omega R}{\sqrt{E_{st}/\rho}}\right)^2 \frac{\left(1 - Kz\left(\frac{L_y}{z}\right)\sqrt{\frac{\omega}{\nu}}\right)^2}{\left[1 - K^2 z^2\left(\frac{L_y}{z}\right)\frac{\omega}{\nu}\right]^2}\right)\frac{1}{\mathrm{Re}^2{}_{st}} \tag{38}$$

Substituting Equation (33) in expressions (36)—(38) for the Reynolds number and neglecting the terms containing the value K^2, we may write the estimate for the critical Reynolds number in the flow near the rotating disk (K~0.5) as:

$$(\mathrm{Re}_{critic})_1 \approx \left(1.5\left(\frac{\omega R}{\sqrt{E_{st}/\rho}}\right)^2\left(1 - 2K\left(\frac{L_y}{z}\right)z\sqrt{\frac{\omega}{\nu}}\right)\right)\mathrm{Re}^2{}_{st} \approx 1.5\left(\frac{\omega R}{\sqrt{E_{st}/\rho}}\right)^2\left(1 - 2\left(\frac{L_y}{\delta}\right)\right)\mathrm{Re}^2{}_{st} \tag{39}$$

For the value of L_y/δ near the wall at the critical point in accordance with [41,49–59], we have

$$\left(\frac{L_y}{\delta}\right) \approx 0.02 \div 0.04 \tag{40}$$

Thus, finally, we have the theoretical solution for the critical Reynolds number for the motion of the flow near the rotating disk

$$(\mathrm{Re}_{critic})_1 \approx 1.3\left(\frac{\omega R}{\sqrt{E_{st}/\rho}}\right)^2 \mathrm{Re}_{st}{}^2 \tag{41}$$

In accordance with [41,77,78], there are the following values for the degree of turbulence observed in the laboratory and the turbulent Reynolds numbers Re_{st} near the wall of the disk: $\left(\omega R/\sqrt{E_{st}/\rho}\right)^{-1} = 0.01 \div 0.02$ and $\mathrm{Re}_{st} = 5 \div 10$. As a result, using Equation (41), we have $\mathrm{Re}_{critic} = 325,000$, which agrees with the data [41]. In the case, when the turbulent Reynolds number is $\mathrm{Re}_{st} = 5 \div 15$, we have $3.25 \times 10^5 \leq \mathrm{Re}_{critic} \leq 7.3 \times 10^6$. Therefore, the defined range for the first critical Reynolds number for the motion near a rotating disk is within the experimental values for the transition mode $2.9 \times 10^5 \leq \mathrm{Re}_{critic} < 7 \times 10^5$ [41].

7. Conclusions

Analytical Formulas (39) and (41), for the critical Reynolds number for the motion of the flow near a rotating disk based on the theory of stochastic equations of continuum laws and the equivalence of measures between random and deterministic motion are presented. Also, analytical Formulas (32) and (33) for the critical point in the case of the motion of the flow near a rotating disk are derived. The results of solutions show the satisfactory correspondence between the values obtained with using the analytical dependences for critical Reynolds number (39) and (41), and the experimental data [41]. For the degree of turbulence observed in the laboratory $\left(\omega R/\sqrt{E_{st}/\rho}\right)^{-1} = 0.01 \div 0.02$ and the turbulent Reynolds numbers Re_{st} near the wall of the disk $\mathrm{Re}_{st} = 5 \div 10$, we have $\mathrm{Re}_{critic} = 325,000$, which agrees with the data [41]. It seems that the obtained dependences for the critical Reynolds number can be useful for estimating of the flow regime in gas or steam turbines.

Funding: This research received no external funding.

Acknowledgments: This work was supported by the program of increasing the competitive ability of National Research Nuclear University MEPhI (agreement with the Ministry of Education and Science of the Russian Federation, 27 August 2013, project no. 02.a03.21.0005).

Conflicts of Interest: The authors declare no conflict of interest.

References

1. Kolmogorov, A.N. Dissipation of energy in locally isotropic turbulence. *Dokl. Akad. Nauk SSSR* **1941**, *32*, 16–18.
2. Kolmogorov, A.N. A new metric invariant of transitive dynamic sets and automorphisms of the Lebesgue spaces. *Dokl. Akad. Nauk SSSR* **1958**, *119*, 861–864.
3. Kolmogorov, A.N. About the entropy per time unit as a metric invariant of automorphisms. *Dokl. Akad. Nauk SSSR* **1958**, *124*, 754–755.
4. Kolmogorov, A.N. Mathematical models of turbulent motion of an incompressible viscous fluid. *Uspekhi Mat. Nauk* **2004**, *59*, 5–10. [CrossRef]
5. Landau, L.D. Toward the problem of turbulence. *Dokl. Akad. Nauk SSSR* **1944**, *44*, 339–342.
6. Lorenz, E.N. Deterministic nonperiodic flow. *J. Atmos. Sci.* **1963**, *20*, 130–141. [CrossRef]
7. Ruelle, D.; Takens, F. On the nature of turbulence. *Commun. Math. Phys.* **1971**, *20*, 167–192. [CrossRef]
8. Feigenbaum, M. The transition to aperiodic behavior in turbulent sets. *Commun. Math. Phys.* **1980**, *77*, 65–86. [CrossRef]
9. Klimontovich, Y.L. Problems of the statistical theory of open sets: Criteria of the relative degree of the ordering of states in the self-organization processes. *Usp. Fiz. Nauk.* **1989**, *158*, 59–91. [CrossRef]
10. Haller, G. *Chaos Near Resonance*; Springer: Berlin/Heidelberg, Germany, 1999. [CrossRef]
11. Struminskii, V.V. Origination of turbulence. *Dokl. Akad. Nauk SSSR* **1989**, *307*, 564–567.
12. Samarskii, A.A.; Mazhukin, V.I.; Matus, P.P.; Mikhailik, I.A. Z/2 conservative schemes for the Korteweg–de Vries equations, Dokl. *Akad. Nauk* **1997**, *357*, 458–461.
13. Orzag, S.A.; Kells, L.C. Transition to turbulence in plane Poiseuille and plane Couette flow. *J. Fluid Mech.* **1980**, *96*, 159–205. [CrossRef]
14. Vishik, M.I.; Zelik, S.V.; Chepyzhov, V.V. Regular attractors and nonautonomous perturbations of them. *Sb. Math.* **2013**, *204*, 3–46. [CrossRef]
15. Carvalho, A.N.; Langa, J.A.; Robinson, J.C.; Su'arez, A. Characterization of non-autonomous attractors of a perturbed infinite-dimensional gradient system. *J. Differ. Equ.* **2007**, *236*, 570–603. [CrossRef]
16. Vishik, M.I.; Chepyzhov, V.V. Trajectory attractors of equations of mathematical physics. *Uspekhi Mat. Nauk.* **2011**, *66*, 3–102. [CrossRef]
17. Vishik, M.I.; Chepyzhov, V.V. Trajectory attractors of equations of mathematical physics. *Russ. Math. Surv.* **2011**, *66*, 637–731. [CrossRef]
18. Ladyzhenskaya, O.A. On a dynamical system generated by Navier–Stokes equations. *J. Sov. Math.* **1975**, *3*, 458–479. [CrossRef]
19. Vishik, M.I.; Zelik, S.V. Attractors for the nonlinear elliptic boundary value problems and their parabolic singular limit. *Commun. Pure Appl. Anal.* **2014**, *13*, 2059–2093. [CrossRef]
20. Landau, L.D.; Lifshits, E.F. *Fluid Mechanics*; Perg. Press Oxford: London, UK, 1959.
21. Constantin, P.; Foais, C.; Temam, R. On dimensions of the attractors in two-dimensional turbulence. *Phys. D Nonlinear Phenom.* **1988**, *30*, 284–296. [CrossRef]
22. Vishik, M.I.; Komech, A.I. Kolmogorov equations corresponding to a two-dimensional stochastic Navier–Stokes system. *Tr. Mosk. Mat. Obs.* **1983**, *46*, 3–43.
23. Packard, N.H.; Crutchfield, J.P.; Farmer, J.D.; Shaw, R.S. Geometry from a time series. *Phys. Rev. Lett.* **1980**, *45*, 712–715. [CrossRef]
24. Malraison, B.; Berge, P.; Dubois, M. Dimension of strange attractors: An experimental determination for the chaotic regime of two convective systems. *J. Phys. Lett.* **1983**, *44*, L897–L902. [CrossRef]
25. Procaccia, I.; Grassberger, P. Characterization of strange attractors. *Phys. Rev. Lett.* **1983**, *50*, 346–349.
26. Procaccia, I.; Grassberger, P. Estimation of the Kolmogorov entropy from a chaotic signal. *Phys. Rev. A* **1983**, *28*, 2591–2593.
27. Grassberger, P.; Procaccia, I. Measuring the strangeness of strange attractors. *Phys. D Nonlinear Phenom.* **1983**, *9*, 189–208. [CrossRef]
28. Grassberger, P.; Procaccia, I. Dimensions and entropies of strange attractors from a fluctuating dynamics approach. *Phys. D Nonlinear Phenom.* **1984**, *13*, 34–54. [CrossRef]
29. Rabinovich, M.I.; Reiman, A.M.; Sushchik, M.M. Correlation dimension of the flow and spatial development of dynamic chaos in the boundary layer. *JETP Lett.* **1987**, *13*, 987.
30. Brandstater, A.; Swift, J.; Swinney, H.L.; Wolf, A.; Farmer, D.J.; Jen, E.; Crutchfield, P.J. Low-dimensional chaos in hydrodynamic system. *Phys. Rev. Lett.* **1983**, *51*, 1442–1446. [CrossRef]
31. Sreenivasan, K.R. Fractals and multifractals in fluid turbulence. *Ann. Rev. Fluid Mech.* **1991**, *23*, 539–600. [CrossRef]
32. Priymak, V.G. Splitting dynamics of coherent structures in a transitional round-pipe flow. *Dokl. Phys.* **2013**, *58*, 457–465. [CrossRef]
33. Mayer, C.S.J.; von Terzi, D.A.; Fasel, H.F. Direct numerical simulation of investigation of complete transition to turbulence via oblique breakdown at Mach 3. *J. Fluid Mech.* **2011**, *674*, 5–42. [CrossRef]
34. Newton, P.K. The fate of random initial vorticity distributions for two-dimensional Euler equations on a sphere. *J. Fluid Mech.* **2016**, *786*, 1–4. [CrossRef]
35. Fursikov, A.V. Moment theory for Navier–Stokes equations with a random right-hand side. *Izv. Ross. Akad. Nauk.* **1992**, *56*, 1273–1315.
36. Davidson, P.A. *Turbulence*; Oxford University Press: Oxford, UK, 2004.
37. Millionshchikov, M.D. *Turbulent Flow in Boundary Layers and in Pipes*; Nauka: Moscow, Russia, 1969.
38. Hinze, J.O. *Turbulence*, 2nd ed.; McGraw-Hill: New York, NY, USA, 1975.

39. Monin, A.S.; Yaglom, A.M. *Statistical Fluid Mechanics*; MIT Press: Cambridge, MA, USA, 1971.
40. Schlichting, H. *Boundary-Layer Theory*, 6th ed.; McGraw-Hill: New York, NY, USA, 1968.
41. Pope, S.B. *Turbulent Flows*; Cambridge University Press: Cambridge, UK, 2000. [CrossRef]
42. Dmitrenko, A.V. *Fundamentals of Heat and Mass Transfer and Hydrodynamics of Single-Phase and Two-Phase Media. Criterial Integral Statistical Methods and Direct Numerical Simulation*; Galleya Print: Moscow, Russia, 2008; Available online: http://search.rsl.ru/ru/catalog/record/6633402 (accessed on 24 December 2020).
43. Dmitrenko, A.V. Calculation of pressure pulsations for a turbulent heterogeneous medium. *Dokl. Phys.* **2007**, *52*, 384–387. [CrossRef]
44. Dmitrenko, A.V. Calculation of the boundary layer of a two-phase medium. *High. Temp.* **2002**, *40*, 706–715. [CrossRef]
45. Dmitrenko, A.V. Heat and mass transfer and friction in injection to a supersonic region of the Laval nozzle. *Heat Transf. Res.* **2000**, *31*, 338–399. [CrossRef]
46. Dmitrenko, A.V. Film cooling in nozzles with large geometric expansion using method of integral relation and second moment closure model for turbulence. In Proceedings of the 33th AIAA/ASME/SAE/ASEE Joint Propulsion Conference and Exhibit, AIAA Paper 97–2911, Seattle, WA, USA, 6–9 July 1997. [CrossRef]
47. Dmitrenko, A.V. Heat and mass transfer in combustion chamber using a second-moment turbulence closure including an influence coefficient of the density fluctuation in film cooling conditions. In Proceedings of the 34th AIAA/ASME/SAE/ASEE Joint Propulsion Conference and Exhibit, AIAA Paper 98–3444, Cleveland, OH, USA, 13–15 July 1998. [CrossRef]
48. Dmitrenko, A.V. Nonselfsimilarity of a boundary-layer flow of a high-temperature gas in a Laval nozzle. *Aviats. Tekh.* **1993**, *1*, 39–42.
49. Dmitrenko, A.V. Computational investigations of a turbulent thermal boundary layer in the presence of external flow pulsations. In Proceedings of the 11th Conference on Young Scientists, Moscow, Physicotechnical Institute, Part. 2, Moscow, Russia, 10 November 1986; pp. 48–52, Deposited at VINITI 08.08.86, No. 5698-B8.
50. Heisenberg, W. Zur statistischen Theorie der Turbulenz. *Z. Phys.* **1948**, *124*, 628–657. [CrossRef]
51. Starikov, F.A.; Kochemasov, G.G.; Kulikov, S.M.; Manachinsky, A.N.; Maslov, N.V.; Ogorodnikov, A.V.; Soldatenkov, I.S. Wavefront reconstruction of an optical vortex by a Hartmann-Shack sensor. *Opt. Lett.* **2007**, *32*, 2291–2293. [CrossRef]
52. Starikov, F.A.; Khokhlov, S.V. Phase correction of laser radiation with the use of adaptive optical systems at the Russian Federal Nuclear Center–Institute of Experimental Physics. *Optoelectron. Instr. Data Proc.* **2012**, *48*, 134–141.
53. Dmitrenko, A.V. Equivalence of measures and stochastic equations for turbulent flows. *Dokl. Phys.* **2013**, *58*, 228–235. [CrossRef]
54. Dmitrenko, A.V. *Regular Coupling between Deterministic (Laminar) and Random (Turbulent) Motions-Equivalence of Measures*; Scientific Discovery Diploma No. 458, Registration No. 583 of December 2; IAASD; Russian Federation: Moscow, Russia, 2013.
55. Dmitrenko, A.V. *Theory of Equivalent Measures and Sets with Repeating Denumerable Fractal Elements. Stochastic Thermodynamics and Turbulence. Determinacy–Randomness Correlator*; Galleya-Print: Moscow, Russia, 2013; Available online: https://search.rsl.ru/ru/record/01006633402 (accessed on 24 December 2020). (In Russian)
56. Dmitrenko, A.V. Some analytical results of the theory of equivalence measures and stochastic theory of turbulence for nonisothermal flows. *Adv. Stud. Theor. Phys.* **2014**, *8*, 1101–1111. [CrossRef]
57. Dmitrenko, A.V. Analytical estimation of velocity and temperature fields in a circular tube on the basis of stochastic equations and equivalence of measures. *J. Eng. Phys. Thermophys.* **2015**, *88*, 1569–1576. [CrossRef]
58. Dmitrenko, A.V. Determination of critical Reynolds numbers for nonisothermal flows using stochastic theory of turbulence and equivalent measures. *Heat Transf. Res.* **2016**, *47*, 41–48. [CrossRef]
59. Dmitrenko, A.V. The theory of equivalence measures and stochastic theory of turbulence for non-isothermal flow on the flat plate. *Int. J. Fluid Mech. Res.* **2016**, *43*, 182–187. [CrossRef]
60. Dmitrenko, A.V. An estimation of turbulent vector fields, spectral and correlation functions depending on initial turbulence based on stochastic equations. The Landau fractal equation. *Int. J. Fluid Mech. Res.* **2016**, *43*, 82–91. [CrossRef]
61. Dmitrenko, A.V. Stochastic equations for continuum and determination of hydraulic drag coefficients for smooth flat plate and smooth round tube with taking into account intensity and scale of turbulent flow. *Contin. Mech. Thermodyn.* **2017**, *29*, 1–9. [CrossRef]
62. Dmitrenko, A.V. Analytical determination of the heat transfer coefficient for gas, liquid and liquidmetal flows in the tube based on stochastic equations and equivalence of measures for continuum. *Contin. Mech. Thermodyn.* **2017**, *29*, 1197–1205. [CrossRef]
63. Dmitrenko, A.V. Determination of the coefficients of heat transfer and friction in supercritical-pressure nuclear reactors with account of the intensity and scale of flow turbulence on the basis of the theory of stochastic equations and equivalence of measures. *J. Eng. Phys. Thermophys.* **2017**, *90*, 1288–1294. [CrossRef]
64. Dmitrenko, A.V. Results of investigations of non-isothermal turbulent flows based on stochastic equations of the continuum and equivalence of measures. *J. Phys. Conf. Ser.* **2018**, *1009*, 012017. [CrossRef]
65. Dmitrenko, A.V. The stochastic theory of the turbulence. *IOP Conf. Ser. Mater. Sci. Eng.* **2018**, *468*, 012021. [CrossRef]
66. Dmitrenko, A.V. Determination of the correlation dimension of an attractor in a pipe based on the theory of stochastic equations and equivalence of measures. *J. Phys.Conf. Ser.* **2019**, *1250*. [CrossRef]
67. Dmitrenko, A.V. Some aspects of the formation of the spectrum of atmospheric turbulence. *JP J. Heat Mass Transf.* **2020**, *19*, 201–208. [CrossRef]
68. Dmitrenko, A.V. The construction of the portrait of the correlation dimension of an attractor in the boundary layer of Earth's atmosphere. *J. Phys. Conf. Ser.* **2019**, *1337*. [CrossRef]
69. Dmitrenko, A.V. The correlationdimension of an attarctor determined on the base of the theory of equivalence of measures and stochastic equations for continuum. *Contin. Mechan. Thermod.* **2020**, *32*, 63–74. [CrossRef]

70. Dmitrenko, A.V. Uncertainty relation in turbulent shear flow based on stochastic equations of the continuum and the equivalence of measures. *Contin. Mech. Thermod.* **2020**, *32*, 161–171. [CrossRef]
71. Dmitrenko, A.V. Formation of the turbulence spectrum in the inertial interval on the basis of the theory of stochastic equations and equivalence of measures. *J. Eng. Phys. Thermophys.* **2020**, *93*, 122–127. [CrossRef]
72. Dmitrenko, A.V. Theoretical solutions for spectral function of the turbulent medium based on the stochastic equations and equivalence of measures. *Contin. Mech. Thermod.* 2020. [CrossRef]
73. Dmitrenko, A.V. The possibility of using low-potential heat based on the organic Rankine cycle and determination of hydraulic characteristics of industrial units based on the theory of stochastic equations. *JP J. Heat Mass Transf.* **2020**, *21*, 125–132. [CrossRef]
74. Dmitrenko, A.V. The theoretical solution for the Reynolds analogy based on the stochastic theory of turbulence. *JP J. Heat Mass Transf.* **2019**, *18*, 463–476. [CrossRef]
75. Dmitrenko, A.V. Determination of critical Reynolds number in the jet based on the theory of stochastic equations and equivalence of measures. *J. Phys. Conf. Ser.* **2020**, *1705*, 012015. [CrossRef]
76. Dmitrenko, A.V. The Spectrum of the turbulence based on theory of stochastic equations and equivalence of measures. *J. Phys. Conf. Ser.* **2020**, *1705*, 012021. [CrossRef]
77. Bunker, R.S. A review of turbine shaped film cooling technology. *J. Heat Transf.* **2005**, *127*, 441–453. [CrossRef]
78. Srinath, E.; Hanb, J.-C. A review of hole geometry and coolant density effect on film cooling. *Front. Heat Mass Transf.* **2015**, *6*. [CrossRef]

fluids

MDPI

Article

Thermal Performance of a Heated Pipe in the Presence of a Metal Foam and Twisted Tape Inserts

K. Papazian [1], Z. Al Hajaj [2] and M. Z. Saghir [1,*]

[1] Department of Mechanical and Industrial Engineering, Ryerson University, Toronto, ON M5B 2K3, Canada; khajag.papazian@ryerson.ca

[2] Department of Mechanical Engineering, Australian College of Kuwait, P.O. Box 1411, Safat-13015, Kuwait; z.hajaj@ack.edu.kw

* Correspondence: zsaghir@ryerson.ca

Received: 6 October 2020; Accepted: 28 October 2020; Published: 30 October 2020

Abstract: To meet the demand for more efficient ways of cooling and heating, new designs and further development of heat exchangers is essential in industry. The present study focuses on the thermal performance of a circular pipe with two inserts. The first insert consists of a porous medium having a porosity of 0.91, and the second one consists of a single twist solid insert. Different ranges of heating conditions have been applied for different flow rates. Water and titanium dioxide (TiO_2) nanofluid 1% vol are the liquid media used for cooling. Laminar flow is assumed for two different Reynolds numbers of 1000 and 2000. The results of the study have shown that the twisted tape insert increases the thermal efficiency of the pipe more than the porous media insert and the plain pipe. In addition, different temperature readings in the cross section of the pipe have indicated that the twisted tape helps mixing up the fluid and provides a constant temperature in the overall volume of the fluid, whereas for the porous media insert and plain pipe the fluid temperature increases in the fluid particles close to the pipe inner surface. TiO_2 nanofluid exhibited an enhancement when compared to water for a plain and porous pipe. However, this enhancement was absent when a twisted insert is used.

Keywords: heat enhancement; nanofluid; circular pipe; twisted tape; porous media; metal foam

1. Introduction

The heat transfer method and technology of fluids have been developed a long time ago. It has been further advanced in the last couple years due to the technological demand for more efficient ways to cool or heat materials in different industrial sectors. There are a great number of studies focused on nanofluid technology as a way to improve cooling and heating efficiency in different industries, such as processor and electronic chip cooling, nuclear reactor cooling, food processing facilities and for use as a replacement of existing cooling fluids in commercial buildings [1–6]. The objective of this paper was to investigate heat enhancement in a circular pipe in the presence of metal foam, twisted tapes, helical coils and a heat exchanger. The novelty is to create mixing without having to embark into a turbulent flow modelling study. The findings will be useful toward a design of new heat exchangers.

1.1. *Nanofluid in Circular Pipe*

Ho et al. [7], conducted an experiment in a circular pipe using water-based suspensions of Al_2O_3 nanoparticles and microencapsulated nanoparticles. The nanofluid is known to enhance heat removal and the MicroEncapsulated Phase Change Material (MEPCM) nanoparticles were used for heat storage. This mixture of both particles in the fluid made the system very attractive for further investigation. The uniqueness of this study is that the authors were able to accurately measure the physical properties of the fluid for different concentrations of each particle type. It was found that the effectiveness of thermal diffusivity of nanofluids appears significantly increased, while the opposite effect is observed if MEPCM in water is used. Additional findings indicated that the forced convective cooling efficiency of incorporating the nanofluid, or the Phase Change Material(PCM) suspension as the heat transfer fluid in the tube heated with a certain heat flux strongly depends on the flow rate and particle fraction. In their investigation, *n*-eicosane was the phase change material used. Saghir et al. [8], further investigated numerically different types of fluid mixtures—binary and ternary—to study heat enhancement and heat storage in a circular pipe. The fluid mixture consisted of Al_2O_3 nanoparticles and MEPCM nanoparticles mixed in water. This new type of fluid shows a good heat enhancement and heat storage capability. It was found that microencapsulated phase change material nanoparticles in water is found to have an excellent heat storage capability when MEPCM is in molten condition. The higher the MEPCM concentration the better the heat storage achieved. A ternary mixture of Al_2O_3 and MEPCM nanoparticles in water is the best fluid for heat storage under any flow rate conditions. It is found that a 3%vol Al_2O_3 nanoparticles mixed with 20% vol MEPCM nanoparticles in water is the recommended fluid mixture. In addition, Sekrani et al. [9], investigated the usefulness of nanofluids in heat enhancement. Their results predicted an enhancement not exceeding 6% assuming that no segregation or sedimentation of nanoparticles take place.

Bianco et al. [10], investigated experimentally the heat transfer behavior on the water and ethylene glycol-Al_2O_3 nanofluids flowing in circular cross-section tubes. The results of the analysis presented that the heat transfer conductivity of the fluid increases with increasing the particle volume concentration, although that was increasing the shear stress value on the wall. They found that the temperature-dependent model increases in the heat transfer coefficient and Nusselt number and decreases the shear stress on the wall.

1.2. *Twisted Tape in Circular Pipe*

Later researchers investigated the insert of twisted tape in the pipe with the aim of creating a strong mixing even in the laminar regime to enhance heat extraction. Eiamsa-Ard et al. [11], obtained results of heat transfer, friction factor and the thermal performance of a uniformly heated microtube which is equipped with a single and double twisted tape. Analysis was performed twice on the double twisted taps; when the blades were rotating in the same direction, and when they were rotating in an opposite direction. The study used two different values for the Reynolds number (5650 and 1700). It was reported that the double tapes models were higher for Nusselt number and thermal performance than the single tape model and the oppositely rotating taps had a higher Nusselt number than the same direction rotated model. The overall thermal performance factor of the micro-fin tube was increased by a factor of 2.03 by using the oppositely rotating taps. Kumar et al. [12], conducted a study for a swirling air jet, running under different conditions, and then compared each one of them to get the best heat transfer characteristics. In the experiments, a bronze twisted tape was added into the jet nozzle in order to create a swirling effect. Four different twist ratios were used, and the distance between the jet and the surface plate varied from 1 to 4. The Reynolds number was maintained in the range of 500 to 3000. In addition, the researchers introduced another dimensionless parameter called swirl number, which is the ratio between the axial flux angular momentum to the axial flux axial momentum. They obtained an equivalent swirl number for each twist ratio used. The temperature distribution on the flat plate is an IR camera used to capture the temperature distribution when the airflow hits the plate. The results of the experiments showed that the best heat transfer characteristic was increasing when the twist ratio is between 1 to 4.5 and it decreased when the twist ratio was between 4 to 7.5. They also found that the heat transfer rate increases with Reynolds number and decreases with increasing distance of the jet to plate spacing.

Maddah et al. [13], compared different heat transfer characteristics, using a shad, tube type heat exchanger with both water and nanofluid flowing in a plain tube, and nanofluid flowing in a twisted insert tube. The nanofluid used had a 0.01% vol concentration of titanium dioxide (TiO_2)

nanoparticles with a water base fluid. The particle size was 30 nm, and the study was performed using various Reynolds numbers greater than 2300 and a counter-current flow. The results showed that by increasing the Reynolds number, the heat transfer coefficient increased for all three models. On the other hand, the nanofluid model in twisted tapes had a heat transfer coefficient which was approximately 20% higher than the base pure water model. The efficiency of the heat exchanger was 30% higher for the nanofluid twisted tape model than for water. The experiment also showed that using a nanofluid in twisted tapes increases the friction factor of the system by 2.5, suggesting that the size of the nanoparticles and the additional surface area for the twist tape has a significant effect.

1.3. Metal Foam in Circular Pipe

Huang et al. [14], performed both experimental and numerical studies. The main objective of both studies was to determine how adding a porous medium to a tube influences the core flow. To do this, porous media with a diameter (marginally smaller than the tube's diameter) were created and placed in the tube's core. The type of porous media was altered, meaning porosities of 0.951, 0.966 and 0.975 were used, respectively, for different tests. Uniform heat flux was also applied to the walls of the tube, as a boundary condition. The type of flow experienced in the tube was also altered, meaning both laminar and fully developed turbulent was used. This was done to see the impact of each and how they relate to the performance of the porous media. In terms of the numerical study, it also analyzed how the pore radius ratio could be used to influence the heat transfer performance of the tube. The results showed that adding porous media influences the heat transfer rate by a factor between 1.6–5.5 (depending on the type of flow). The flow resistance also increased because the area of contact between the porous media and the fluid flow increased. For example, the heat transfer enhancement was overshadowed greatly by the flow resistance, deeming the addition of porous media in that particular case to be poor. Ghosh et al. [15], conducted a study using an insert of metal foam in a circular pipe They did the analysis using a cubic shape structure from the metal foam, attached to an isothermal surface once and placed in between isothermal plates for another attempt. The analysis was done in order to gain a better understanding on the effect of the porosity and the density in the heat transfer characteristic of the metal foam. The analysis results concluded that the metal foam sink solution is more suitable for high Nusselt number and high flow applications.

Additional studies were done in a helical pipe [16–19]. The investigations focused toward the development of correlation to calculate the friction factor of laminar and turbulent flow. Nanofluids has been used in the analysis and it was found that the heat transfer decreases as the twist ratio increases.

According to our literature investigation, it was found that there are not many studies comparing the performance of nanofluids flowing in a twisted tube with metal foam and plain pipes. This provides an opportunity for more studies to highlight nanofluid technology with the arrangement of twist tape and metal foam in order to improve the performance of the nanofluid heat transfer characteristics and the cooling capability. In this current paper, an attempt is made to investigate the effect of the presence of inserts in a circular pipe. One may mention that the findings should be beneficial for the design of new heat exchangers for engineering applications. Section 2 will present the problem description. In Section 3, we present the finite element formulation. Section 4 presents the analysis results such as the temperature changes, Nusselt number, friction factor and the thermal efficacy for different conditions. Finally, Section 5 presents the conclusions.

2. Problem Description

In this study, we present a detailed analysis of a circular cross-section pipe with different metal inserts. The aim is to determine which one is the most efficient to be used in the heat exchanger as the heat observer component, that will result in an increase in the heat exchanger's overall efficiency. The problem consists of a circular pipe having a length L equal to 85 mm and an inner diameter equal to 6.8 mm and an outer diameter equal to 8 mm. The pipe is composed of three parts where the first part known as L_1, has a length 27.2 mm, followed by the heated section having a length L_2

equal to 37.4 mm and finally the third part insulated section L_3 equal to 20.4 mm. Part 1 and part 3 external surface is insulated and part 2 is heated externally with a heat flux q″ having an independent value of 50,000 W/m^2 and 75,000 W/m^2. Figure 1 presents the model in question. Flow enters the pipe at a constant flow rate Q and an inlet temperature T_{in} equal to 18 °C. Three different cases were investigated. The aim is to select the best scenario leading to heat enhancement and the best thermal efficiency. In the first case, the inside the pipe is free allowing the flow to circulate along the length of the pipe. In the second case an insert made of aluminum oxide metallic foam having a porosity of 0.91 and a permeability equivalent to 10 PPI (i.e., 9.5478×10^{-7} m^2) is investigated. In the third case, a metallic a single twist insert made of copper is studied in details. Calculated temperatures were investigated on the surface of the internal pipe in the heated part only. The pipe is made of copper with a thickness of 0.006 mm. Figure 1 displays in details all components of the numerical setup.

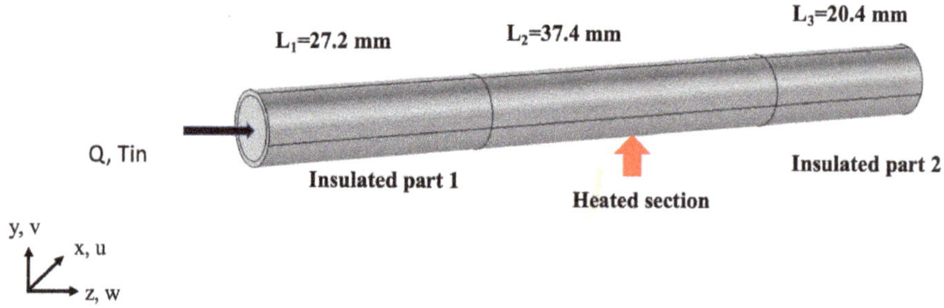

Figure 1. Pipe configuration.

3. Finite Element Formulation

The copper pipe's heat transfer rate and thermal efficiency problem presented in this paper required solving the full Navier-Stokes equation, continuity equation and the energy equation. These equations were solved using COMSOL software (version 4.3a, COMSOL, Stockholm, Sweden) [20], which utilizes a finite element method. A three-dimensional model has been created and three velocity vectors u, v, w, were added in the x, y, z directions, respectively. Two consistent stabilization methods were used which are the streamline diffusion and the crosswind diffusion. The pressure was the 4th parameter used in solving the problem. For the free flow in the model, the formulation adopted are as follows:

X-direction momentum equation:

$$\rho_f\left(u\frac{\partial u}{\partial x}+v\frac{\partial u}{\partial y}+w\frac{\partial u}{\partial z}\right)=-\frac{\partial p}{\partial x}+\mu_f\left(\frac{\partial^2 u}{\partial x^2}+\frac{\partial^2 u}{\partial y^2}+\frac{\partial^2 u}{\partial z^2}\right) \quad (1)$$

Y-direction momentum equation:

$$\rho_f\left(u\frac{\partial v}{\partial x}+v\frac{\partial v}{\partial y}+w\frac{\partial v}{\partial z}\right)=-\frac{\partial p}{\partial y}+\mu_f\left(\frac{\partial^2 v}{\partial x^2}+\frac{\partial^2 v}{\partial y^2}+\frac{\partial^2 v}{\partial z^2}\right)+\rho_f g \quad (2)$$

Z-direction momentum equation:

$$\rho_f\left(u\frac{\partial w}{\partial x}+v\frac{\partial w}{\partial y}+w\frac{\partial w}{\partial z}\right)=-\frac{\partial p}{\partial z}+\mu_f\left(\frac{\partial^2 w}{\partial x^2}+\frac{\partial^2 w}{\partial y^2}+\frac{\partial^2 w}{\partial z^2}\right) \quad (3)$$

Continuity equation:

$$\left(\frac{\partial u}{\partial x}+\frac{\partial v}{\partial y}+\frac{\partial w}{\partial z}\right)=0 \quad (4)$$

Energy conservation equation:

$$(\rho Cp)_f\left(u\frac{\partial T}{\partial x}+v\frac{\partial T}{\partial y}+w\frac{\partial T}{\partial z}\right)=k_f\left(\frac{\partial^2 T}{\partial x^2}+\frac{\partial^2 T}{\partial y^2}+\frac{\partial^2 T}{\partial z^2}\right) \tag{5}$$

The density of the fluid is ρ_f, the dynamic viscosity is μ_f, the pressure is p, and g is the gravity vector. The three-dimensional velocities are u, v and w in the x, y and z direction, respectively. The specific heat of the fluid is Cp_f and the conductivity is k_f. Heat conduction formulation is used to study the heat in the copper pipe.

When an insert of porous medium is used, one needs to solve in conjunction with the Navier stokes equation, the Darcy-Brinkman formulation. The formulation in this case becomes:

X-direction Darcy-Brinkman:

$$\frac{\mu_f}{\kappa}u=-\frac{\partial p}{\partial x}+\mu_f\left(\frac{\partial^2 u}{\partial x^2}+\frac{\partial^2 u}{\partial y^2}+\frac{\partial^2 u}{\partial z^2}\right) \tag{6}$$

Y-direction Darcy-Brinkman:

$$\frac{\mu_f}{\kappa}v=-\frac{\partial p}{\partial y}+\mu_f\left(\frac{\partial^2 v}{\partial x^2}+\frac{\partial^2 v}{\partial y^2}+\frac{\partial^2 v}{\partial z^2}\right)+\rho_f\,g \tag{7}$$

Z-direction Darcy-Brinkman:

$$\frac{\mu_f}{\kappa}w=-\frac{\partial p}{\partial z}+\mu_f\left(\frac{\partial^2 w}{\partial x^2}+\frac{\partial^2 w}{\partial y^2}+\frac{\partial^2 w}{\partial z}\right) \tag{8}$$

and the energy equation used was the following:

$$(\rho Cp)_{eff}\left(u\frac{\partial T}{\partial x}+v\frac{\partial T}{\partial y}+w\frac{\partial T}{\partial z}\right)=k_{eff}\left(\frac{\partial^2 T}{\partial x^2}+\frac{\partial^2 T}{\partial y^2}+\frac{\partial^2 T}{\partial z^2}\right) \tag{9}$$

The effective heat capacity takes into consideration the heat capacity in the liquid and in the solid matrix. Similarly, the effective conductivity combines the conductivity of the fluid and the conductivity of the metallic foam. Readers are invited to consult [21], for more details.

3.1. Boundary Conditions

Two different heat fluxes were applied at the external wall of the middle section of the pipe as shown in Figure 1. The first applied heat flux is 50,000 W/m^2, another equal to 75,000 W/m^2 was applied uniformly around the external wall. In order to remain in the laminar regime two Reynolds number were selected (1000 and 2000) leading to water flow rate of Q = 5.361 cm^3/s and Q = 10.722 cm^3/s respectively. If nanofluid is used, the flow rate is identical to the water flow rate because the kinematic viscosity of water and TiO_2 nanofluid is the same. The differences in physical properties between the two fluids are the thermal conductivity and the specific heat. The inlet fluid's temperature is set equal to T_{in} = 18 °C, and the study runs in a steady state condition. At the outlet, open boundary is applied which means that normal stresses are set equal to zero. Because the inlet Reynolds number is set constant for different fluids, the mass rate will change from one case to another with changing fluid. Table 1 presents the physical properties used in our calculation. The twisted tape used is a 1 mm thick full 360° rotated tape with the same length as the pipe itself.

Table 1. Physical Property of the material used in the study.

Material	Thermal Conductivity K (W/m·k)	Density (kg·m^3)	Heat Capacity Cp (J/kg·k)
Copper	385	8960	376.812
Water	0.613	998.2	4182
Copper Porous Media	82.2713 (effective)	216	895 (effective)

3.2. *Mesh Sensitivity*

Before running different study cases on COMSOL, the mesh sensitivity was examined whether it was accurate enough to record the numerical results from the model. In the Table 2, we demonstrated different mesh sensitivity that we can select using COMSOL software. The mesh levels that COMSOL supports and the elements numbers for each mesh level are shown above. The average Nusselt number was evaluated at the inner wall of the pipe for each mesh level, and the results are represented in the Figure 2a. It is evident that a coarse level will be suitable to be used in the COMSOL model. Figure 2b, presents the finite element mesh used in our simulation. In addition, the model has been tested by comparing the numerical data with some experimental results [8]. The model consisted of the same boundary condition to the current one.

Table 2. Mesh information for different level of meshing.

Mesh Type [20]	Number and Types of Elements Used
Extremely Coarse	11,361 domain elements, 2722 boundary elements, 444 edge elements
Extra coarse	25,662 domain elements, 5034 boundary elements, 624 edge elements
Coarser	51,376 domain elements, 8634 boundary elements, 804 edge elements
Coarse	130,257 domain elements, 17620 boundary elements, 1168 edge elements

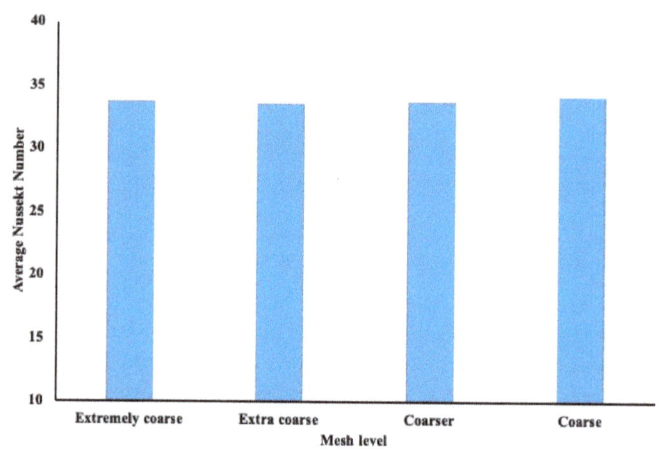

(**a**) Mesh sensitivity analysis

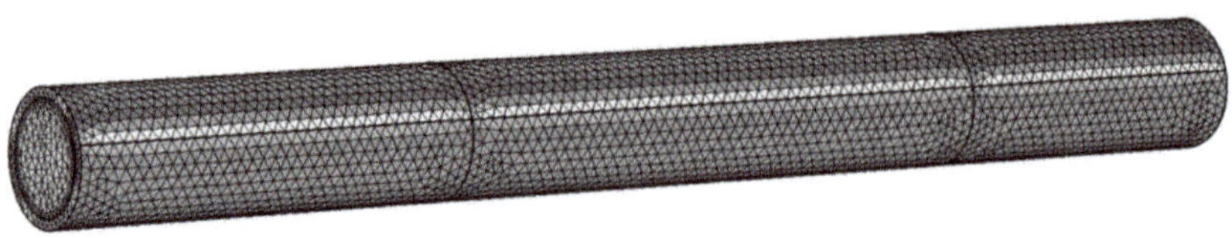

(**b**) Finite element model

Figure 2. Finite element analysis.

3.3. *Pipe Inserts*

In order to investigate the importance of flow behavior inside the pipe, two different types of insert have been included in the modelling. The first one consists of a metal foam porous medium. The porous insert is made of copper and its physical properties are presented in Table 1. Figure 3a presents the twisted insert in the pipe and Figure 3b presents the 360° single twist tape having a 2 mm thickness. The length (between two twists) is 40.8 mm and the diameter of the insert is identical to the internal pipe diameter of 6.8 mm. Thus, in this particular case the ratio of the twisted length to the diameter is equal to 6.

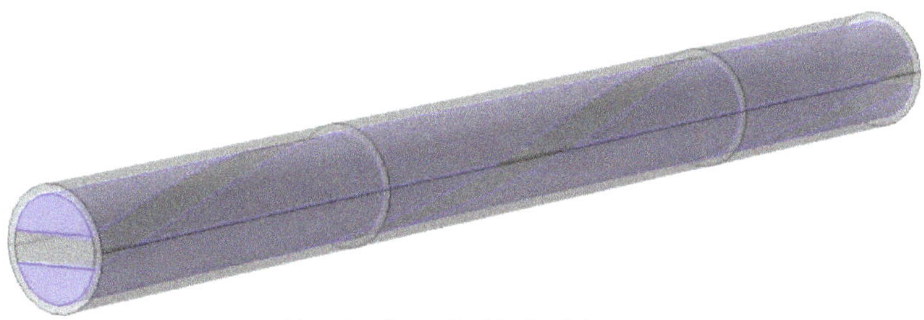

(**a**) twisted tape inside the tube

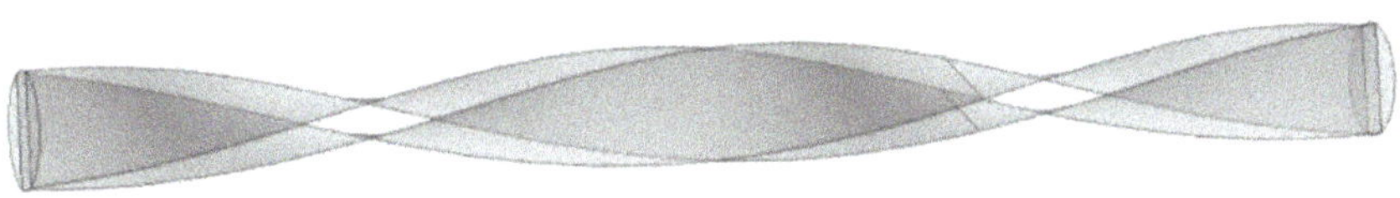

(**b**) Twisted tape

Figure 3. Twisted shape insert.

4. Results and Discussion

This paper focuses on examining the best pipe model to be used for heat enhancement and heat extraction. As indicated previously, three different types of pipe model were investigated. The first pipe is a hollow cylinder heated in one external surface and insulated at the remaining surface using internal fluid circulation to extract heat. The second configuration is identical to the previous configuration with a porous insert. The third configuration is identical to the first configuration but with a 360° single twist tape insert. In order to keep the investigation under the laminar regime valid in all the scenarios two different Reynolds numbers of 1000 and 2000 were adopted. This translates into flow rates of 5.361 cm^3/s and 10.722 cm^3/s for both water and TiO_2 nanofluid. The inlet temperature is always maintained constant at 18 °C.

4.1. *Temperature and Nusselt Number Variation*

The temperature has been measured at the wall of the internal pipe along the heated section only. Figures 4–6 highlight the temperature variation for the three configurations. Figure 4a presents the temperature variation at the external heated wall. A high temperature is observed and the temperature profile is expected. In Figure 4b the temperature variation is non-linear mainly near the end of the heated section. As the Reynolds number increases, the flow rate leads to more cooling of the internal wall. The temperature profile is identical for the cases, however, the Nusselt number is evaluated as well to examine the heat enhancement. By definition, the local Nusselt number is calculated as follows:

$$Nu = \frac{h\,D}{k_w} \tag{10}$$

where h is the heat convection coefficient, D is the pipe diameter and k_W is the water thermal conductivity. The heat convection coefficient h is calculated as follows:

$$h = \frac{q''}{(T - Tin)} \tag{11}$$

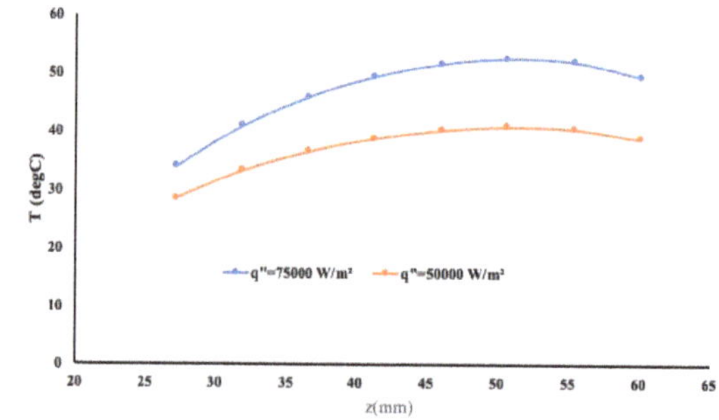

(**a**) Temperature variation at the external heated wall

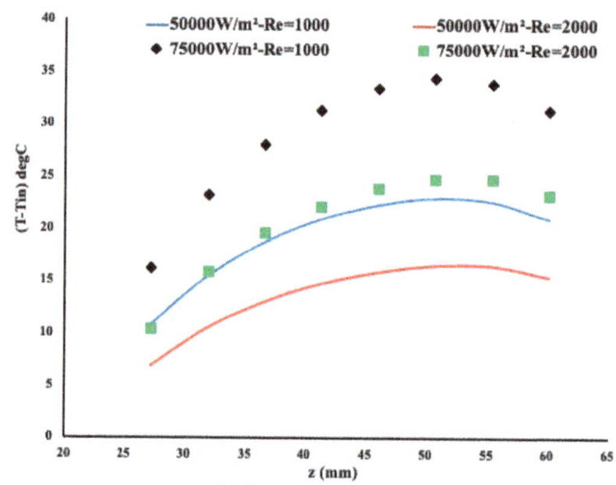

(**b**) Temperature variation

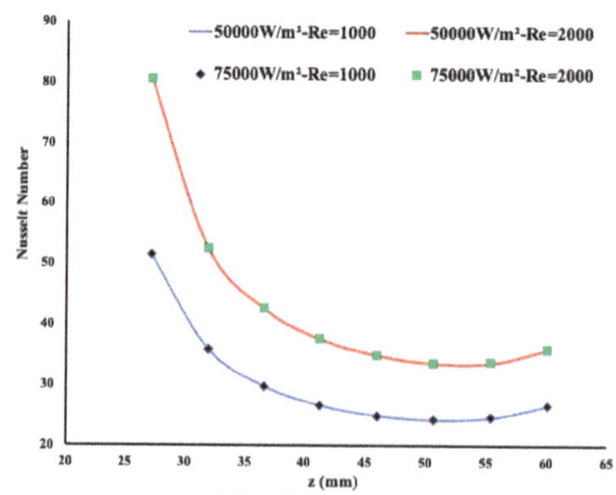

(**c**) Local Nusselt number

Figure 4. Temperature and the Nusselt number at the inner and external wall surface (no insert).

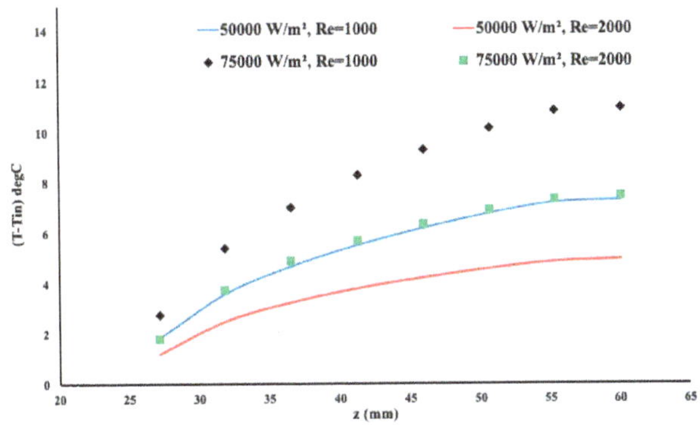

(**a**) Temperature variation

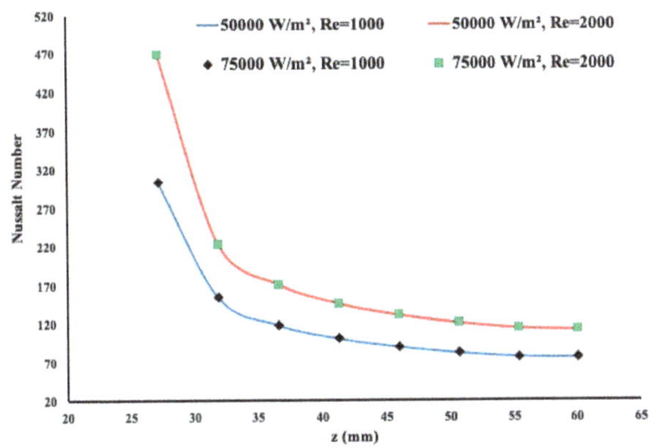

(**b**) Local Nusselt number

Figure 5. Temperature and the Nusselt number at the inner wall surface with porous media insert.

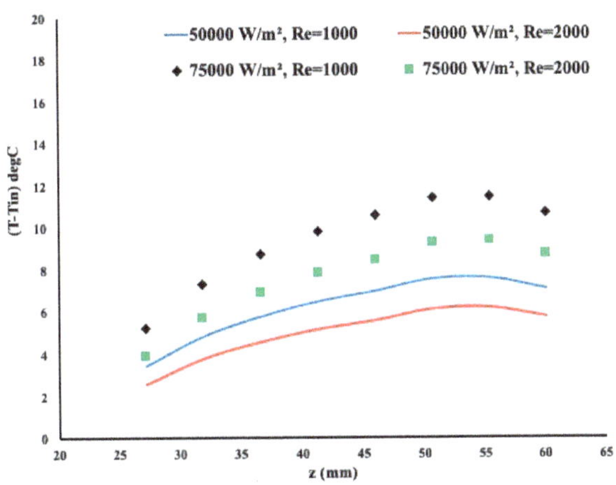

(**a**) Temperature variation

Figure 6. *Cont.*

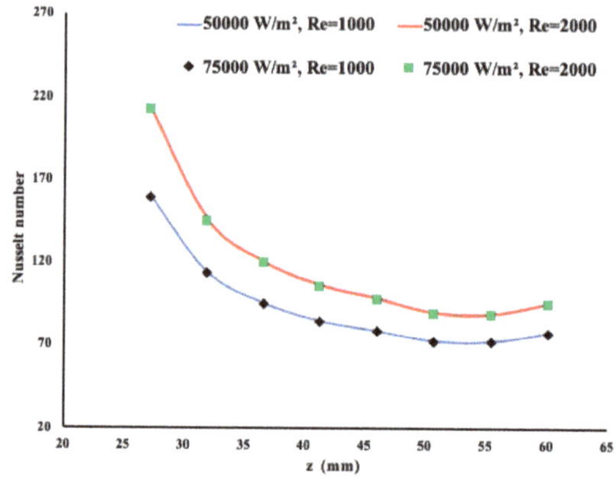

(**b**) Local Nusselt number

Figure 6. Temperature and the Nusselt number at the inner wall surface with twisted tape insert.

Figure 4c presents the Nusselt number variation for the two conditions. Due to the laminar regime, the Nusselt number is independent of the heating intensity and provide a limited heat enhancement for a constant Reynolds number. However, as the Reynolds number increases the heat enhancement is more pronounced.

Similarly, the heating and the flow rate are applied when a porous insert is implemented in the model. By examining Figure 5a, a slower flow rate leads to more heat extraction as observed in the previous case. In addition, the presence of a porous medium forces the flow to exert more contact with the pore and thus more heat is extracted. The temperature variation is almost linear contrary to the previous case where the linearity is removed towards the end of the heated section. Figure 5b presents the local Nusselt number variation. It is evident from this graph that heat extraction is more pronounced in the presence of a porous insert as the local Nusselt number is larger in magnitude.

Figure 6 showed the temperature variation and the local Nusselt number when a twisted tape insert is implemented. Interestingly enough one may judge the shape of the temperature as between that of a hollow pipe displayed in Figure 4 and a porous configuration as shown in Figure 5.

The waviness of the temperature is due to the measurement of temperature. In some locations, the insert touches the internal pipe wall creating a drop of the temperature due to metal contact. However, this waviness is very small and unnoticeable. Figure 6b displays the local Nusselt number. It is obvious that the heat extraction is more pronounced compared to a regular pipe but the porous insert provides a better heating enhancement. Although for a hollow pipe the flow is fully developed in the case of twisted insert (Figure 7) the flow structure inside the pipe was observed, forcing the fluid to extract the best possible of heat from the pipe. The amount of heat extraction is constant regardless of the heating conditions due to the laminar regime.

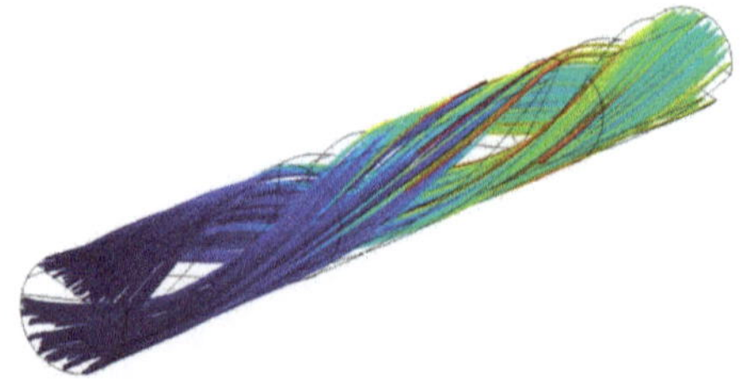

Figure 7. Flow structure with the twisted tape insert.

4.2. Friction Factor Variation

Friction factor is an important parameter to be investigated. Although water is the working fluid but in the presence of the insert it may affect the pumping power. By definition, the friction factor is defined as:

$$f = \frac{2 * \Delta p}{\frac{L}{D} * \rho w^2} \quad (12)$$

where Δp is the pressure change between the pipe inlet and outlet in kPa, D is the diameter of the pipe, L is the length of the pipe and ρ_f is the fluid density and w is the fluid velocity on the z direction. Figure 8 presents the friction coefficient along the entire length of the pipe for a heat flux of q″ = 75,000 W/m^2 and for two different Reynolds numbers. It is evident that the flow in the plain pipe exhibited the lowest friction coefficient when compared to the porous insert case or the twisted tape case. However, as observed earlier in the presence of a twisted tape case or porous case, the heat extraction showed better performance at the expense of a higher friction coefficient. As the Reynolds number increases, the friction coefficient increases correspondingly. The variation pattern along the flow direction is the same for the case of plain pipe and twisted insert. However, fluctuation in the pattern exists for the porous insert due to the non-uniform flow circulation between the pores.

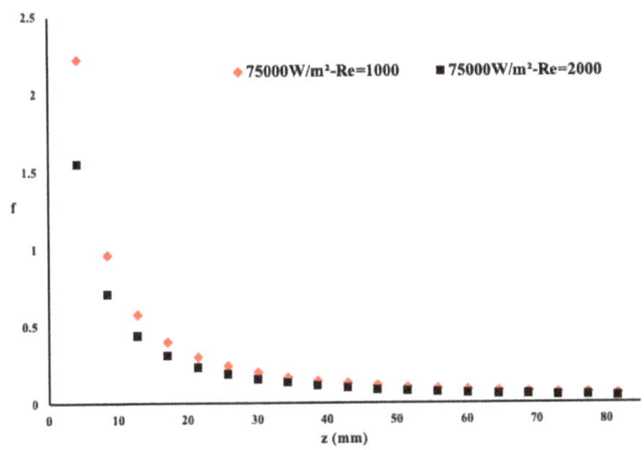

(**a**) No insert

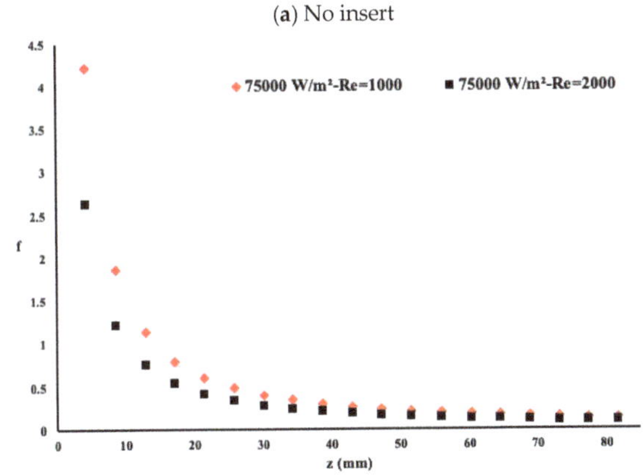

(**b**) Porous insert

Figure 8. *Cont.*

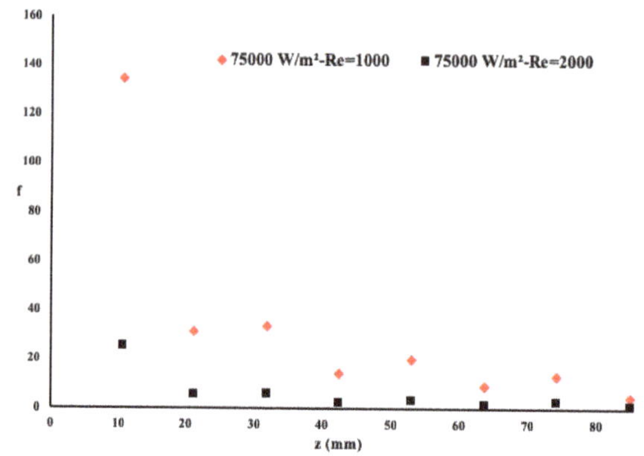

(c) Twisted tape insert

Figure 8. Friction factor along the pipe length.

4.3. Thermal Efficiency Ratio

Thermal efficiency is the most accurate parameter to identify the best configuration related to heat extraction. It has been demonstrated that some configurations may provide a good heat enhancement but at the expense of a high friction coefficient and therefore a pressure drop. Other configurations provide a low heat enhancement at the expense of low pressure drop. By definition, the thermal efficiency is defined as follows:

$$\text{Thermal efficiency} = \frac{Nu_{average} * z}{f * D} \tag{13}$$

In the current comparison, Figure 9 presents the thermal efficiency ratio of the case of Reynolds number equal to 2000 over the case where the Reynolds number is equal to 1000. This coefficient is calculated along the entire pipe. The heat flux is set equal to 75,000 W/m^2 and the inlet temperature is maintained at 18 °C.

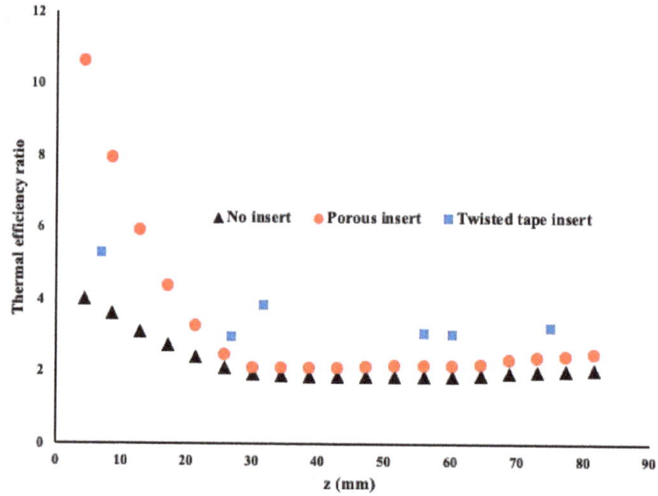

Figure 9. Thermal efficiency ratio for different configurations.

As shown in Figure 9, it is evident that a twisted insert provided the best thermal efficiency followed by the porous insert. The flow mixing as well as the flow through a porous structure is evident for the good performance of such pipe as a heat exchanger. Because the twisted tape touches the inside diameter of the pipe at different location, the point shown in Figure 9 are at the wall of the internal pipe diameter where no contact exists between the twisted tape and the pipe.

4.4. *Effectiveness in Using Nanofluid as Working Fluid*

In the present work, the distilled water working fluid showed that the insert of a twist pipe provided the best thermal efficiency when compared to a porous insert or plain pipe. The reason is due to the mixing phenomenon happening in the entire pipe. Two factors may have influenced the performance of the heat enhancement, the low friction coefficient and the high heat extraction provided by high Nusselt number. In order the study the importance of working fluid in heat extraction, the model was repeated using a nanofluid. The nanofluid consist of 1% vol TiO_2 nanoparticles in distilled water. The metallic nanoparticles have a diameter of 50 nm and are uniformly distributed in the solution.

Table 3 presents the physical properties of the nanofluid. The model was computed for the three configurations for two different Reynolds number of 1000 and 2000, respectively. The heat flux was maintained constant at 75,000 W/m^2 with a constant inlet temperature of 18 °C. In this section the ratio of the thermal efficiency of the nanofluid to distilled water is discussed with the aim to highlight the usefulness of using nanofluid.

Table 3. 1% vol TiO_2 nanofluid physical properties.

ρ_f (kg/m^3)	μ_f (kg/m/s)	k_f (W/m.K)	Cp_f (J/kg/K)
1030	0.001019	0.835	4040

Figure 10 presents the study case for no insert. As shown, the nanofluid provided a better performance when compared to distilled water regardless of the Reynolds number. A very large heat enhancement with an average of 22% is noticeable for a Reynolds number of 2000 and 15% enhancement for a Reynolds number of 1000. Both fluids had very small pressure drops but the high conductivity of the nanofluid and nearly similar viscosity led to a better performance of the nanofluid over water.

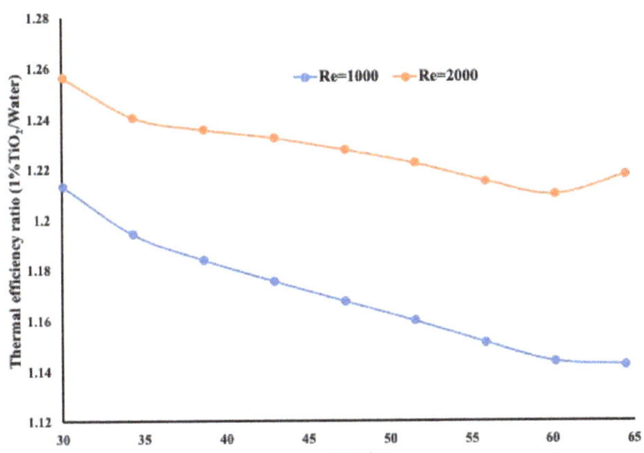

Figure 10. Thermal efficiency ratio between nanofluid and water (no insert).

Figure 11 presents the same scenario but with a porous insert. The insert is identical to the previous model, having a porosity of 0.91. This copper metallic insert helps heat extraction regardless of whether the working fluid is a nanofluid or distilled water. In this figure, one may notice that the nanofluid outperforms distilled water but with a heat enhancement of an average 2% when the Reynolds number is equal to 2000 and 0.5% when the Reynolds number is equal to 1000. The difference in performance for the two different Reynolds numbers is due to higher flow rate when the Reynolds number is equal to 2000. Nerveless the nanofluid demonstrated here its capability to improve the heat enhancement when compared to water.

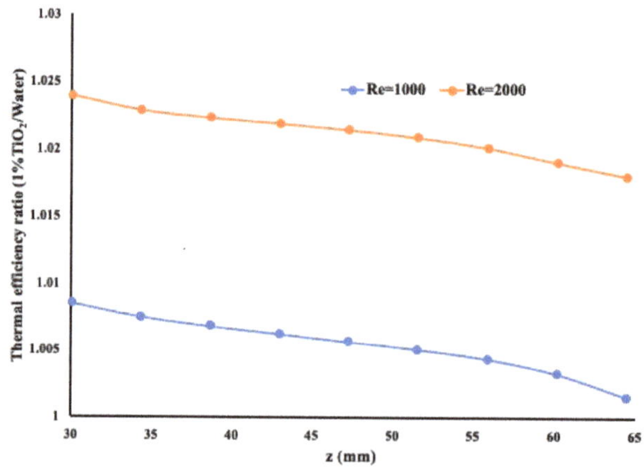

Figure 11. Thermal efficiency ratio between nanofluid and water (porous insert).

Finally, the study has been extended to a twisted tape insert as shown in Figure 12. It is evident from the results presented in Figure 12 that the nanofluid did not perform as expected. It is strongly believed that the mixing effect, while in laminar regime, provides the best heat extraction regardless of whether a nanofluid is used or not. It is also important to mention that the twisted tape is attached at some of its part to the internal wall of the pipe thus making it difficult to identify a location to measure the flow velocity.

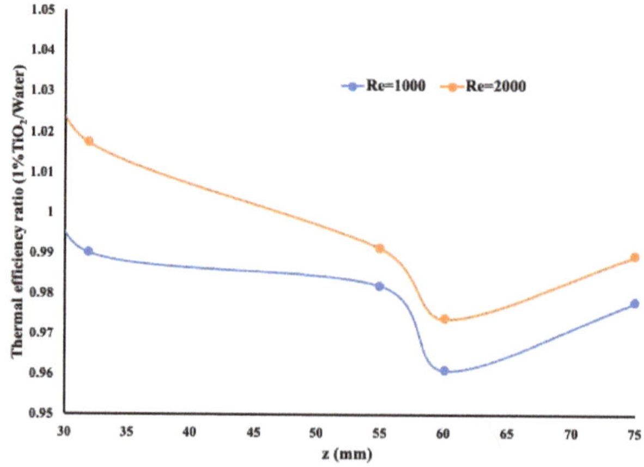

Figure 12. Thermal efficiency ratio between nanofluid and water (twisted tape inserts).

5. Conclusions

The findings of the current study of the heat enhancement for three different pipe assemblies showed the following:

(1) The temperature increases by increasing the heat flux and decreases by increasing the Reynolds number. As the working fluid changes from water to a nanofluid, the heat extraction by the nanofluid is in general more pronounced due to its higher conductivity.
(2) The Nusselt number value is higher in the presence of a porous insert but at the expense of a higher friction coefficient.
(3) The thermal efficiency of the twisted tape insert pipe is higher than that of a porous medium insert pipe and a plain pipe. This is mainly due to the strong mixing effect.
(4) A nanofluid provided a better heat enhancement in a plain pipe and with a porous insert when compared to water via thermal efficiency. However, for the twisted pipe, the nanofluid performance was weak compared to water as working fluid. Thus, one can simply avoid using a nanofluid and simply use a twisted pipe insert.

Author Contributions: Conceptualization, M.Z.S.; methodology, K.P. and M.Z.S.; software, K.P. and Z.A.H. validation, K.P.; formal analysis, K.P., Z.A.H. and M.Z.S.; investigation, K.P., Z.A.H. and M.Z.S.; resources, M.Z.S.; data curation, M.Z.S.; writing—original draft preparation, K.P., M.Z.S. and Z.A.H.; writing—review and editing, K.P., M.Z.S. and Z.A.H.; funding acquisition, M.Z.S. All authors have read and agreed to the published version of the manuscript.

Funding: The authors acknowledge the full financial support of the National Science and Engineering Research Council Canada (NSERC), Ryerson University for the financial support and Qatar National Research Foundation (NPRP12S-0123-190011).

Conflicts of Interest: The authors declare no conflict of interest.

Nomenclature

u, v, w	Velocity of the fluid in m/s
T_{in}	Inlet temperature (°C)
T	Measured temperature (°C)
x, y, z	Coordinates in mm
P	Pressure in Pa
q″	Heat flux in W/m^2
L	Length of the pipe in mm
D	Pipe diameter in mm
ρ_f	Density in kg/m^3
μ_f	Viscosity in kg/m s
Cp_f	Specific heat in J/kg K
k_f	Conductivity
k_w	Conductivity of water

References

1. Wang, X.-Q.; Mujumdar, A.S. Heat transfer characteristics of nanofluids: A review. *Int. J. Therm. Sci.* **2007**, *46*, 1–19. [CrossRef]
2. Wang, X.; Xu, X.; Choi, S.U.S. Thermal conductivity of nanoparticle-fluid mixture. *J. Thermophys. Heat Transf.* **1999**, *13*, 474–480. [CrossRef]
3. Maıgga, S.E.B.; Nguyen, C.T.; Galanis, N.; Roy, G. Heat transfer behaviours of nanofluids in a uniformly heated tube. *Superlattices Microstruct.* **2004**, *35*, 543–557. [CrossRef]
4. Halelfadl, S.; Estellé, P.; Aladag, B.; Doner, N.; Maré, T. Viscosity of carbon nanotubes water-based nanofluids: Influence of concentration and temperature. *Int. J. Therm. Sci.* **2013**, *71*, 111–117. [CrossRef]
5. Halelfadl, S.; Maré, T.; Estellé, P. Efficiency of carbon nanotubes water based nanofluids as coolants. *Exp. Therm. Fluid Sci.* **2014**, *53*, 104–110. [CrossRef]

6. Das, S.K.; Putra, N.; Thiesen, P.; Roetzel, W. Temperature dependence of thermal conductivity enhancement for nanofluids. *J. Heat Transf.* **2003**, *125*, 567–574. [CrossRef]
7. Ho, C.J.; Chang, C.Y.; Cheng, C.Y.; Cheng, S.J.; Guo, Y.W.; Hsu, S.T.; Yan, W.-M. Laminar forced convection effectiveness of Al_2O_3-water nanofluid flow in a circular tube at various operation temperatures: Effects of temperature-dependent proper-ties. *Int. J. Heat Mass Transf.* **2016**, *100*, 464–481. [CrossRef]
8. Saghir, M.Z.; Bayomy, A. Heat enhancement and heat storage for a ternary mixture in a circular pipe. *Therm. Sci. Eng. Prog.* **2018**, *5*, 32–43. [CrossRef]
9. Sekrani, G.; Poncet, S.; Proulx, P. Modeling of convective turbulent heat transfer of water-based Al_2O_3 nanofluids in an uniformly heated pipe. *Chem. Eng. Sci.* **2018**, *176*, 205–219. [CrossRef]
10. Bianco, V.; Manca, O.; Nardini, S. Numerical investigation of transient single phase forced convection of nanofluids in circular tubes. *WIT Trans. Eng. Sci.* **2008**, *61*, 3–12. [CrossRef]
11. Eiamsa-Ard, S.; Wongcharee, K. Heat transfer characteristics in micro-fin tube equipped with double twisted tapes: Effect of twisted tape and micro-fin tube arrangements. *J. Hydrodyn.* **2013**, *25*, 205–214. [CrossRef]
12. Kumar, S.; Hindasageri, V.; Prabhu, S. Local heat transfer distribution on a flat plate impinged by a swirling jet generated by a twisted tape. *Int. J. Therm. Sci.* **2017**, *111*, 351–368. [CrossRef]
13. Maddah, H.; Aghayari, R.; Farokhi, M.; Jahanizadeh, S.; Ashtary, K. Effect of Twisted-Tape Turbulators and Nanofluid on Heat Transfer in a Double Pipe Heat Exchanger. *J. Eng.* **2014**, *2014*, 1–9. [CrossRef]
14. Huang, Z.; Nakayama, A.; Yang, K.; Yang, C.; Liu, W. Enhancing heat transfer in the core flow by using porous medium insert in a tube. *Int. J. Heat Mass Transf.* **2010**, *53*, 1164–1174. [CrossRef]
15. Ghosh, I. Heat-Transfer Analysis of High Porosity Open-Cell Metal Foam. *J. Heat Transf.* **2008**, *130*, 034501. [CrossRef]
16. Zhao, H.; Li, X.; Wu, Y.; Wu, X. Friction factor and Nusselt number correlations for forced convection in helical tubes. *Int. J. Heat Mass Transf.* **2020**, *155*. [CrossRef]
17. Pathipakka, G.; Sivashanmugam, P. Heat transfer behaviour of nanofluids in a uniformly heated circular tube fitted with helical inserts in laminar flow. *Superlattices Microstruct.* **2010**, *47*, 349–360. [CrossRef]
18. Prabhanjan, D.; Raghavan, G.; Rennie, T. Comparison of heat transfer rates between a straight tube heat exchanger and a helically coiled heat exchanger. *Int. Commun. Heat Mass Transf.* **2002**, *29*, 185–191. [CrossRef]
19. Xin, R.C.; Ebadian, M.A. The Effects of Prandtl Numbers on Local and Average Convective Heat Transfer Characteristics in Helical Pipes. *J. Heat Transf.* **1997**, *119*, 467–473. [CrossRef]
20. COMSOL. *Software User Manual*; COMSOL: Newton, MA, USA, 2020.
21. Alhajaj, Z.; Bayomy, A.; Saghir, M. A comparative study on best configuration for heat enhancement using nanofluid. *Int. J. Thermofluids* **2020**, *8*, 100041. [CrossRef]

Publisher's Note: MDPI stays neutral with regard to jurisdictional claims in published maps and institutional affiliations.

© 2020 by the authors. Licensee MDPI, Basel, Switzerland. This article is an open access article distributed under the terms and conditions of the Creative Commons Attribution (CC BY) license (http://creativecommons.org/licenses/by/4.0/).

MDPI
St. Alban-Anlage 66
4052 Basel
Switzerland
Tel. +41 61 683 77 34
Fax +41 61 302 89 18
www.mdpi.com

Fluids Editorial Office
E-mail: fluids@mdpi.com
www.mdpi.com/journal/fluids

www.ingramcontent.com/pod-product-compliance
Lightning Source LLC
Chambersburg PA
CBHW040744250926
51850CB00007BB/1405